Springer Series in Reliability Engineering

More information about this series at http://www.springer.com/series/6917

Xufeng Zhao · Toshio Nakagawa

Advanced Maintenance Policies for Shock and Damage Models

Springer

Xufeng Zhao
Nanjing University of Aeronautics
and Astronautics
Nanjing, Jiangsu
China

Toshio Nakagawa
Aichi Institute of Technology
Toyota, Aichi
Japan

ISSN 1614-7839 ISSN 2196-999X (electronic)
Springer Series in Reliability Engineering
ISBN 978-3-319-88939-9 ISBN 978-3-319-70456-2 (eBook)
https://doi.org/10.1007/978-3-319-70456-2

Printed on acid-free paper

This Springer imprint is published by Springer Nature
The registered company is Springer International Publishing AG
The registered company address is: Gewerbestrasse 11, 6330 Cham, Switzerland

Preface

The number of aged plants and infrastructures has been greatly increasing in advanced nations [1]. In order to conduct safe and economical maintenance strategies, modeling and analysis of wear or damage lurked within operating units in analytical ways play important roles in reliability theory and engineering. The damage models have been studied for decades, and some of which were summarized in the book *Shock and Damage Models in Reliability Theory* [2]. In this book, literatures of the past and our latest research results are surveyed systematically, and some examples in the book *Stochastic Process with Applications to Reliability Theory* [3] are cited to build the bridge between theory and practice.

We recently have proposed the models of *replacement first*, *replacement last*, *replacement middle*, and *replacement overtime* in maintenance theory [4–14], which were also surveyed in books *Random Maintenance Policies* [15] and *Maintenance Overtime Policies in Reliability Theory* [16]. These new models would be more effective in maintaining production systems with random working cycles and computer systems with continuous processing times. We have also noticed that these new models would be applicable to damage models [17–21]. We will compare the damage models with approaches of *replacement first*, *replacement last*, *replacement middle*, and *replacement overtime* with the standard model in the book [2] and show that our theoretical damage models can be applied to defragmentation and backup schemes for database management in computer systems.

Nine chapters with appendix, which are based on our original works, are included in this book: In Chap. 1, we take the reliability systems with repairs as examples to introduce stochastic processes, e.g., Poisson process, renewal process, and cumulative process. Formulations of damage models such as cumulative damage model, independent damage model, etc., are given without detailed explanations and full proofs.

In Chap. 2, we review the standard replacement model for cumulative damage process, in which shocks for an operating unit occur randomly and an amount of damage due to shocks is additive, causing the unit to fail when the total damage exceeds a failure threshold K. The unit is supposed to be replaced correctively after failure K and preventively before K at planned time T, at shock number N, or at damage level Z, whichever occurs first. We name this standard replacement model as *replacement first*, as it is formulated under the classical approach of *whichever triggering event occurs first*. Several combinational models of replacement policies with T, N and Z are optimized analytically, when shocks occur at a renewal process and at a Poisson process. In addition, extended replacement models, e.g., the level of failure threshold K is a random variable and the unit fails when the total number of shocks reaches N, are obtained.

In Chaps. 3 and 4, we center on discussions of the models with new approaches of *whichever triggering event occurs last*, *replacing over a planned measure*, and *whichever triggering event occurs middle*, which are named as *replacement last*, *replacement overtime*, and *replacement middle*, respectively:

1. Replacement Last: The unit is replaced preventively at time T, at shock N, or at damage Z, whichever occurs last.
2. Replacement Overtime: The unit is replaced preventively at the forthcoming shock over time T and at the next shock over damage Z.
3. Replacement Middle: Denoting t_N and t_Z be the respective replacement times at shock N and at damage Z, the unit is replaced preventively, e.g., at planned time T for $\{t_N < T \leq t_Z\}$ and $\{t_Z < T \leq t_N\}$.

In Chaps. 5 and 6, minimal repairs, to fix the failures with probability $p(x)$ when the total damage is x at some shock, and minimal maintenance, to preserve an operating unit when the total damage has exceeded a failure threshold K, are introduced into the modified models of replacement first, last, and middle. In Chap. 5, replacement overtime is modeled into the discussed policies, which are named as *replacement overtime first* and *replacement overtime last*. In Chap. 6, replacement models with shock numbers and failure numbers are surveyed, respectively.

In Chap. 7, it is assumed that an operating unit, degrading with *additive damage* produced by shocks, is also suffered for *independent damage* that occurs at a nonhomogeneous Poisson process. Corrective replacement is done when the total additive damage exceeds K, and minimal repair is made for the independent damage to let the unit return to operation. When the unit is replaced preventively at time T and number N of independent damages, the modified models of replacement first and replacement last are obtained. Furthermore, replacement overtime first and replacement overtime last for independent and additive damages are modeled and discussed, respectively. In addition, both number N of shocks and number M of independent damages are considered simultaneously for the modified replacement first, last, and middle, and their expected cost rates are obtained for further discussions.

In Chap. 8, the new approaches discussed in the above chapters are applied to database maintenance models. We suppose that a database system updates in large volumes at a stochastic process, and the fragmentation, which refers to the non-contiguous regions and should be freed back into contiguous areas, and the updated data files, which should be copied to a safer storage system, arise with respective amounts of random variables. We formulate several kinds of defragmentation and backup models, by replacing the random *shocks* with *database updates* in large volumes, and the amount of *damage* with the volumes of *fragmentation* and *updated data*.

Finally, in Chap. 9, we present compactly other damage models and their maintenance policies, such as follows:

1. Replacement policies for the periodic damage model where the damage produced by shocks is measured exactly at periodic times.
2. Periodic and sequential maintenance policies that are imperfectly conducted for periodic damage models.
3. Inspection policies for the continuous damage model where the total damage increases continuously with time.
4. Inspection and maintenance policies for the Markov chain model where the total damage transits among several states.

An interesting study throughout this book is that we compare models of new approaches with the standard model given in Chap. 2, and critical solutions of comparisons are found analytically and computed numerically. In Chap. 3, models of replacement last are compared with replacement first to find in what cases which model is better from the point of cost rates. In order to compare replacement overtime with replacement first, costs for preventive replacement policies are modified and a new policy of *replacement overtime first* is first modeled in Chap. 4. For the replacement middle policies, a new approach of *whichever triggering event occurs middle* is proposed for modeling and numerical examples of comparisons are conducted. In Chap. 5, replacement first and replacement last are compared for their optimum times T with given shock N and optimum shocks N with given T, replacement overtime first is compared with replacement overtime last for their optimum times T with given shock N, and the replacement policy done over time T is compared with the standard replacement and the policy done at shock N. Similar comparisons are also made in the following chapters.

We would like to express our sincere appreciations to Prof. Hoang Pham for providing us the opportunity to write this book and to Editor Anthony Doyle and the Springer staff for their editorial work.

Nanjing, China Xufeng Zhao
Toyota, Japan Toshio Nakagawa

Contents

Chapter 1
Introduction

1.1 Stochastic Processes

We take the reliability systems with repairs as examples to introduce the following stochastic processes briefly [3, 23]. Consider a one-unit system with repair action, i.e., the operating unit is repaired only at failure, and it becomes as good as new to start the operation again after repair is completed.

1. When the repair time is assumed to be negligible and failures occur exponentially, e.g., the failure time of the operating unit has an exponential distribution $F(t) = 1 - \mathrm{e}^{-\lambda t}$ for time $t \geq 0$ and $0 < \lambda < \infty$, the occurrence of failures forms a *Poisson process*, and the unit fails during any time interval $[t, t + \mathrm{d}t]$ with constant probability $\lambda \mathrm{d}t$. A Poisson process is the simplest stochastic process in reliability and its theoretical properties can be easily investigated.
2. When the failure probability of the unit increases or sometimes decreases with its age, e.g., the probability that the unit fails in $[t, t + \mathrm{d}t]$ increases or decreases with time t, a *renewal process* is formed, which has the property of self-renewing aggregates. Obviously, the above Poisson process is one particular case of renewal process with exponential failure times. The renewal process plays a major role in analysis of probability models with sums of independent and nonnegative random variables, and also, is a basic tool in reliability theory.
3. When the repair time is non-negligible and two types of repairs are done to fix the minor and major failures, the system forms a *Markov process* with three states of operation, repair for minor failure, and repair for major failure. In this case, the process transits among the states, which follows a *Markov property* claiming that the future behavior only depends on the present state and is independent of the past history. The Markov process becomes a renewal process when there is only one failure state to be fixed. Furthermore, if the durations of transitions between states are discrete times, such as days, weeks, months, and etc., then the process becomes a *Markov chain*.

X. Zhao and T. Nakagawa, *Advanced Maintenance Policies for Shock and Damage Models*, Springer Series in Reliability Engineering,
https://doi.org/10.1007/978-3-319-70456-2_1

4. Suppose that shocks occur randomly according to a stochastic process and produce variable damages to an operating unit, and each amount of damage due to shocks is additive. This forms a *cumulative process*, or *cumulative damage process*, which depends on the rate of shocks and the total amount of damage accumulated. In particular, when the total damage is classified into several states without counting its quantity, the cumulative process becomes a Markov process mentioned above. In this book, a nonhomogeneous Poisson process with varying rate of shocks, e.g., $\lambda(t)$, and a homogeneous Poisson process with constant rate of shocks, e.g., λ, are usually supposed for shock arrivals. The cumulative process makes the discussions more difficult for reliability systems, as we should observe the two stochastic processes of shocks and damages simultaneously.

1.1.1 Poisson Process

A Poisson process is the simplest stochastic process that arises in applications of events arriving randomly in time. We firstly investigate the properties of failure times with an exponential distribution for an operating unit, and then demonstrate how a Poisson process is defined from a counting process and applied for the reliability systems.

(1) Exponential Distribution

Suppose that an operating unit is replaced immediately with a new one at each failure, where the time for replacement is negligible. The successive operating units have their failure times X_j $(j = 1, 2, \ldots)$ from the beginning of operations to respective failures. It is assumed that X_j are independent random variables with an identical distribution $F(t) \equiv \Pr\{X_j \le t\} = 1 - \mathrm{e}^{-\lambda t}$ $(0 < \lambda < \infty)$ and a density function $f(t) \equiv \mathrm{d}F(t)/\mathrm{d}t = \lambda \mathrm{e}^{-\lambda t}$ for $t \ge 0$. Then, the statistical mean and variance of X_j are

$$E\{X_j\} = \int_0^\infty t\lambda \mathrm{e}^{-\lambda t}\mathrm{d}t = \frac{1}{\lambda},$$

$$V\{X_j\} = \int_0^\infty t^2\lambda \mathrm{e}^{-\lambda t}\mathrm{d}t - \frac{1}{\lambda^2} = \frac{1}{\lambda^2},$$

and its LS (Laplace-Stieltjes) transform is

$$F^*(s) \equiv E\{\mathrm{e}^{-sX_j}\} = \int_0^\infty \mathrm{e}^{-st}\mathrm{d}F(t) = \int_0^\infty \mathrm{e}^{-st}\lambda \mathrm{e}^{-\lambda t}\mathrm{d}t = \frac{\lambda}{s+\lambda}$$

for any $Re(s) > 0$, where note that $Re(s) > 0$ is omitted throughout this book.

Denoting $S_n \equiv \sum_{j=1}^n X_j$ $(j = 1, 2, \ldots)$, from the assumption that random variables X_j are independent and identical,

$$E\{S_n\} = nE\{X_j\} = \frac{n}{\lambda},$$
$$V\{S_n\} = nV\{X_j\} = \frac{n}{\lambda^2},$$
$$E\{\mathrm{e}^{-sS_n}\} = \left(E\{\mathrm{e}^{-sX_j}\}\right)^n = \left(\frac{\lambda}{s+\lambda}\right)^n.$$

The distribution of S_n follows

$$G_n(t) \equiv \Pr\{S_n \le t\} = \int_0^t \frac{\lambda(\lambda u)^{n-1}}{(n-1)!}\mathrm{e}^{-\lambda u}\mathrm{d}u = \sum_{j=n}^{\infty} \frac{(\lambda t)^j}{j!}\mathrm{e}^{-\lambda t}$$
$$=1 - \sum_{j=0}^{n-1} \frac{(\lambda t)^j}{j!}\mathrm{e}^{-\lambda t} \quad (n = 1, 2, \ldots), \tag{1.1}$$

which is the formulation of a *gamma distribution* with order n. A density function of S_n is

$$g_n(t) \equiv \frac{\mathrm{d}G_n(t)}{\mathrm{d}t} = \frac{\lambda(\lambda t)^{n-1}}{(n-1)!}\mathrm{e}^{-\lambda t} \quad (n = 1, 2, \ldots),$$

and is generalized as

$$g_\alpha(t) \equiv \frac{\lambda(\lambda t)^{\alpha-1}}{\Gamma(\alpha)}\mathrm{e}^{-\lambda t}$$

for $\alpha > 0$, where $\Gamma(\alpha) \equiv \int_0^\infty x^{\alpha-1}\mathrm{e}^{-x}\mathrm{d}x$ for $\alpha > 0$, which is called a *gamma function* with order α (Problem 1.1).

The conditional probability that the unit fails during $(t, t+u]$ $(t, u \ge 0)$, given that it is operating at time t, is

$$\Pr\{t < X \le t+u | X > t\} = \frac{F(t+u) - F(t)}{\overline{F}(t)}, \tag{1.2}$$

where $\overline{\phi}(t) \equiv 1 - \phi(t)$ for any function $\phi(t)$. Thus, the *failure rate* is defined as [1, 22]

$$h(t) \equiv \frac{1}{\overline{F}(t)} \lim_{u\to 0} \frac{F(t+u) - F(t)}{u} = \frac{f(t)}{\overline{F}(t)}.$$

When $F(t) = 1 - \mathrm{e}^{-\lambda t}$, the failure rate is

$$h(t) = \frac{\lambda \mathrm{e}^{-\lambda t}}{\mathrm{e}^{-\lambda t}} = \lambda,$$

which is irrespective of time t. Thus, if the failure rate can be estimated from the actual life data, then MTTF (Mean Time to Failure) is obtained by taking the reciprocal of λ.

In addition, the conditional probability in (1.2) is

$$\frac{F(t+u)-F(t)}{\overline{F}(t)} = \frac{\mathrm{e}^{-\lambda t}-\mathrm{e}^{-\lambda(t+u)}}{\mathrm{e}^{-\lambda t}} = 1-\mathrm{e}^{-\lambda u},$$

which is irrespective of time t. This means that the age of t has no effect on the operating time u after t, which is called the *memoryless property* [3, 23].

Next, letting $N(t) \equiv \max\{n;\ S_n \le t\}$ denote the number of failures during the time interval $[0, t]$, we have the relation

$$\{S_n \le t\} \Leftrightarrow \{N(t) \ge n\}. \tag{1.3}$$

When $F(t) = 1 - \mathrm{e}^{-\lambda t}$, from (1.1),

$$\Pr\{N(t) \ge n\} = \Pr\{S_n \le t\} = \sum_{j=n}^{\infty} \frac{(\lambda t)^j}{j!}\mathrm{e}^{-\lambda t},$$

and

$$\begin{aligned}\Pr\{N(t) = n\} &= \Pr\{S_n \le t\} - \Pr\{S_{n+1} \le t\} \\ &= \frac{(\lambda t)^n}{n!}\mathrm{e}^{-\lambda t} \quad (n = 0, 1, 2, \ldots).\end{aligned} \tag{1.4}$$

The mean and variance of $N(t)$ are

$$E\{N(t)\} = \sum_{n=0}^{\infty} n\frac{(\lambda t)^n}{n!}\mathrm{e}^{-\lambda t} = \lambda t,$$

$$V\{N(t)\} = \sum_{n=0}^{\infty} n^2\frac{(\lambda t)^n}{n!}\mathrm{e}^{-\lambda t} - (\lambda t)^2 = \lambda t.$$

Note that a probability function given in (1.4) is called a *Poisson distribution* with mean λt.

(2) Poisson Process

A *counting process* is a stochastic process $N(t)$ for $t \ge 0$ with values that are positive, integer, and increasing, and $N(t) - N(s)$ is the number of events occurred during the time interval $[s, t]$ for $s < t$, where the events can be considered as failures for reliability systems. The examples of a counting process include a Poisson process and a renewal process. Next, we define a Poisson process as follows [3, 23]:

1. $N(0) = 0$.
2. The counting process $N(t)$ has independent increment.
3. The probability that n events occur during any interval Δt is

$$\Pr\{N(t+\Delta t) - N(t) = n\} = \frac{(\lambda \Delta t)^n}{n!} \mathrm{e}^{-\lambda \Delta t} \quad (n = 0, 1, 2, \ldots).$$

It can be clearly shown that

$$\Pr\{N(t) = n\} = \frac{(\lambda t)^n}{n!} \mathrm{e}^{-\lambda t} \quad (n = 0, 1, 2, \ldots),$$

which agrees with (1.4).

In order to introduce a nonhomogeneous Poisson process, we firstly take up the following concepts of minimal repair and failure rate in reliability. The failed unit undergoes minimal repair and resumes operation when the repair is completed, where the time for repair is negligible. Let $0 \equiv S_0 \le S_1 \le \cdots \le S_{j-1} \le S_j \le \cdots$ be the successive failure times of the unit and $X_j \equiv S_j - S_{j-1}$ $(j = 1, 2, \ldots)$ be the times between failures with an identical distribution $F(t) \equiv \Pr\{X_j \le t\}$. Then, we define minimal repair [1] as: The unit undergoes minimal repair at failures if and only if

$$\Pr\{X_j \le u | S_{j-1} = t\} = \frac{F(t+u) - F(t)}{\overline{F}(t)} \quad (j = 2, 3, \ldots) \tag{1.5}$$

for $t, u \ge 0$, where (1.5) is also called *failure rate*, representing the probability that the unit surviving at time t fails during interval $(t, t+u]$. This definition means that the failure rate remains undisturbed by any minimal repair, i.e., the unit restored after minimal repair has the same failure rate as it does before failure.

Suppose that the failure distribution $F(t)$ has a density function $f(t)$ and a failure rate $h(t) = f(t)/\overline{F}(t)$. The *cumulative hazard function* is defined as $H(t) \equiv \int_0^t h(u)\mathrm{d}u$ and satisfies $F(t) = 1 - \mathrm{e}^{-H(t)}$. Then, we have the distribution of the nth failure time S_n (Problem 1.2),

$$\begin{aligned} G_n(t) \equiv \Pr\{S_n \le t\} &= \sum_{j=n}^{\infty} \frac{H(t)^j}{j!} \mathrm{e}^{-H(t)} \\ &= 1 - \sum_{j=0}^{n-1} \frac{H(t)^j}{j!} \mathrm{e}^{-H(t)} \quad (n = 1, 2, \ldots), \end{aligned}$$

and the means of S_n and X_n are

$$E\{S_n\} = \int_0^\infty \overline{G}_n(t)\mathrm{d}t = \sum_{j=0}^{n-1} \int_0^\infty \frac{H(t)^j}{j!}\mathrm{e}^{-H(t)}\mathrm{d}t,$$

$$E\{X_n\} = E\{S_n\} - E\{S_{n-1}\} = \int_0^\infty \frac{H(t)^{n-1}}{(n-1)!}\mathrm{e}^{-H(t)}\mathrm{d}t.$$

In addition, if $h(t)$ increases, then $E\{X_n\}$ decreases with n and converges to $1/h(\infty)$ as $n \to \infty$ [1, 3], and from (1.3),

$$\Pr\{N(t) = n\} = \Pr\{S_n \le t < S_{n+1}\} = \frac{H(t)^n}{n!}\mathrm{e}^{-H(t)} \quad (n = 0, 1, 2, \ldots), \tag{1.6}$$

and the mean and variance of $N(t)$ become

$$E\{N(t)\} = V\{N(t)\} = H(t).$$

We next define a nonhomogeneous Poisson process as follows [3, 23]:

1. $N(0) = 0$.
2. A counting process $\{N(t)\}$ has independent increment.
3. The probability that n events occur during any interval t is given in (1.6) for $t \ge 0$.

In reliability systems, the events in the above definition can be considered as failures that occur with varying rates. When $H(t) = \lambda t$, a nonhomogeneous Poisson process degrades into a Poisson process with failure rate λ.

1.1.2 Renewal Process

In *renewal theory*, a failed unit is replaced with a new one, i.e., an operating unit is always renewed at each failure, forming a *renewal process*. We next observe the properties of a renewal process as follows: Consider generally a sequence of independent and nonnegative random variables $\{X_1, X_2, \ldots\}$ with an identical distribution $F(t)$, a density $f(t) \equiv \mathrm{d}F(t)/\mathrm{d}t$ and finite mean $\mu \equiv \int_0^\infty \overline{F}(t)\mathrm{d}t < \infty$. Denoting $S_n \equiv \sum_{j=1}^n X_j$ $(n = 1, 2, \ldots)$ with $S_0 \equiv 0$, $N(t) \equiv \max\{n; S_n \le t\}$ that represents the number of failures or renewals during the interval $[0, t]$.

Letting $F^{(n)}(t) \equiv \int_0^t F^{(n-1)}(t-u)\mathrm{d}F(u)$ $(n = 1, 2, \ldots)$ with $F^{(0)}(t) \equiv 1$ for $t \ge 0$ be the nth Stieltjes convolution of $F(t)$, the probability that n renewals occurred during $[0, t]$ is

$$\Pr\{N(t) = n\} = \Pr\{S_n \le t < S_{n+1}\} = F^{(n)}(t) - F^{(n+1)}(t) \quad (n = 0, 1, 2, \ldots). \tag{1.7}$$

We define a *renewal function* $M(t) \equiv E\{N(t)\}$ as the expected number $M(t) \equiv E\{N(t)\}$ of renewals during $[0, t]$ with a renewal density $m(t) \equiv \mathrm{d}M(t)/\mathrm{d}t$. Then, from (1.7),

$$M(t) = \sum_{n=1}^{\infty} n \Pr\{N(t) = n\} = \sum_{n=1}^{\infty} F^{(n)}(t). \tag{1.8}$$

From the notation of convolution, we obtain a *renewal equation* as

$$\begin{aligned} M(t) &= F(t) + \sum_{n=1}^{\infty} \int_0^t F^{(n)}(t-u)\mathrm{d}F(u) \\ &= \int_0^t [1 + M(t-u)]\mathrm{d}F(u). \end{aligned} \tag{1.9}$$

The renewal equation has appeared frequently in analysis of reliability models in several types, as most systems are assumed to be renewed after perfect maintenance.

The Laplace-Stieltjes transform of $M(t)$ is [3, 23]

$$M^*(s) \equiv \int_0^{\infty} \mathrm{e}^{-st}\mathrm{d}M(t) = \frac{F^*(s)}{1 - F^*(s)}, \tag{1.10}$$

where $\phi^*(s) \equiv \int_0^{\infty} \mathrm{e}^{-st}\mathrm{d}\phi(t)$, and $\int_0^{\infty} \mathrm{e}^{-st}\mathrm{d}F^{(n)}(t) = [F^*(s)]^n$ $(n = 0, 1, 2, \ldots)$. Clearly, by inverting (1.10),

$$M(t) = F(t) + F(t) * F(t) + F(t) * F(t) * F(t) + \cdots,$$

where the asterisk represents the pairwise convolution.

We summarize the following limiting theorems of a renewal theory [1, 3, 22]:

1. $M(t)/t \to 1/\mu$ as $t \to \infty$.
2. If $\mu_2 \equiv \int_0^{\infty} t^2 \mathrm{d}F(t)$ and $\sigma^2 = \mu_2 - \mu^2$, then

$$M(t) = \frac{t}{\mu} + \frac{1}{2}\left(\frac{\sigma^2}{\mu^2} - 1\right) + o(1) \tag{1.11}$$

as $t \to \infty$, where the function $f(h)$ is denoted as $o(h)$ if $\lim_{h \to 0} f(h)/h = 0$.

3. When $F(t)$ is IFR (Increasing Failure Rate), i.e., the failure rate $h(t) = f(t)/\overline{F}(t)$ increases with t,

$$\frac{t}{\mu} - 1 \le \frac{t}{\int_0^t \overline{F}(u)\mathrm{d}u} - 1 \le M(t) \le \frac{tF(t)}{\int_0^t \overline{F}(u)\mathrm{d}u} \le \frac{t}{\mu}. \tag{1.12}$$

4. The central limit theorem for a renewal process is

$$\lim_{t\to\infty}\Pr\left\{\frac{N(t)-t/\mu}{\sqrt{\sigma^2 t/\mu^3}}\le x\right\}=\frac{1}{\sqrt{2\pi}}\int_{-\infty}^{x}\mathrm{e}^{-u^2/2}\mathrm{d}u. \tag{1.13}$$

From the above results, $M(t)$ is approximately given by

$$M(t)\approx\frac{t}{\mu}+\frac{1}{2}\left(\frac{\sigma^2}{\mu^2}-1\right),$$

and for $\sigma \ll \mu$,

$$M(t)\approx\frac{t}{\mu}-\frac{1}{2}.$$

In addition, an alternating renewal process combining two renewal processes $\{X_j\}$ and $\{Y_j\}$ $(j=1,2,\ldots)$, a geometric renewal process $\{X_1, X_2/a, \cdots, X_k/a^{k-1}, \cdots\}$, and a discrete renewal process with discrete times [3], can be extended from the renewal process studied above.

1.1.3 Cumulative Process

An operating unit suffered for damage due to shocks fails when the total damage exceeds a failure threshold of its mechanical strength. We suppose that random variables of X_j and W_j $(j=1,2,\ldots)$ are respective sequences of interarrival times between shocks and damages due to shocks, where X_j are independent of W_j and $X_0=W_0=0$. A *cumulative process* or *cumulative damage process* is formed, when the shock process $\{X_j\}$ is compounded with the damage process $\{W_j\}$.

It is assumed that both X_j and W_j $(j=1,2,\ldots)$ are independent variables and have general distributions $F(t)\equiv\Pr\{X_j\le t\}$ with finite mean $\mu\equiv\int_0^\infty\overline{F}(t)\mathrm{d}t<\infty$ and $G(x)\equiv\Pr\{W_j\le x\}$ with finite mean $1/\omega\equiv\int_0^\infty\overline{G}(x)\mathrm{d}x<\infty$, respectively. Let the random variable $N(t)$ be the total number of shocks during $[0,t]$. Then, from (1.7), the probability that n shocks occur during $[0,t]$ is

$$\Pr\{N(t)=n\}=F^{(n)}(t)-F^{(n+1)}(t)\quad(n=0,1,2,\ldots),$$

and the distribution of the total damage accumulated at time t for n shocks is

$$\Pr\left\{\sum_{j=0}^{n}W_j\le x|N(t)=n\right\}\Pr\{N(t)=n\}=G^{(n)}(x)[F^{(n)}(t)-F^{(n+1)}(t)].$$

Denoting $W(t)$ as the total damage accumulated at time t for $N(t)$ shocks, i.e.,

$$W(t) \equiv \sum_{j=0}^{N(t)} W_j \quad (N(t) = 0, 1, 2, \ldots),$$

the distribution of $W(t)$ is

$$\Pr\{W(t) \le x\} = \sum_{n=0}^{\infty} \Pr\left\{\sum_{j=0}^{n} W_j \le x | N(t) = n\right\} \Pr\{N(t) = n\}$$
$$= \sum_{n=0}^{\infty} G^{(n)}(x)[F^{(n)}(t) - F^{(n+1)}(t)], \tag{1.14}$$

and the survival distribution is

$$\Pr\{W(t) > x\} = \sum_{n=0}^{\infty} [G^{(n)}(x) - G^{(n+1)}(x)] F^{(n+1)}(t). \tag{1.15}$$

Thus, the total expected damage at time t is

$$E\{W(t)\} = \int_0^{\infty} x \mathrm{d} \Pr\{W(t) \le x\}$$
$$= \sum_{n=0}^{\infty} F^{(n+1)}(t) \int_0^{\infty} [G^{(n)}(x) - G^{(n+1)}(x)] \mathrm{d}x$$
$$= \frac{1}{\omega} \sum_{n=1}^{\infty} F^{(n)}(t) = \frac{1}{\omega} M_F(t), \tag{1.16}$$

where $M_F(t) \equiv E\{N(t)\} = \sum_{n=1}^{\infty} F^{(n)}(t)$ is a renewal function of $F(t)$ and represents the expected number of shocks during $[0, t]$. From (1.11), $E\{W(t)\}$ is approximately given as

$$E\{W(t)\} \approx \frac{1}{\omega} \left(\frac{t}{\mu} + \frac{\sigma_F^2 - \mu^2}{2\mu^2} \right), \tag{1.17}$$

and from (1.13), $W(t)$ has the following asymptotic distribution [2, 3]:

$$\lim_{t \to \infty} \Pr \left\{ \frac{W(t) - t/(\mu\omega)}{\sqrt{(t/\mu)[(\sigma_F/\mu\omega)^2 + \sigma_G^2]}} \le x \right\} = \frac{1}{\sqrt{2\pi}} \int_{-\infty}^{x} \mathrm{e}^{-u^2/2} \mathrm{d}u, \tag{1.18}$$

where

$$\sigma_F^2 \equiv V\{X_j\} = \int_0^\infty t^2 \mathrm{d}F(t) - \mu^2,$$
$$\sigma_G^2 \equiv V\{W_j\} = \int_0^\infty x^2 \mathrm{d}G(x) - \frac{1}{\omega^2}.$$

1.2 Damage Models

Suppose that an operating unit fails when the damage exceeds a failure threshold K $(0 < K < \infty)$ of its mechanical strength. Then, we consider cumulative damage model, independent damage model, continuous damage model, and Markov chain model in the following sections.

1.2.1 Cumulative Damage Model

An operating unit with cumulative damage process has been introduced in Sect. 1.1.3. Using the same notations such as $F(t)$, $G(x)$, $W(t)$, $N(t)$, and etc., we next suppose that the damaged unit fails when the total damage exceeds a failure threshold K. Denoting Y as the first-passage time to failure K, i.e., $Y \equiv \min\{t; W(t) > K\}$, its distribution is, from (1.15),

$$\Pr\{Y \leq t\} = \Pr\{W(t) > K\} = \sum_{n=0}^{\infty} [G^{(n)}(K) - G^{(n+1)}(K)] F^{(n+1)}(t), \tag{1.19}$$

and its LS transform is

$$\int_0^\infty \mathrm{e}^{-st} \mathrm{d}\Pr\{Y \leq t\} = \sum_{n=0}^{\infty} [G^{(n)}(K) - G^{(n+1)}(K)][F^*(s)]^{n+1}. \tag{1.20}$$

Thus, the mean time to failure K is

$$\begin{aligned} E\{Y\} &= \sum_{n=0}^{\infty} [G^{(n)}(K) - G^{(n+1)}(K)] \int_0^\infty t \mathrm{d}F^{(n+1)}(t) \\ &= \mu \sum_{n=0}^{\infty} G^{(n)}(K) = \mu[1 + M_G(K)]. \end{aligned} \tag{1.21}$$

where $M_G(x) \equiv \sum_{j=1}^{\infty} G^{(j)}(x)$. This obviously means from (1.21) that the mean time to failure is given by the product of mean time between shocks and the expected number of shock over K. Similarly, from (1.11), $E\{Y\}$ is approximately given

$$E\{Y\} \approx \mu\left(\omega K + \frac{\omega^2\sigma_G^2 + 1}{2}\right). \tag{1.22}$$

In addition, from (1.12) and $\omega K - 1 < M_G(K) \leq \omega K$ when $G(x)$ is IFR,

$$\mu\omega K < E\{Y\} \leq \mu(\omega K + 1).$$

Letting p_{n+1} $(n = 0, 1, 2, \ldots)$ denote the probability that the unit fails at the $(n+1)$th shock,

$$\begin{aligned} p_{n+1} &\equiv \Pr\{W_1 + W_2 + \cdots + W_n \leq K \text{ and } W_1 + W_2 + \cdots + W_{n+1} > K\} \\ &= G^{(n)}(K) - G^{(n+1)}(K) \quad (n = 0, 1, 2, \ldots), \end{aligned} \tag{1.23}$$

and the surviving probability of the first n shocks is

$$G^{(n)}(K) = \Pr\{W_1 + W_2 + \cdots + W_n \leq K\} = \sum_{j=n+1}^{\infty} p_j \quad (n = 0, 1, 2, \ldots).$$

Thus, the probability that the unit surviving at the nth shock fails at the $(n+1)$th shock is

$$r_{n+1}(K) \equiv \frac{p_{n+1}}{\sum_{j=n+1}^{\infty} p_j} = \frac{G^{(n)}(K) - G^{(n+1)}(K)}{G^{(n)}(K)} \quad (n = 0, 1, 2, \ldots), \tag{1.24}$$

which is called the discrete failure rate of a probability function $\{p_n\}_{n=1}^{\infty}$. In particular, when $G(x) = 1 - e^{-\omega x}$,

$$r_{n+1}(K) = \frac{(\omega K)^n/n!}{\sum_{j=n}^{\infty}[(\omega K)^j/j!]}, \tag{1.25}$$

which increases strictly with n from $e^{-\omega K}$ to 1 (Problem 1.3).

Furthermore, suppose that shocks occur at a nonhomogeneous Poisson process with cumulative hazard function $H(t)$. From (1.6),

$$\Pr\{N(t) = n\} = \frac{H(t)^n}{n!} e^{-H(t)} \quad (n = 0, 1, 2, \ldots).$$

Replacing $F^{(n)}(t)$ with $\sum_{j=n}^{\infty}[H(t)^j/j!]e^{-H(t)}$ formally, the following results are obtained:

$$\Pr\{W(t) \le K\} = \sum_{n=0}^{\infty} G^{(n)}(K)\frac{H(t)^n}{n!}e^{-H(t)},$$
$$E\{W(t)\} = \frac{1}{\omega}H(t),$$
$$E\{Y\} = \sum_{n=0}^{\infty} G^{(n)}(K)\int_0^{\infty} \frac{H(t)^n}{n!}e^{-H(t)}dt.$$

1.2.2 Independent Damage Model

Suppose that the damage due to shocks is independent with each other and has no effect on the operating unit unless its amount exceeds a failure threshold K of the mechanical strength. In addition, damages produced by shocks are not additive to the current level, i.e., the unit only fails when some damage exceeds K for the first time. This is called *independent damage*, and its typical examples are the fracture of brittle materials such as glass and semiconductor part which fails due to over-current or over-voltage [2, 24].

We use the same notations of $F(t)$, $G(x)$, $W(t)$, $N(t)$, and etc. in Sect. 1.1.3 for discussions. In this case, the probability that the unit fails exactly at the $(n+1)$th $(n = 0, 1, 2, \ldots)$ shock is, from (1.23),

$$p_{n+1} = G(K)^n - G(K)^{n+1}.$$

Thus, the distribution of time to failure is

$$\Pr\{Y \le t\} = \sum_{n=0}^{\infty}[G(K)^n - G(K)^{n+1}]F^{(n+1)}(t), \tag{1.26}$$

and its LS transform is

$$\int_0^{\infty} e^{-st}d\Pr\{Y \le t\} = \frac{\overline{G}(K)F^*(s)}{1 - G(K)F^*(s)}. \tag{1.27}$$

MTTF is

$$E\{Y\} = \frac{\mu}{\overline{G}(K)}, \tag{1.28}$$

and the failure rate is

$$r_{n+1}(K) = p_1 = \overline{G}(K), \tag{1.29}$$

which is constant for any n.

When shocks occur at a nonhomogeneous Poisson process,

$$\Pr\{Y \le t\} = \sum_{j=0}^{\infty}[1 - G(K)^j]\frac{H(t)^j}{j!}\mathrm{e}^{-H(t)} = 1 - \mathrm{e}^{-\overline{G}(K)H(t)},$$

$$E\{Y\} = \int_0^{\infty} \mathrm{e}^{-\overline{G}(K)H(t)}\mathrm{d}t,$$

and the failure rate is

$$r(t) = \overline{G}(K)h(t).$$

1.2.3 Continuous Damage Model

Suppose that the total damage is not accumulated by shocks but increases continuously and swayingly at a stochastic path $W(t)$ with time t from $W(0) \equiv 0$. It is assumed that $W(t) = A_t t + B_t$, in which $A_t \ge 0$, B_t is a standard Brownian Motion [3], and the unit fails when $W(t)$ exceeds a failure threshold K.

The reliability $R(t) \equiv \Pr\{Y > t\}$ of the unit at time t is

$$R(t) = \Pr\{W(t) \le K\} = \Pr\{A_t t + B_t \le K\}. \tag{1.30}$$

We give two examples of (1.30), and extensive discussions of the continuous damage model and its replacement policies will be addressed in Chap. 9. When $A_t \equiv \omega$ and B_t is normally distributed with mean 0 and variance $\sigma^2 t$,

$$R(t) = \Pr\{B_t \le K - \omega t\} = \Phi\left(\frac{K - \omega t}{\sigma\sqrt{t}}\right), \tag{1.31}$$

where $\Phi(x)$ is a standard normal distribution with mean 0 and variance 1, i.e., $\Phi(x) = (1/\sqrt{2\pi})\int_{-\infty}^{x} \mathrm{e}^{-u^2/2}\mathrm{d}u$. When $A_t \equiv \omega$ and B_t is exponentially distributed with mean $\sigma\sqrt{t}$, i.e., $\Pr\{B_t \le t\} = 1 - \mathrm{e}^{-t/\sigma\sqrt{t}}$,

$$R(t) = \Pr\{B_t \le K - \omega t\} = 1 - \exp\left(-\frac{K - \omega t}{\sigma\sqrt{t}}\right). \tag{1.32}$$

1.2.4 Markov Chain Model

In either situation, the total damage accumulated by shocks in Sect. 1.2.1 and the continuously increased damage in Sect. 1.2.3 can be inspected at periodic times kT ($k = 1, 2, \ldots; 0 < T < \infty$). It is assumed that the increment of damage Z_k for the interval $[(k-1)T, kT]$ $(k = 1, 2, \ldots)$ is independent with each other and has an identical distribution $G(x) \equiv \Pr\{Z_k \le x\}$.

We suppose that the total damage is 0 at time 0, i.e., $Z_0 = 0$ when the unit starts operation, and becomes Z_1 at some inspection and reaches a threshold Z_n that makes the unit fail at the following inspections, where $0 < Z_1 < Z_n < \infty$. Further, the increment of damage between Z_1 and Z_n is divided into $n-1$ different levels such as $0 \equiv Z_0 < Z_1 < Z_2 < \cdots < Z_{n-1} < Z_n$ $(n = 2, 3, \ldots)$.

To formulate a Markov model, we define the following states of an operating unit [25]:

State 0: The total damage is less than Z_1.
State j: The total damage is between Z_j and Z_{j+1} $(j = 1, 2, \ldots, n-1)$.
State n: The total damage reaches Z_n.

It is assumed that the process remains in State j if the total damage does not exceed Z_{j+1} at some inspections. The above states forms a Markov chain with an absorbing State n. Then, one-step transition probabilities Q_{ij} are

$$Q_{ii} = G(Z_{i+1} - Z_i) \quad (i = 0, 1, \ldots, n-1),$$
$$Q_{ij} = G(Z_{j+1} - Z_i) - G(Z_j - Z_i) \quad (j = i+1, \ldots, n), \tag{1.33}$$

where $Z_0 = 0$ and $Z_{n+1} = \infty$, and $\sum_{j=i}^{n} Q_{ij} \equiv 1$.

The expected number I_i of inspections from State i to State n is

$$I_i = \sum_{j=i}^{n-1} Q_{ij}(1 + I_j) + Q_{in}.$$

Solving the above equation for I_i,

$$I_i = \frac{\sum_{j=i+1}^{n-1} Q_{ij} I_j + 1}{1 - Q_{ii}} \quad (i = 0, 1, \ldots, n-1), \tag{1.34}$$

where $\sum_{n}^{n-1} \equiv 0$.

In particular, when $G(x) = 1 - \mathrm{e}^{-\omega x}$,

$$Q_{ii} = 1 - \mathrm{e}^{-\omega(Z_{i+1} - Z_i)},$$
$$Q_{ij} = \mathrm{e}^{-\omega(Z_j - Z_i)} - \mathrm{e}^{-\omega(Z_{j+1} - Z_i)} \quad (j = i+1, \ldots, n-1),$$
$$Q_{i,n} = \mathrm{e}^{-\omega(Z_n - Z_i)}.$$

Thus, from (1.34),

$$I_i \mathrm{e}^{-\omega Z_{i+1}} = \mathrm{e}^{-\omega Z_i} + \sum_{j=i+1}^{n-1} (\mathrm{e}^{-\omega Z_j} - \mathrm{e}^{-\omega Z_{j+1}}) I_j. \tag{1.35}$$

By computing I_i successively for $i = n-1, n-2, \ldots, 1, 0$ (Problem 1.4),

$$I_0 = \sum_{j=0}^{n-1} e^{\omega(Z_{j+1}-Z_j)} - (n-1). \tag{1.36}$$

Therefore, MTTF is

$$E\{Y\} = T\left[\sum_{j=0}^{n-1} e^{\omega(Z_{j+1}-Z_j)} - (n-1)\right]. \tag{1.37}$$

1.3 Problem 1

1.1 When the density function of a gamma distribution is generalized as

$$g_\alpha(t) \equiv \frac{\lambda(\lambda t)^{\alpha-1}}{\Gamma(\alpha)} e^{-\lambda t},$$

derive its mean, variance and LS transform.

1.2 Derive $G_n(t)$, $E\{S_n\}$ and $E\{X_n\}$.

1.3 Prove that when $G(x) = 1 - e^{-\omega x}$ for $x > 0$,

$$r_{n+1}(x) = \frac{(\omega x)^n/n!}{\sum_{j=n}^{\infty}[(\omega x)^j/j!]} \quad (n = 0, 1, 2, \ldots)$$

increases strictly with n from $e^{-\omega x}$ to 1, and decreases strictly with x from 1 to 0.

1.4 Derive (1.36).

Chapter 2
Standard Replacement Policies

To begin with, this chapter gives three standard replacement models that have been obtained as basic replacement policies for an operating unit with shock and damage [2]. Here, the so-called standard models are formulated under the classical approach of *whichever triggering event occurs first* [22] in reliability theory. That is, the unit is replaced preventively at some thresholds or planned measurements such as operating time, usage number, damage level, repair cost, number of faults or repairs, etc., or at failure, *whichever occurs first*, which is named as *replacement first*.

Replacement first is absolutely reasonable when a single preventive replacement scenario is planned to avoid catastrophic failure, and models based on this approach have been surveyed and extended for decades. However, it may typically cause frequent and unnecessary replacement actions when several compound preventive replacement scenarios are scheduled. The features of models for replacement policies acting on the approach of first have been observed in [7, 9, 12, 19]. We name the models in this chapter as replacement first, as it will be used to compare with *replacement last* in Chap. 3, and *replacement middle* and *replacement overtime* in Chap. 4, which are based on the newly proposed approaches of *whichever triggering event occurs last*, *whichever triggering event occurs middle*, and *replacing over a planned measure*, respectively.

In this chapter, we consider an operating unit suffered for cumulative damage due to random shocks should operate over an infinite time span, in which it is of great importance to make suitable replacement plans to avoid catastrophic failure when the total damage has exceeded a failure threshold K. That is, we focus mainly on replacement policies for an operating unit with the failure mode of cumulative damage. As discussed in [2], we give the following three preventive replacement actions: The most easy way is to replace an operating unit at planned ages without monitoring its shock and damage; however, using monitoring equipment, we could count the number of shocks and investigate the amount of total damage at shock times, and more precise replacement plans can be done at a pre-specified number of shocks

X. Zhao and T. Nakagawa, *Advanced Maintenance Policies for Shock and Damage Models*, Springer Series in Reliability Engineering,
https://doi.org/10.1007/978-3-319-70456-2_2

or damage level before failure, despite the relatively rough. So that the operating unit is supposed to be replaced correctively after failure K and preventively before failure at planned time T, at shock number N, or at damage level Z, whichever occurs first [2].

The planned replacement actions can be optimized in order to balance failure losses and replacement costs. It was shown from numerical examples [2] that optimum preventive replacement policies acting on shock and damage conditions show more superiority than the policy acting on a planned time, that is, from the cost saving point, replacement done at Z is better than those at T and N, and replacement done at N is better than that at T in most cases. However, replacement at T is much easier to perform but additional monitoring equipment should be provided for those at N and Z, of which monitoring costs may not be neglected in practice. Additional comparisons of replacement policies for cumulative damage models have been studied in [26]. Further, it seems to be a waste of cost for replacement first, as we know that, replacement should be done as soon as possible before failure when any policy at T, N, or Z is first triggered. However, we will find the cases when replacement first saves cost for a long run, by comparing to the proposed replacement policies in the following chapters.

In Sect. 2.1, we obtain the expected cost rate of three combined preventive replacement policies planned at time T, at shock N, and at damage Z. In Sects. 2.1.1 and 2.1.2, we derive analytically optimum policies which minimize the expected cost rates for each of three policies and for two combinations of three ones. It would be of great interest to show theoretically that when all preventive replacement costs of three policies are the same, the best policy among three ones is replacement with damage Z, the next one is replacement with shock N, and the last one is replacement with time T. In Sect. 2.1.3, when shocks occur at a Poisson process and each amount of damage due to shocks is exponential, optimum policies in Sects. 2.1.1 and 2.1.2 are computed numerically and compared with each other. Sections 2.2 and 2.3 consider several extended replacement models, of which the level of failure threshold K is a random variable with an estimated probability distribution, and the unit also fails when the total number of shocks reaches to a certain value of N.

2.1 Three Replacement Policies

A new unit with damage level 0 begins to operate at time 0 and degrades with damage produced by shocks. It is assumed that shocks occur at a renewal process according to an identical distribution $F(t)$ with a density function $f(t) \equiv \mathrm{d}F(t)/\mathrm{d}t$ and finite mean $\mu \equiv \int_0^\infty \overline{F}(t)\mathrm{d}t$, where $\overline{F}(t) \equiv 1 - F(t)$. When $F(t)$ has a density function $f(t)$, $h(t) \equiv f(t)/\overline{F}(t)$ is assumed to increase from $h(0) \equiv \lim_{t\to 0} h(t)$ to $h(\infty) \equiv \lim_{t\to\infty} h(t)$. Clearly, when $F(t) = 1 - \mathrm{e}^{-\lambda t}$, $h(t) = \lambda$ for any $t \geq 0$. An amount W_j $(j = 1, 2, \cdots)$ of damage due to the jth shock has an identical distribution $G(x) \equiv \Pr\{W_j \leq x\}$ with finite mean $1/\omega \equiv \int_0^\infty \overline{G}(x)\mathrm{d}x$ and is additive to the current damage level.

Let $N(t)$ denote the number of shocks during the interval $[0, t]$. Then, the probability that j shocks occur exactly in $[0, t]$ is

$$\Pr\{N(t) = j\} = F^{(j)}(t) - F^{(j+1)}(t) \quad (j = 0, 1, 2, \cdots),$$

and the distribution of the total damage $W(t)$ at time t is

$$\Pr\{W(t) \leq x\} = \sum_{j=0}^{\infty} [F^{(j)}(t) - F^{(j+1)}(t)] G^{(j)}(x),$$

where $\Phi^{(j)}(t)$ denotes the j-fold Stieltjes convolution of any function $\Phi(t)$ with itself, i.e., $\Phi^{(j)}(t) = \int_0^t \Phi^{(j-1)}(t-u)\mathrm{d}\Phi(u)$ $(j = 1, 2, \cdots)$ and $\Phi^{(0)}(t) \equiv 1$ for $t \geq 0$.

The unit fails when the total damage has exceeded a failure threshold K $(0 < K < \infty)$ at some shock, and its failure is detected and *corrective replacement* (CR) is done immediately. *Preventive replacement* (PR) times are scheduled before failure at planned time T $(0 < T \leq \infty)$, at shock number N $(N = 1, 2, \cdots)$, or at damage level Z $(0 < Z \leq K)$, whichever occurs first, which is called *replacement first* (RF). In addition, the unit is supposed to be replaced at damage K or Z rather than at shock N, when the total damage has exceeded K or Z at shock N. Furthermore, it is assumed that both CR and PR remove all damage perfectly, and the unit becomes as good as new after any replacement.

It can be understood for RF that the replacement should be done as soon as possible before CR when any PR is first triggered. Then, the probability that the unit is replaced at time T is

$$\sum_{j=0}^{N-1} [F^{(j)}(T) - F^{(j+1)}(T)] G^{(j)}(Z), \tag{2.1}$$

the probability that it is replaced at shock N is

$$F^{(N)}(T) G^{(N)}(Z), \tag{2.2}$$

the probability that it is replaced at damage Z is

$$\sum_{j=0}^{N-1} F^{(j+1)}(T) \int_0^Z [G(K-x) - G(Z-x)]\mathrm{d}G^{(j)}(x), \tag{2.3}$$

and the probability that it is replaced at failure K is

$$\sum_{j=0}^{N-1} F^{(j+1)}(T) \int_0^Z \overline{G}(K-x)\mathrm{d}G^{(j)}(x), \tag{2.4}$$

where note that (2.1) + (2.2) + (2.3) + (2.4) = 1. Thus, the mean time to replacement is (Problem 2.1)

$$
\begin{aligned}
&T\sum_{j=0}^{N-1}[F^{(j)}(T)-F^{(j+1)}(T)]G^{(j)}(Z)+G^{(N)}(Z)\int_0^T t\mathrm{d}F^{(N)}(t)\\
&+\sum_{j=0}^{N-1}\int_0^T t\mathrm{d}F^{(j+1)}(t)\int_0^Z[G(K-x)-G(Z-x)]\mathrm{d}G^{(j)}(x)\\
&+\sum_{j=0}^{N-1}\int_0^T t\mathrm{d}F^{(j+1)}(t)\int_0^Z\overline{G}(K-x)\mathrm{d}G^{(j)}(x)\\
&=\sum_{j=0}^{N-1}G^{(j)}(Z)\int_0^T[F^{(j)}(t)-F^{(j+1)}(t)]\mathrm{d}t. \qquad (2.5)
\end{aligned}
$$

Therefore, the expected replacement cost rate is

$$
C_F(T,N,Z)=\frac{\begin{array}{l}c_K-(c_K-c_T)\sum_{j=0}^{N-1}[F^{(j)}(T)-F^{(j+1)}(T)]G^{(j)}(Z)\\-(c_K-c_N)F^{(N)}(T)G^{(N)}(Z)\\-(c_K-c_Z)\sum_{j=0}^{N-1}F^{(j+1)}(T)\int_0^Z[G(K-x)-G(Z-x)]\mathrm{d}G^{(j)}(x)\end{array}}{\sum_{j=0}^{N-1}G^{(j)}(Z)\int_0^T[F^{(j)}(t)-F^{(j+1)}(t)]\mathrm{d}t}, \qquad (2.6)
$$

where c_T = replacement cost at time T, c_N = replacement cost at shock N, c_Z = replacement cost at damage Z, and c_K = replacement cost at failure K, where $c_K > c_T$, $c_K > c_N$, and $c_K > c_Z$.

In particular, when the unit is replaced only after failure K, i.e., when $T \to \infty$, $N \to \infty$ and $Z \to K$,

$$
C \equiv \lim_{\substack{T\to\infty\\N\to\infty\\Z\to K}} C_F(T,N,Z)=\frac{c_K}{\mu[1+M_G(K)]}, \qquad (2.7)
$$

where $M_G(x)\equiv\sum_{j=1}^{\infty}G^{(j)}(x)$ is a renewal function of $G(x)$ and represents the expected number of shocks at damage level x. Further, if $Z \to 0$, then N is equal to 1, and

$$
\lim_{\substack{T\to\infty\\Z\to 0}} C_F(T,N,Z)=\lim_{\substack{T\to\infty\\Z\to 0}} C_F(T,1,Z)=\frac{1}{\mu}[c_K-(c_K-c_Z)G(K)]. \qquad (2.8)
$$

2.1.1 Optimum Policies with One Variable

We should avoid the high failure cost without any PR in (2.7). Meanwhile, it would be unreasonable to make frequent replacement actions to waste PR cost, such as an extreme case in (2.8). In order to minimize the respective cost rates of the above three replacement policies, we next obtain analytically optimum time T^*, shock N^*, and damage Z^*, respectively.

(1) Optimum T^*

Suppose that the unit is replaced preventively only at time T $(0 < T \leq \infty)$. Then, putting that $N \to \infty$ and $Z \to K$ in (2.6),

$$C(T) \equiv \lim_{\substack{N\to\infty \\ Z\to K}} C_F(T, N, Z) = \frac{c_K - (c_K - c_T)\sum_{j=0}^{\infty}[F^{(j)}(T) - F^{(j+1)}(T)]G^{(j)}(K)}{\sum_{j=0}^{\infty} G^{(j)}(K)\int_0^T [F^{(j)}(t) - F^{(j+1)}(t)]\mathrm{d}t}. \tag{2.9}$$

It can be easily seen that $\lim_{T\to 0} C(T) = \infty$ and $\lim_{T\to\infty} C(T) = C$ in (2.7). We find optimum T^* to minimize $C(T)$. Differentiating $C(T)$ with respect to T and setting it equal to zero,

$$Q_1(T)\sum_{j=0}^{\infty} G^{(j)}(K)\int_0^T [F^{(j)}(t) - F^{(j+1)}(t)]\mathrm{d}t - \sum_{j=0}^{\infty} F^{(j+1)}(T)[G^{(j)}(K) - G^{(j+1)}(K)] = \frac{c_T}{c_K - c_T}, \tag{2.10}$$

where

$$Q_1(T, N) \equiv \frac{\sum_{j=0}^{N-1} f^{(j+1)}(T)[G^{(j)}(K) - G^{(j+1)}(K)]}{\sum_{j=0}^{N-1}[F^{(j)}(T) - F^{(j+1)}(T)]G^{(j)}(K)}, \tag{2.11}$$

$$Q_1(T) \equiv \lim_{N\to\infty} Q_1(T, N) = \frac{\sum_{j=0}^{\infty} f^{(j+1)}(T)[G^{(j)}(K) - G^{(j+1)}(K)]}{\sum_{j=0}^{\infty}[F^{(j)}(T) - F^{(j+1)}(T)]G^{(j)}(K)},$$

and $f^{(j+1)}(t) \equiv \mathrm{d}F^{(j+1)}(t)/\mathrm{d}t$. Note that if $h(t)$ increases with t, i.e., $F(t)$ has an IFR (Increasing Failure Rate) property, its convolution is also IFR, and so that, $f^{(j+1)}(t)/[F^{(j)}(t) - F^{(j+1)}(t)]$ increases with t [22]. In particular, when $F(t) = 1 - \mathrm{e}^{-\lambda t}$, $f^{(j+1)}(t)/[F^{(j)}(t) - F^{(j+1)}(t)] = \lambda$ for any $t \geq 0$.

If $Q_1(T)$ increases strictly with T to $Q_1(\infty) \equiv \lim_{T\to\infty} Q_1(T)$, then the left-hand side of (2.10) increases strictly to (Problem 2.2)

$$\mu Q_1(\infty)[1 + M_G(K)] - 1.$$

Therefore, if

$$Q_1(\infty)[1 + M_G(K)] > \frac{c_K}{\mu(c_K - c_T)},$$

then there exists a finite and unique T^* $(0 < T^* < \infty)$ which satisfies (2.10), and the resulting cost rate is

$$C(T^*) = (c_K - c_T)Q_1(T^*). \tag{2.12}$$

Conversely, if

$$Q_1(\infty)[1 + M_G(K)] \leq \frac{c_K}{\mu(c_K - c_T)},$$

then $T^* = \infty$, i.e., the unit is replaced only at failure, and the resulting cost rate is given in (2.7).

(2) Optimum N^*

Suppose that the unit is replaced preventively only at shock N $(N = 1, 2, \cdots)$. Then, putting that $T \to \infty$ and $Z \to K$ in (2.6),

$$C(N) \equiv \lim_{\substack{T\to\infty \\ Z\to K}} C_F(T, N, Z) = \frac{c_K - (c_K - c_N)G^{(N)}(K)}{\mu \sum_{j=0}^{N-1} G^{(j)}(K)} \quad (N = 1, 2, \cdots). \tag{2.13}$$

In particular, when $N = 1$, i.e., the unit is always replaced at the first shock, the expected cost rate is given in (2.8) when $c_N = c_Z$.

We find optimum N^* to minimize $C(N)$. Forming the inequality $C(N + 1) - C(N) \geq 0$ (Problem 2.3),

$$r_{N+1}(K) \sum_{j=0}^{N-1} G^{(j)}(K) - [1 - G^{(N)}(K)] \geq \frac{c_N}{c_K - c_N}, \tag{2.14}$$

where for $0 < x \leq K$,

$$r_{N+1}(x) \equiv \frac{G^{(N)}(x) - G^{(N+1)}(x)}{G^{(N)}(x)} \quad (N = 1, 2, \cdots).$$

If $r_{N+1}(x)$ increases strictly with N, i.e., $G^{(N+1)}(x)/G^{(N)}(x)$ decreases strictly with N, then the left-hand side of (2.14) increases strictly with N to $r_\infty(K)[1 + M_G(K)] - 1$ (Problem 2.4), where $r_\infty(K) \equiv \lim_{N\to\infty} r_{N+1}(K) \leq 1$. Therefore, if

$$r_\infty(K)[1+M_G(K)] > \frac{c_K}{c_K - c_N},$$

then there exists a finite and unique minimum N^* $(1 \le N^* < \infty)$ which satisfies (2.14), and the resulting cost rate is

$$(c_K - c_N)r_{N^*}(K) < \mu C(N^*) \le (c_K - c_N)r_{N^*+1}(K), \tag{2.15}$$

whose cost rate is given in (2.8) when $N^* = 1$. Conversely, if

$$r_\infty(K)[1+M_G(K)] \le \frac{c_K}{c_K - c_N},$$

then $N^* = \infty$, i.e., the unit is replaced only at failure, and the expected cost rate is given in (2.7).

Note that $r_{N+1}(K)$ represents the probability that the unit surviving at the Nth shock will fail at the next shock, which might increase with N to 1.

(3) Optimum Z^*

Suppose that the unit is replaced preventively only at damage Z $(0 < Z \le K)$. Then, putting that $T \to \infty$ and $N \to \infty$ in (2.6),

$$C(Z) \equiv \lim_{\substack{T\to\infty \\ N\to\infty}} C_F(T, N, Z) = \frac{c_K - (c_K - c_Z)[G(K) - \int_0^Z \overline{G}(K-x)\mathrm{d}M_G(x)]}{\mu[1+M_G(Z)]}. \tag{2.16}$$

When $Z \to 0$, (2.16) agrees with (2.8).

We find optimum Z^* to minimize $C(Z)$. Differentiating $C(Z)$ with respect to Z and setting it equal to zero,

$$\int_{K-Z}^{K} [1+M_G(K-x)]\mathrm{d}G(x) = \frac{c_Z}{c_K - c_Z}, \tag{2.17}$$

whose left-hand side increases strictly with Z from 0 to $M_G(K)$ (Problem 2.5). Therefore, if $M_G(K) > c_Z/(c_K - c_Z)$, then there exists a finite and unique Z^* $(0 < Z^* < K)$ which satisfies (2.17), and the resulting cost rate is

$$\mu C(Z^*) = (c_K - c_Z)\overline{G}(K - Z^*). \tag{2.18}$$

Conversely, if $M_G(K) \le c_Z/(c_K - c_Z)$, then $Z^* = K$, i.e., the unit is replaced only at failure, and the resulting cost rate is given in (2.7).

2.1.2 Optimum Policies with Two Variables

Our next concerns are (i) how to compare respective replacement policies done at time T, at shock N, and at damage Z, and (ii) what are optimum policies with two variables to minimize their expected cost rates in (2.6). We answer for (i) and (ii) in the following sections, and further discussions for (i) will be addressed in Chap. 3.

(1) Optimum T_F^* and N_F^*

Suppose that the unit is replaced preventively at time T $(0 < T \leq \infty)$ or at shock N $(N = 1, 2, \cdots)$, whichever occurs first. When $c_T = c_N$, putting that $Z \to K$ in (2.6), the expected cost rate is

$$C_F(T, N) = \frac{c_T + (c_K - c_T)\sum_{j=0}^{N-1} F^{(j+1)}(T)[G^{(j)}(K) - G^{(j+1)}(K)]}{\sum_{j=0}^{N-1} G^{(j)}(K)\int_0^T [F^{(j)}(t) - F^{(j+1)}(t)]\mathrm{d}t}. \tag{2.19}$$

We find optimum (T_F^*, N_F^*) to minimize $C_F(T, N)$. Forming the inequality $C_F(T, N-1) - C_F(T, N) > 0$,

$$\frac{r_N(K)F^{(N)}(T)}{\int_0^T [F^{(N-1)}(t) - F^{(N)}(t)]\mathrm{d}t}\sum_{j=0}^{N-1} G^{(j)}(K)\int_0^T [F^{(j)}(t) - F^{(j+1)}(t)]\mathrm{d}t$$
$$-\sum_{j=0}^{N-1} F^{(j+1)}(T)[G^{(j)}(K) - G^{(j+1)}(K)] < \frac{c_T}{c_K - c_T}, \tag{2.20}$$

where $r_N(K)$ is given in (2.14).

Differentiating $C_F(T, N)$ with respect to T and setting it equal to zero,

$$Q_1(T, N)\sum_{j=0}^{N-1} G^{(j)}(K)\int_0^T [F^{(j)}(t) - F^{(j+1)}(t)]\mathrm{d}t$$
$$-\sum_{j=0}^{N-1} F^{(j+1)}(T)[G^{(j)}(K) - G^{(j+1)}(K)] = \frac{c_T}{c_K - c_T}, \tag{2.21}$$

where $Q_1(T, N)$ is given in (2.11).

In addition, because both T_F^* and N_F^* have to satisfy (2.20) and (2.21), (2.21) is rewritten as, by substituting (2.20) for (2.21),

$$Q_1(T, N) > \frac{r_N(K)F^{(N)}(T)}{\int_0^T [F^{(N-1)}(t) - F^{(N)}(t)]\mathrm{d}t}. \tag{2.22}$$

Thus, if the inequality (2.22) does not hold for any N, there does not exist any finite T_F^* which satisfies (2.21), i.e., the optimum policy is $(T_F^* = \infty, N_F^* = N^*)$, where N^* is given in (2.14).

For example, when $F(t) = 1 - \mathrm{e}^{-\lambda t}$ and $r_N(x)$ increases strictly with N, (2.22) is

$$\frac{\sum_{j=0}^{N-1}[(\lambda T)^j/j!][G^{(j)}(K) - G^{(j+1)}(K)]}{\sum_{j=0}^{N-1}[(\lambda T)^j/j!]G^{(j)}(K)} > r_N(K),$$

whose left-hand side increases with T from $\overline{G}(K)$ to $r_N(K)$ as $T \to \infty$ (Problem 2.2). That is, the above inequality does not hold for $0 < T \leq \infty$. This means that when optimum N_F^* satisfies (2.20), the left-hand side of (2.21) is less than $c_T/(c_K - c_T)$, i.e., $C_F(T, N_F^*)$ decreases with T, and hence, $T_F^* = \infty$. This concludes that if the inequality (2.22) does not hold, then the optimum policy which minimizes $C_F(T, N)$ is $(T_F^* = \infty, N_F^* = N^*)$.

In other words, when $c_T \geq c_N$ is supposed for the replacement policies done at time T and at shock N, its optimum policy degrades into the case in which only finite N^* for $T = \infty$ can be found in (2.14).

Next, we obtain optimum T_F^* for given N when $F(t) = 1 - \mathrm{e}^{-\lambda t}$ and $c_T = c_N$. In this case, (2.21) is rewritten as

$$\widetilde{Q}_1(T, N)\sum_{j=0}^{N-1} F^{(j+1)}(T)G^{(j)}(K) - \sum_{j=0}^{N-1} F^{(j+1)}(T)[G^{(j)}(K) - G^{(j+1)}(K)] = \frac{c_T}{c_K - c_T}, \tag{2.23}$$

where

$$\widetilde{Q}_1(T, N) \equiv \frac{\sum_{j=0}^{N-1}[(\lambda T)^j/j!][G^{(j)}(K) - G^{(j+1)}(K)]}{\sum_{j=0}^{N-1}[(\lambda T)^j/j!]G^{(j)}(K)}.$$

When $r_N(x)$ increases strictly with N, it is approved that $\widetilde{Q}_1(T, N) = \overline{G}(K)$ for $N = 1$ and increases strictly with T for $N \geq 2$ from $\overline{G}(K)$ to $r_N(K)$ (Problem 2.2). Thus, the left-hand side of (2.23) increases strictly with T from 0 to

$$r_N(K)\sum_{j=0}^{N-1} G^{(j)}(K) - [1 - G^{(N)}(K)] < r_{N+1}(K)\sum_{j=0}^{N} G^{(j)}(K) - [1 - G^{(N+1)}(K)]$$

$$= r_{N+1}(K)\sum_{j=0}^{N-1} G^{(j)}(K) - [1 - G^{(N)}(K)],$$

which agrees with that of (2.14). Thus, if $N > N^*$ in (2.14), then there exists a finite and unique T_F^* $(0 < T_F^* < \infty)$ which satisfies (2.23), and the resulting cost rate is

$$\frac{C_F(T_F^*, N)}{\lambda} = (c_K - c_T)\widetilde{Q}_1(T_F^*, N). \tag{2.24}$$

Conversely, if $N \le N^*$, then $T_F^* = \infty$.

(2) Optimum T_F^* and Z_F^*

Suppose that the unit is replaced preventively at time T $(0 < T \le \infty)$ or at damage Z $(0 < Z \le K)$, whichever occurs first. When $c_T = c_Z$, putting that $N \to \infty$ in (2.6), the expected cost rate is

$$C_F(T, Z) = \frac{c_T + (c_K - c_T)\sum_{j=0}^{\infty} F^{(j+1)}(T)\int_0^Z \overline{G}(K - x)\mathrm{d}G^{(j)}(x)}{\sum_{j=0}^{\infty} G^{(j)}(Z)\int_0^T [F^{(j)}(t) - F^{(j+1)}(t)]\mathrm{d}t}. \tag{2.25}$$

We find optimum (T_F^*, Z_F^*) to minimize $C_F(T, Z)$. Differentiating $C_F(T, Z)$ with respect to Z and setting it equal to zero,

$$Q_2(T, Z)\sum_{j=0}^{\infty} G^{(j)}(Z)\int_0^T [F^{(j)}(t) - F^{(j+1)}(t)]\mathrm{d}t$$
$$- \sum_{j=0}^{\infty} F^{(j+1)}(T)\int_0^Z \overline{G}(K - x)\mathrm{d}G^{(j)}(x) = \frac{c_T}{c_K - c_T}, \tag{2.26}$$

where

$$Q_2(T, Z) \equiv \frac{\overline{G}(K - Z)\sum_{j=0}^{\infty} g^{(j)}(Z)F^{(j+1)}(T)}{\sum_{j=0}^{\infty} g^{(j)}(Z)\int_0^T [F^{(j)}(t) - F^{(j+1)}(t)]\mathrm{d}t},$$

and $g^{(j)}(x) \equiv \mathrm{d}G^{(j)}(x)/\mathrm{d}x$ $(j = 1, 2, \cdots)$ and $g^{(0)}(x) \equiv 0$.

Differentiating $C_F(T, Z)$ with respect to T and setting it equal to zero,

$$Q_3(T, Z)\sum_{j=0}^{\infty} G^{(j)}(Z)\int_0^T [F^{(j)}(t) - F^{(j+1)}(t)]\mathrm{d}t$$
$$- \sum_{j=0}^{\infty} F^{(j+1)}(T)\int_0^Z \overline{G}(K - x)\mathrm{d}G^{(j)}(x) = \frac{c_T}{c_K - c_T}, \tag{2.27}$$

where

$$Q_3(T,Z) \equiv \frac{\sum_{j=0}^{\infty} f^{(j+1)}(T)\int_0^Z \overline{G}(K-x)\mathrm{d}G^{(j)}(x)}{\sum_{j=0}^{\infty}[F^{(j)}(T)-F^{(j+1)}(T)]G^{(j)}(Z)}.$$

Substituting (2.26) for (2.27),

$$Q_3(T,Z) = Q_2(T,Z). \tag{2.28}$$

Thus, if $Q_3(T,Z) < Q_2(T,Z)$ for any Z, then there dose not exist any finite T_F^* which satisfies (2.27), i.e., the optimum policy is ($T_F^* = \infty$, $Z_F^* = Z^*$), where Z^* is given in (2.17).

For example, when $F(t) = 1 - \mathrm{e}^{-\lambda t}$, it is proved that $Q_2(T,Z) = \lambda\overline{G}(K-Z)$, and (Problem 2.6)

$$\frac{\sum_{j=0}^{\infty}[(\lambda T)^j/j!]\int_0^Z \overline{G}(K-x)\mathrm{d}G^{(j)}(x)}{\sum_{j=0}^{\infty}[(\lambda T)^j/j!]G^{(j)}(Z)} < \overline{G}(K-Z)$$

for any Z. That is, (2.28) does not hold for $0 < T \le \infty$. This means that when optimum Z_F^* satisfies (2.26), the left-hand side of (2.27) is less than $c_T/(c_K - c_T)$, i.e., $C_F(T, Z_F^*)$ decreases with T, and hence, $T_F^* = \infty$. This concludes that if $Q_3(T,Z) < Q_2(T,Z)$ and $c_T \ge c_Z$, then the optimum policy which minimizes $C_F(T,Z)$ is ($T_F^* = \infty$, $Z_F^* = Z^*$).

In other words, when $c_T \ge c_Z$ is supposed for the replacement policies done at time T and at damage Z, its optimum policy degrades into the case in which only Z^* for $T = \infty$ can be found in (2.17).

Next, we obtain optimum T_F^* for given Z when $F(t) = 1 - \mathrm{e}^{-\lambda t}$ and $c_T = c_Z$. In this case, (2.27) is rewritten as

$$\widetilde{Q}_3(T,Z)\sum_{j=0}^{\infty} F^{(j+1)}(T)G^{(j)}(Z) - \sum_{j=0}^{\infty} F^{(j+1)}(T)\int_0^Z \overline{G}(K-x)\mathrm{d}G^{(j)}(x) = \frac{c_T}{c_K - c_T}, \tag{2.29}$$

where

$$\widetilde{Q}_3(T,Z) \equiv \frac{\sum_{j=0}^{\infty}[(\lambda T)^j/j!]\int_0^Z \overline{G}(K-x)\mathrm{d}G^{(j)}(x)}{\sum_{j=0}^{\infty}[(\lambda T)^j/j!]G^{(j)}(Z)}.$$

If $\widetilde{Q}_3(T,Z)$ increases strictly with T to $\overline{G}(K-Z)$ (Problem 2.6), then the left-hand side of (2.29) increases strictly with T from 0 to

$$\int_{K-Z}^{K}[1+M_G(K-x)]\mathrm{d}G(x),$$

which agrees with that of (2.17). Thus, if $Z > Z^*$ in (2.17), then there exists a finite and unique T_F^* $(0 < T_F^* < \infty)$ which satisfies (2.29). Conversely, if $Z \le Z^*$, then $T_F^* = \infty$.

(3) Optimum N_F^* and Z_F^*

Suppose that the unit is replaced preventively at shock N $(N = 1, 2, \cdots)$ or at damage Z $(0 < Z \le K)$, whichever occurs first. When $c_N = c_Z$, putting that $T \to \infty$ in (2.6), the expected cost rate is

$$C_F(N,Z) = \frac{c_N + (c_K - c_N)\sum_{j=0}^{N-1}\int_0^Z \overline{G}(K-x)\mathrm{d}G^{(j)}(x)}{\mu\sum_{j=0}^{N-1}G^{(j)}(Z)}. \tag{2.30}$$

We find optimum (N_F^*, Z_F^*) to minimize $C_F(N,Z)$. Differentiating $C_F(N,Z)$ with respect to Z and setting it equal to zero,

$$\sum_{j=0}^{N-1}\int_0^Z[\overline{G}(K-Z)-\overline{G}(K-x)]\mathrm{d}G^{(j)}(x) = \frac{c_N}{c_K - c_N}. \tag{2.31}$$

Forming the inequality $C_F(N+1,Z) - C_F(N,Z) \ge 0$,

$$\frac{\sum_{j=0}^{N-1}G^{(j)}(Z)}{G^{(N)}(Z)}\int_0^Z \overline{G}(K-x)\mathrm{d}G^{(N)}(x) - \sum_{j=0}^{N-1}\int_0^Z \overline{G}(K-x)\mathrm{d}G^{(j)}(x) \ge \frac{c_N}{c_K - c_N}. \tag{2.32}$$

Substituting (2.31) for (2.32),

$$\int_0^Z[\overline{G}(K-Z)-\overline{G}(K-x)]\mathrm{d}G^{(N)}(x) \le 0, \tag{2.33}$$

which dose not hold for any Z (Problem 2.7). Thus, there does not exist any finite N_F^* which satisfies (2.32).

When optimum Z_F^* satisfies (2.31), the left-hand side of (2.32) is less than $c_N/(c_K - c_N)$, which means $C_F(N, Z_F^*)$ decreases strictly with N, and hence, $N_F^* = \infty$. This concludes that if $c_N \ge c_Z$, then the optimum policy which minimizes $C_F(N,Z)$ is $(N_F^* = \infty, Z_F^* = Z^*)$, where Z^* is given in (2.17).

Next, we obtain optimum N_F^* for given Z when $G(x) = 1 - \mathrm{e}^{-\omega x}$ and $c_N = c_Z$. In this case, (2.32) is

$$e^{-\omega(K-Z)}\left[r_{N+1}(Z)\sum_{j=0}^{N-1}G^{(j)}(Z)+G^{(N)}(Z)-1\right]\geq\frac{c_Z}{c_K-c_Z}, \tag{2.34}$$

whose left-hand side increases strictly with N to (Problem 2.4)

$$\omega Z e^{-\omega(K-Z)},$$

which agrees with that of (2.17). Thus, if $Z > Z^*$ in (2.17), then there exists a finite and unique minimum N_F^* $(1 \leq N_F^* < \infty)$ which satisfies (2.34). Conversely, if $Z \leq Z^*$, then $N_F^* = \infty$.

The above optimum results show under the suitable conditions, e.g., different PR costs for $c_T \geq c_N \geq c_Z$, the policy with damage Z is the best among three ones, the next one is the policy with shock N, and the third one is the policy with time T. However, the order of working load for each preventive replacement is usually damage Z, shock N and time T, because we have to investigate the amount of total damage at each shock for damage Z, count the number of shocks for shock N, and record only the passed time for time T which is the easiest work load among three policies. Therefore, the case of $c_T < c_N < c_Z$ should be investigated to compare the above three policies, which will be discussed numerically in **(2)** of Sect. 2.1.3.

2.1.3 Poisson Shock Times

When shocks occur at a Poisson process with rate λ, and each amount of damage due to shocks is exponential with parameter ω, i.e., $F(t) = 1 - e^{-\lambda t}$ and $G(x) = 1 - e^{-\omega x}$,

$$F^{(j)}(t)=\sum_{i=j}^{\infty}\frac{(\lambda t)^i}{i!}e^{-\lambda t},\quad G^{(j)}(x)=\sum_{i=j}^{\infty}\frac{(\omega x)^i}{i!}e^{-\omega x}\quad (j=0,1,2,\cdots).$$

The expected cost rate in (2.6) is rewritten as

$$\frac{C_F(T,N,Z)}{\lambda}=\frac{\begin{array}{l}c_K-(c_K-c_T)\sum_{j=0}^{N-1}[(\lambda T)^j/j!]e^{-\lambda T}G^{(j)}(Z)\\ -(c_K-c_N)F^{(N)}(T)G^{(N)}(Z)\\ -(c_K-c_Z)(e^{-\omega Z}-e^{-\omega K})\sum_{j=0}^{N-1}F^{(j+1)}(T)[(\omega Z)^j/j!]\end{array}}{\sum_{j=0}^{N-1}F^{(j+1)}(T)G^{(j)}(Z)}. \tag{2.35}$$

We survey again optimum policies with one variable T^*, N^* and Z^* discussed in Sect. 2.1.1 and with two variables T_F^*, N_F^* and Z_F^* discussed in Sect. 2.1.2 for different PR costs c_T, c_N and c_Z.

(1) Optimum Policies with One Variable

We obtain optimum T^*, N^* and Z^* to minimize $C(T)$, $C(N)$ and $C(Z)$, respectively.

(a) Optimum T^*

From (2.9) and (2.35),

$$\frac{C(T)}{\lambda} = \frac{c_T + (c_K - c_T)\sum_{j=0}^{\infty} F^{(j+1)}(T)[(\omega K)^j/j!]e^{-\omega K}}{\sum_{j=0}^{\infty} F^{(j+1)}(T)G^{(j)}(K)}. \tag{2.36}$$

Differentiating $C(T)$ with respect to T and setting it equal to zero,

$$\widetilde{Q}_1(T)\sum_{j=0}^{\infty} F^{(j+1)}(T)G^{(j)}(K) - \sum_{j=0}^{\infty} F^{(j+1)}(T)\frac{(\omega K)^j}{j!}e^{-\omega K} = \frac{c_T}{c_K - c_T}, \tag{2.37}$$

where $\widetilde{Q}_1(T) \equiv \lim_{N\to\infty} \widetilde{Q}_1(T, N)$ is given in (2.23). The left-hand side of (2.37) increases strictly from 0 to ωK because $\widetilde{Q}_1(T)$ increase strictly with T from $e^{-\omega K}$ to 1. Therefore, if $\omega K > c_T/(c_K - c_T)$, then there exists a finite and unique T^* $(0 < T^* < \infty)$ which satisfies (2.37), and the resulting cost rate is

$$\frac{C(T^*)}{\lambda} = (c_K - c_T)\widetilde{Q}_1(T^*). \tag{2.38}$$

(b) Optimum N^*

From (2.13) and (2.35),

$$\frac{C(N)}{\lambda} = \frac{c_N + (c_K - c_N)[1 - G^{(N)}(K)]}{\sum_{j=0}^{N-1} G^{(j)}(K)} \quad (N = 1, 2, \cdots). \tag{2.39}$$

Forming the inequality $C(N+1) - C(N) \geq 0$,

$$r_{N+1}(K)\sum_{j=0}^{N-1} G^{(j)}(K) - 1 + G^{(N)}(K) \geq \frac{c_N}{c_K - c_N}, \tag{2.40}$$

where

$$r_{N+1}(K) \equiv \frac{(\omega K)^N/N!}{\sum_{j=N}^{\infty}[(\omega K)^j/j!]} \quad (N = 0, 1, 2, \cdots)$$

increases strictly with N from $e^{-\omega K}$ to 1 (Problem 2.4). Thus, the left-hand side of (2.40) increases strictly with N to ωK. Therefore, if $\omega K > c_N/(c_K - c_N)$, then there exists a finite and unique minimum N^* $(1 \leq N^* < \infty)$ which satisfies (2.40), and the resulting cost rate is

$$(c_K - c_N) r_{N^*}(K) < \frac{C(N^*)}{\lambda} \leq (c_K - c_N) r_{N^*+1}(K). \tag{2.41}$$

(c) Optimum Z^*

From (2.16) and (2.35),

$$\frac{C(Z)}{\lambda} = \frac{c_Z + (c_K - c_Z)\mathrm{e}^{-\omega(K-Z)}}{1 + \omega Z}. \tag{2.42}$$

Differentiating $C(Z)$ with respect to Z and setting it equal to zero,

$$\omega Z \mathrm{e}^{-\omega(K-Z)} = \frac{c_Z}{c_K - c_Z}, \tag{2.43}$$

whose left-hand side increases strictly with Z from 0 to ωK. Therefore, if $\omega K > c_Z/(c_K - c_Z)$, then there exists a finite and unique Z^* $(0 < Z^* < K)$ which satisfies (2.43), and the resulting cost rate is

$$\frac{C(Z^*)}{\lambda} = (c_K - c_Z)\mathrm{e}^{-\omega(K-Z^*)}. \tag{2.44}$$

It is of great interest that if $\omega K > c_i/(c_K - c_i)$ $(i = T, N, Z)$, then finite T^*, N^* and Z^* always exist uniquely.

(2) Optimum Policies with Two Variables

We obtain optimum (T_F^*, N_F^*), (T_F^*, Z_F^*) and (N_F^*, Z_F^*) to minimize $C_F(T, N)$, $C_F(T, Z)$ and $C_F(N, Z)$ for $c_T < c_N < c_Z$, respectively.

(a) Optimum T_F^* and N_F^*

Putting that $Z = K$ in (2.35),

$$\frac{C_F(T, N)}{\lambda} = \frac{\begin{array}{l} c_T + (c_K - c_T)\sum_{j=0}^{N-1} F^{(j+1)}(T)[(\omega K)^j/j!]\mathrm{e}^{-\omega K} \\ +(c_N - c_T)F^{(N)}(T)G^{(N)}(K) \end{array}}{\sum_{j=0}^{N-1} F^{(j+1)}(T)G^{(j)}(K)}. \tag{2.45}$$

In particular, when $N = 1$, $T_F^* = \infty$, and

$$\frac{C_F(\infty, 1)}{\lambda} = c_N + (c_K - c_N)\mathrm{e}^{-\omega K}.$$

We find optimum T_F^* for $N \geq 2$ to minimize $C_F(T, N)$. Differentiating $C_F(T, N)$ with respect to T and setting it equal to zero,

$$(c_K - c_T)\left[\widetilde{Q}_1(T,N)\sum_{j=0}^{N-1}F^{(j+1)}(T)G^{(j)}(K) - \sum_{j=0}^{N-1}F^{(j+1)}(T)\frac{(\omega K)^j}{j!}e^{-\omega K}\right]$$
$$+ (c_N - c_T)\left[Q_4(T,N)\sum_{j=0}^{N-1}F^{(j+1)}(T)G^{(j)}(K) - F^{(N)}(T)G^{(N)}(K)\right] = c_T, \tag{2.46}$$

where $\widetilde{Q}_1(T, N)$ is given in (2.23) and increases strictly with T to $r_N(K)$ in (2.40), and

$$Q_4(T,N) \equiv \frac{[(\lambda T)^{N-1}/(N-1)!]G^{(N)}(K)}{\sum_{j=0}^{N-1}[(\lambda T)^j/j!]G^{(j)}(K)},$$

which increases strictly with T from 0 to $G^{(N)}(K)/G^{(N-1)}(K)$ (Problem 2.8). Thus, the left-hand side of (2.46) increases strictly with T from 0 to

$$(c_K - c_T)\left[r_N(K)\sum_{j=0}^{N-1}G^{(j)}(K) + G^{(N)}(K) - 1\right]$$
$$+ (c_N - c_T)\left\{[1 - r_N(K)]\sum_{j=0}^{N-1}G^{(j)}(K) - G^{(N)}(K)\right\}.$$

Therefore, if

$$(c_K - c_N)\left[r_N(K)\sum_{j=0}^{N-1}G^{(j)}(K) + G^{(N)}(K)\right]$$
$$+ (c_N - c_T)\sum_{j=0}^{N-1}G^{(j)}(K) > c_K, \tag{2.47}$$

then there exists a finite T_F^* $(0 < T_F^* < \infty)$ which satisfies (2.46).

In addition, letting $L(N)$ be the left-hand side of (2.47),

$$L(N+1) - L(N) > (c_K - c_N)[r_{N+1}(K) - r_N(K)]\sum_{j=0}^{N-1}G^{(j)}(K) > 0,$$
$$\lim_{N\to\infty} L(N) = (c_K - c_T)(1 + \omega K).$$

Therefore, if $\omega K \le c_T/(c_K - c_T)$, then $T_F^* = \infty$. If $\omega K > c_T/(c_K - c_T)$, then a finite T_F^* might exist. On the other hand, if

$$L(2) = (c_K - c_T)G(K) - (c_K - c_N)\frac{G^{(2)}(K)}{G(K)} > c_T,$$

then a finite T_F^* $(0 < T_F^* < \infty)$ always exists for $N \geq 2$, and the resulting cost rate is

$$\frac{C_F(T_F^*, N)}{\lambda} = (c_K - c_T)\widetilde{Q}_1(T_F^*, N) + (c_N - c_T)Q_4(T_F^*, N). \tag{2.48}$$

Because $L(2) > (c_N - c_T)G(K)$, if $G(K) \geq c_T/(c_N - c_T)$, then a finite T_F^* exists.

In general, it is very difficult to derive optimum N_F^* for given T analytically (Problem 2.9). So that, we compute numerically T_F^* which satisfies (2.46) and the resulting cost rate $C_F(T_F^*, N)$ in (2.48) for N. Comparisons for different N will be shown in numerical examples.

(b) Optimum T_F^* and Z_F^*

Putting that $N \to \infty$ in (2.35),

$$\frac{C_F(T, Z)}{\lambda} = \frac{\begin{array}{c} c_T + (c_K - c_Z)\sum_{j=0}^{\infty} F^{(j+1)}(T)[(\omega Z)^j/j!]e^{-\omega K} \\ +(c_Z - c_T)\sum_{j=0}^{\infty} F^{(j+1)}(T)[(\omega Z)^j/j!]e^{-\omega Z} \end{array}}{\sum_{j=0}^{\infty} F^{(j+1)}(T)G^{(j)}(Z)}. \tag{2.49}$$

We find optimum T_F^* and Z_F^* to minimize $C_F(T, Z)$. Differentiating $C_F(T, Z)$ with respect to T and setting it equal to zero,

$$[(c_K - c_Z)e^{-\omega(K-Z)} + (c_Z - c_T)]\left[Q_5(T, Z)\sum_{j=0}^{\infty} F^{(j+1)}(T)G^{(j)}(Z) - \sum_{j=0}^{\infty} F^{(j+1)}(T)\frac{(\omega Z)^j}{j!}e^{-\omega Z}\right] = c_T, \tag{2.50}$$

where

$$Q_5(T, Z) \equiv \frac{\sum_{j=0}^{\infty}[(\lambda T)^j/j!][(\omega Z)^j/j!]e^{-\omega Z}}{\sum_{j=0}^{\infty}[(\lambda T)^j/j!]G^{(j)}(Z)}.$$

Note that $Q_5(T, K) = \widetilde{Q}_1(T)$ in (2.37), and $Q_5(T, Z)$ increases strictly with T from $e^{-\omega Z}$ to 1. Thus, the left-hand side of (2.50) increases strictly with T from 0 to

$$\omega Z[(c_K - c_Z)e^{-\omega(K-Z)} + (c_Z - c_T)].$$

Therefore, if

$$\omega Z[(c_K - c_Z)\mathrm{e}^{-\omega(K-Z)} + (c_Z - c_T)] > c_T, \tag{2.51}$$

then there exists a finite and unique T_F^* $(0 < T_F^* < \infty)$ which satisfies (2.50) for given Z, and the resulting cost rate is

$$\frac{C_F(T_F^*, Z)}{\lambda} = [(c_K - c_Z)\mathrm{e}^{-\omega(K-Z)} + (c_Z - c_T)]Q_5(T_F^*, Z). \tag{2.52}$$

Clearly, if $\omega K > c_T/(c_K - c_T)$, then a finite T_F^* might exist.

Next, differentiating $C_F(T, Z)$ with respect to Z and setting it equal to zero,

$$[(c_K - c_Z)\mathrm{e}^{-\omega(K-Z)} + (c_Z - c_T)]\sum_{j=1}^{\infty} F^{(j)}(T)G^{(j)}(Z)$$

$$= c_T + (c_Z - c_T)Q_6(T, Z)\sum_{j=0}^{\infty} F^{(j+1)}(T)G^{(j)}(Z), \tag{2.53}$$

where

$$Q_6(T, Z) \equiv \frac{\sum_{j=0}^{\infty} F^{(j+1)}(T)[(\omega Z)^j/j!]}{\sum_{j=0}^{\infty} F^{(j+2)}(T)[(\omega Z)^j/j!]} \geq 1,$$

which increases strictly with Z from $F(T)/F^{(2)}(T)$ to $Q_6(T, K)$ (Problem 2.10). Thus, the left-hand side of (2.53) increases strictly with Z from 0 to

$$(c_K - c_T)\sum_{j=1}^{\infty} F^{(j)}(T)G^{(j)}(K),$$

and its right-hand side also increase strictly with Z from

$$c_T + (c_Z - c_T)\frac{F(T)^2}{F^{(2)}(T)}$$

to

$$c_T + (c_Z - c_T)Q_6(T, K)\sum_{j=0}^{\infty} F^{(j+1)}(T)G^{(j)}(K).$$

Therefore, if

$$(c_K - c_T)\sum_{j=1}^{\infty} F^{(j)}(T)G^{(j)}(K)$$
$$> c_T + (c_Z - c_T)Q_6(T, K)\sum_{j=0}^{\infty} F^{(j+1)}(T)G^{(j)}(K),$$

then a finite and unique Z_F^* $(0 < Z_F^* < K)$ to satisfy (2.53) for given T exists. Clearly, if $\omega K > c_Z/(c_K - c_Z)$, then a finite Z_F^* might exist, because $Q_6(T, K)$ decreases with T to 1.

(c) Optimum N_F^* and Z_F^*

Putting that $T \to \infty$ in (2.35),

$$\frac{C_F(N, Z)}{\lambda} = \frac{c_Z - (c_Z - c_N)G^{(N)}(Z) + (c_K - c_Z)\mathrm{e}^{-\omega(K-Z)}[1 - G^{(N)}(Z)]}{\sum_{j=0}^{N-1} G^{(j)}(Z)}. \tag{2.54}$$

In particular, when $N = 1$, $Z_F^* = K$, and

$$\frac{C_F(1, K)}{\lambda} = c_N + (c_K - c_N)\mathrm{e}^{-\omega K}.$$

We find optimum N_F^* and Z_F^* to minimize $C_F(N, Z)$. Differentiating $C_F(N, Z)$ with respect to Z and setting it equal to zero for $N \geq 2$,

$$(c_K - c_Z)\mathrm{e}^{-\omega(K-Z)}\sum_{j=1}^{N} G^{(j)}(Z)$$
$$= c_Z + (c_Z - c_N)\left[\widetilde{r}_N(Z)\sum_{j=0}^{N-1} G^{(j)}(Z) - G^{(N)}(Z)\right], \tag{2.55}$$

where

$$\widetilde{r}_{N+1}(x) \equiv \frac{(\omega x)^N/N!}{\sum_{j=0}^{N-1}[(\omega x)^j/j!]} \quad (N = 1, 2, \cdots).$$

The left-hand side of (2.55) increases strictly with Z from 0 to

$$(c_K - c_Z)\sum_{j=1}^{N} G^{(j)}(K),$$

and its right-hand side increases strictly with Z from c_Z to

$$c_Z + (c_Z - c_N)\left[\widetilde{r}_N(K)\sum_{j=0}^{N-1} G^{(j)}(K) - G^{(N)}(K)\right],$$

because $\widetilde{r}_N(Z)$ increases strictly with Z from 0 and decreases strictly with N to 0 (Problem 2.4). Thus, if

$$(c_K - c_Z)\sum_{j=1}^{N} G^{(j)}(K) > c_Z + (c_Z - c_N)\left[\widetilde{r}_N(K)\sum_{j=0}^{N-1} G^{(j)}(K) - G^{(N)}(K)\right],$$

then there exists a unique Z_F^* $(0 < Z_F^* < K)$ to satisfy (2.55), and the expected cost rate is

$$\frac{C_F(N, Z_F^*)}{\lambda} = (c_K - c_Z)\mathrm{e}^{-\omega(K-Z_F^*)} - (c_Z - c_N)\widetilde{r}_N(Z_F^*). \tag{2.56}$$

Clearly, the left-hand side of (2.55) goes to $(c_K - c_Z)\omega K$ as $Z \to K$ and $N \to \infty$, and its right-hand side goes to c_Z as $N \to \infty$. Thus, if $\omega K > c_Z/(c_K - c_Z)$, then a finite Z_F^* $(0 < Z_F^* < K)$ might exist.

Forming the inequality $C_F(N+1, Z) - C_F(N, Z) \geq 0$,

$$[(c_K - c_Z)\mathrm{e}^{-\omega(K-Z)} + (c_Z - c_N)]\left[r_{N+1}(Z)\sum_{j=0}^{N-1} G^{(j)}(Z) + G^{(N)}(Z) - 1\right] \geq c_N, \tag{2.57}$$

whose left-hand side increases strictly with N. Thus, if

$$\omega Z[(c_K - c_Z)\mathrm{e}^{-\omega(K-Z)} + (c_Z - c_N)] > c_N,$$

then there exists a finite and unique minimum N_F^* $(1 \leq N_F^* < \infty)$ to satisfy (2.57) for given Z. Clearly, if $\omega K > c_N/(c_K - c_N)$, then a finite N_F^* $(1 \leq N_F^* < \infty)$ might exist.

(3) Numerical Examples

When $F(t) = 1 - \mathrm{e}^{-\lambda t}$ and $G(x) = 1 - \mathrm{e}^{-\omega x}$, i.e., $F^{(j)}(t) = \sum_{i=j}^{\infty}[(\lambda t)^i/i!]\mathrm{e}^{-\lambda t}$ and $G^{(j)}(x) = \sum_{i=j}^{\infty}[(\omega x)^i/i!]\mathrm{e}^{-\omega x}$ $(j = 0, 1, 2, \cdots)$, we compute optimum λT^*, N^*, ωZ^* and $(\lambda T_F^*, N_F^*)$, $(\lambda T_F^*, \omega Z_F^*)$, $(N_F^*, \omega Z_F^*)$, which minimize the expected cost rates $C(T)$ in (2.36), $C(N)$ in (2.39), $C(Z)$ in (2.42), and $C_F(T, N)$ in (2.45), $C_F(T, Z)$ in (2.49), $C_F(N, Z)$ in (2.54), respectively.

From (2.37), optimum T^* satisfies

$$\frac{\sum_{j=0}^{\infty}[(\lambda T)^j/j!][(\omega K)^j/j!]}{\sum_{j=0}^{\infty}[(\lambda T)^j/j!]\sum_{i=j}^{\infty}[(\omega K)^i/i!]}\sum_{j=0}^{\infty}\left[\sum_{i=j}^{\infty}\frac{(\lambda T)^{i+1}}{(i+1)!}\mathrm{e}^{-\lambda T}\sum_{i=j}^{\infty}\frac{(\omega K)^i}{i!}\mathrm{e}^{-\omega K}\right]$$
$$-\sum_{j=0}^{\infty}\left[\sum_{i=j}^{\infty}\frac{(\lambda T)^{i+1}}{(i+1)!}\mathrm{e}^{-\lambda T}\right]\frac{(\omega K)^j}{j!}\mathrm{e}^{-\omega K}=\frac{c_T}{c_K-c_T},$$

and from (2.38), the resulting cost rate is

$$\frac{C(T^*)}{\lambda}=(c_K-c_T)\frac{\sum_{j=0}^{\infty}[(\lambda T^*)^j/j!][(\omega K)^j/j!]}{\sum_{j=0}^{\infty}[(\lambda T^*)^j/j!]\sum_{i=j}^{\infty}[(\omega K)^i/i!]}.$$

From (2.40), optimum N^* satisfies

$$\frac{(\omega K)^N/N!}{\sum_{j=N}^{\infty}[(\omega K)^j/j!]}\sum_{j=0}^{N-1}\sum_{i=j}^{\infty}\frac{(\omega K)^i}{i!}\mathrm{e}^{-\omega K}-\sum_{j=0}^{N-1}\frac{(\omega K)^j}{j!}\mathrm{e}^{-\omega K}\geq\frac{c_N}{c_K-c_N},$$

and from (2.39), the resulting cost rate is

$$\frac{C(N^*)}{\lambda}=\frac{c_K-(c_K-c_N)\sum_{j=N^*}^{\infty}[(\omega K)^j/j!]\mathrm{e}^{-\omega K}}{\sum_{j=0}^{N^*-1}\sum_{i=j}^{\infty}[(\omega K)^i/i!]\mathrm{e}^{-\omega K}}.$$

Optimum ωZ^* and its resulting cost rate $C(Z^*)/\lambda$ satisfy (2.43) and (2.44), respectively.

Tables 2.1, 2.2 and 2.3 presents optimum λT^*, N^* and ωZ^*, and their cost rates $C(T^*)/(\lambda c_T)$, $C(N^*)/(\lambda c_N)$ and $C(Z^*)/(\lambda c_Z)$ for ωK and c_K/c_T, c_K/c_N and c_K/c_Z. All of optimum values λT^*, N^* and ωZ^* increase with ωK and decrease with c_K/c_i $(i=T,N,Z)$, and their cost rates decrease with ωK and increase with c_K/c_i. When $\omega K=5.0$, 10.0, optimum λT^*, N^* and ωZ^* are almost the same, and when $\omega K=20.0$, $\lambda T^*<N^*<\omega Z^*$. It can be understood that when a failure level K becomes smaller or cost c_K for failure becomes larger, early replacement should be decided and it's replacement cost rate becomes higher.

When the same values c_K/c_i $(i=T,N,Z)$ are assigned, it can be found from Tables 2.1, 2.2 and 2.3 that $C(T^*)>C(N^*)>C(Z^*)$. As is known, unit failure is caused by total cumulative damage, that is, if more precise damage or shock could be observed and monitored, more effective replacement actions can be done. However, monitoring costs should not be neglected in practice, it is more reasonable to suppose $c_T<c_N<c_Z$ when replacement polices are valued, which will be shown in Tables 2.4, 2.5 and 2.6.

Table 2.1 Optimum λT^* and its cost rate $C(T^*)/(\lambda c_T)$

c_K/c_T	$\omega K = 5.0$		$\omega K = 10.0$		$\omega K = 20.0$	
	λT^*	$C(T^*)/(\lambda c_T)$	λT^*	$C(T^*)/(\lambda c_T)$	λT^*	$C(T^*)/(\lambda c_T)$
5	3.328	0.617	5.750	0.260	11.835	0.106
10	2.221	0.910	4.378	0.334	9.971	0.124
15	1.804	1.130	3.805	0.382	9.137	0.134
20	1.567	1.317	3.460	0.420	8.617	0.142
30	1.292	1.637	3.040	0.479	7.962	0.153
50	1.019	2.162	2.594	0.564	7.235	0.168

Table 2.2 Optimum N^* and its cost rate $C(N^*)/(\lambda c_N)$

c_K/c_N	$\omega K = 5.0$		$\omega K = 10.0$		$\omega K = 20.0$	
	N^*	$C(N^*)/(\lambda c_N)$	N^*	$C(N^*)/(\lambda c_N)$	N^*	$C(N^*)/(\lambda c_N)$
5	3	0.508	6	0.213	13	0.089
10	2	0.684	5	0.253	12	0.100
15	2	0.786	5	0.283	11	0.105
20	2	0.887	4	0.299	10	0.110
30	2	1.090	4	0.325	10	0.115
50	1	1.330	4	0.377	9	0.122

Table 2.3 Optimum ωZ^* and its cost rate $C(Z^*)/(\lambda c_Z)$

c_K/c_Z	$\omega K = 5.0$		$\omega K = 10.0$		$\omega K = 20.0$	
	ωZ^*	$C(Z^*)/(\lambda c_Z)$	ωZ^*	$C(Z^*)/(\lambda c_Z)$	ωZ^*	$C(Z^*)/(\lambda c_Z)$
5	2.642	0.378	6.710	0.149	15.850	0.063
10	2.073	0.482	6.009	0.166	15.089	0.066
15	1.783	0.561	5.632	0.178	14.675	0.068
20	1.591	0.628	5.374	0.186	14.389	0.069
30	1.340	0.746	5.019	0.199	13.994	0.071
50	1.055	0.948	4.585	0.218	13.505	0.074

Tables 2.4, 2.5 and 2.6 presents optimum $(\lambda T_F^*, N_F^*)$, $(\lambda T_F^*, \omega Z_F^*)$ and $(N_F^*, \omega Z_F^*)$, and their cost rates $C_F(T_F^*, N_F^*)/\lambda c_Z$, $C_F(T_F^*, Z_F^*)/\lambda c_Z$ and $C_F(N_F^*, \omega Z_F^*)/\lambda c_Z$ for ωK and c_K/c_Z when $\omega K = 10$ and $c_T : c_N : c_Z = 1 : 2 : 3$.

For optimum $(\lambda T_F^*, N_F^*)$ and $C_F(T_F^*, N_F^*)/\lambda c_Z$:

1. When $N = 1$, $T_F^* = \infty$,

$$\frac{C_F(\infty, 1)}{\lambda c_Z} = \frac{c_N}{c_Z} + \left(\frac{c_K}{c_Z} - \frac{c_N}{c_Z}\right) \mathrm{e}^{-\omega K}.$$

Table 2.4 Optimum $(\lambda T_F^*, N_F^*)$ and its cost rate $C_F(T_F^*, N_F^*)/(\lambda c_Z)$ when $c_T : c_N : c_Z = 1 : 2 : 3$

c_K/c_Z	$\omega K = 10.0$			$\omega K = 20.0$		
	λT_F^*	N_F^*	$C_F(T_F^*, N_F^*)/(\lambda c_Z)$	λT_F^*	N_F^*	$C_F(T_F^*, N_F^*)/(\lambda c_Z)$
5	3.887	8	0.127	9.199	17	0.045
10	3.331	6	0.157	8.204	14	0.051
15	2.933	6	0.177	7.722	13	0.054
20	2.912	5	0.191	7.337	13	0.057
30	2.553	5	0.216	6.959	12	0.061
50	2.527	4	0.247	6.523	11	0.065

Table 2.5 Optimum $(\lambda T_F^*, \omega Z_F^*)$ and its cost rate $C_F(T_F^*, Z_F^*)/(\lambda c_Z)$ when $c_T : c_N : c_Z = 1 : 2 : 3$

c_K/c_Z	$\omega K = 10.0$			$\omega K = 20.0$		
	λT_F^*	ωZ_F^*	$C_F(T_F^*, Z_F^*)/(\lambda c_Z)$	λT_F^*	ωZ_F^*	$C_F(T_F^*, Z_F^*)/(\lambda c_Z)$
5	4.932	7.878	0.111	11.008	17.471	0.040
10	4.502	7.136	0.127	10.474	16.692	0.042
15	4.274	6.735	0.137	10.185	16.268	0.043
20	4.119	6.459	0.145	9.986	15.976	0.044
30	3.909	6.079	0.158	9.712	15.571	0.046
50	3.656	5.611	0.176	9.374	15.070	0.048

Table 2.6 Optimum $(N_F^*, \omega Z_F^*)$ and its cost rate $C_F(N_F^*, Z_F^*)/(\lambda c_Z)$ when $c_T : c_N : c_Z = 1 : 2 : 3$

c_K/c_Z	$\omega K = 10.0$			$\omega K = 20.0$		
	N_F^*	ωZ_F^*	$C_F(N_F^*, Z_F^*)/(\lambda c_Z)$	N_F^*	ωZ_F^*	$C_F(N_F^*, Z_F^*)/(\lambda c_Z)$
5	7	7.435	0.140	15	16.920	0.056
10	7	6.522	0.158	14	16.195	0.059
15	6	6.309	0.168	14	15.706	0.061
20	6	5.962	0.177	14	15.367	0.063
30	6	5.486	0.190	13	15.078	0.064
50	5	5.217	0.209	13	14.496	0.067

2. For $N = 2, 3, 4, \cdots$, compute λT_N to satisfy (2.46).
3. Compute and compare $C_F(\infty, 1)$ and $C_F(T_N, N)$ $(N = 2, 3, \cdots)$ to determine optimum $T_F^* = T_N$ and $N_F^* = N$, where

$$\frac{C_F(T_F^*, N_F^*)}{\lambda c_Z} = \left(\frac{c_K}{c_Z} - \frac{c_T}{c_Z}\right) \widetilde{Q}_1(T_F^*, N_F^*) + \left(\frac{c_N}{c_Z} - \frac{c_T}{c_Z}\right) Q_4(T_F^*, N_F^*).$$

For optimum $(\lambda T_F^*, \omega Z_F^*)$ and $C_F(T_F^*, Z_F^*)/\lambda c_Z$:

1. Compute T_Z for $0 < Z \leq K$ to satisfy (2.50), and compute Z_T for $0 < T \leq \infty$ to satisfy (2.53).
2. Let $Z = K$ and compute T_{Z1}, compute Z_{T1} for T_{Z1}, compute T_{Z2} for $Z_{T1}, \cdots$, to determine $\lambda T_F^* = T_{Zi}$ and $\omega Z_F^* = Z_{Ti}$ $(i = 1, 2, \cdots)$, where

$$\frac{C_F(T_F^*, Z_F^*)}{\lambda c_Z} = \left[\left(\frac{c_K}{c_Z} - 1\right) \mathrm{e}^{-\omega(K - Z_F^*)} + \left(1 - \frac{c_T}{c_Z}\right)\right] Q_5(T_F^*, Z_F^*).$$

For optimum $(N_F^*, \omega Z_F^*)$ and $C_F(N_F^*, Z_F^*)/\lambda c_Z$:

1. When $N = 1$, $Z_F^* = K$, and

$$\frac{C_F(1, K)}{\lambda c_Z} = \frac{c_N}{c_Z} + \left(\frac{c_K}{c_Z} - \frac{c_N}{c_Z}\right) \mathrm{e}^{-\omega K}.$$

2. For $N = 2, 3, \cdots$, compute Z_N to satisfy (2.55).
3. Compute and compare $C_F(1, K)$ and $C_F(N, Z_N)$ $(N = 2, 3, \cdots)$ to determine optimum $Z_F^* = Z_N$ and $N_F^* = N$, where

$$\frac{C_F(N_F^*, Z_N)}{\lambda c_Z} = \left(\frac{c_K}{c_Z} - 1\right) \mathrm{e}^{-\omega(K - Z_F^*)} - \left(1 - \frac{c_N}{c_Z}\right) \frac{(\omega Z_F^*)^{N_F^* - 1}/(N_F^* - 1)!}{\sum_{j=0}^{N_F^* - 2} [(\omega Z_F^*)^j / j!]}.$$

These tables indicate that when $c_T : c_N : c_Z = 1 : 2 : 3$, $\lambda T_F^* < N_F^*$ and $\lambda T_F^* < \omega Z_F^*$, but N_F^* and ωZ_F^* are almost the same. As the scale of c_T, c_N and c_Z is roughly given, we cannot compare optimum policies $(\lambda T_F^*, N_F^*)$, $(\lambda T_F^*, \omega Z_F^*)$ and $(N_F^*, \omega Z_F^*)$ exactly. However, we may conclude: (i) Optimum policy $(\lambda T_F^*, N_F^*)$ saves more cost rate than $(\lambda T_F^*, \omega Z_F^*)$ does, though c_Z is greater than c_N. (ii) Optimum policy $(\lambda T_F^*, \omega Z_F^*)$ is better than $(N_F^*, \omega Z_F^*)$, which may be due to $c_T < c_N$. (iii) Either $(\lambda T_F^*, N_F^*)$ or $(N_F^*, \omega Z_F^*)$ would be better, though c_T is much less than c_Z. The reasons of (i)–(iii) should be explored further, and different scales of costs c_T, c_N and c_Z will be discussed in Chap. 4.

2.2 Random Failure Levels

Units operating in different degradation environments exhibit varying levels of toughness, in which case, the above defined failure threshold K could not be determined as constant values; however, its probability distribution can be estimated statistically from historical failure data. In this section, a level of failure threshold K is supposed to be a random variable and has a general distribution $L(x) \equiv \Pr\{K \leq x\}$ with finite mean $1/\theta$ and $L(0) = 0$ [2].

When the unit is replaced preventively at time T $(0 < T \leq \infty)$ or at shock N $(N = 1, 2, \cdots)$, whichever occurs first, the expected cost rate is [2]

$$C_R(T, N; L) = \frac{\begin{array}{c} c_K - (c_K - c_N)F^{(N)}(T) \int_0^\infty G^{(N)}(x)\mathrm{d}L(x) \\ -(c_K - c_T)\sum_{j=0}^{N-1}[F^{(j)}(T) - F^{(j+1)}(T)] \int_0^\infty G^{(j)}(x)\mathrm{d}L(x) \end{array}}{\sum_{j=0}^{N-1} \int_0^T [F^{(j)}(t) - F^{(j+1)}(t)]\mathrm{d}t \int_0^\infty G^{(j)}(x)\mathrm{d}L(x)}. \tag{2.58}$$

When $L(x) = 1 - \mathrm{e}^{-\theta x}$ and the unit is replaced preventively only at time T, the expected cost rate is

$$\begin{aligned} C_R(T; \theta) &\equiv \lim_{N\to\infty} C_R(T, N; L) \\ &= \frac{c_K - (c_K - c_T)\sum_{j=0}^{\infty}[G^*(\theta)]^j[F^{(j)}(T) - F^{(j+1)}(T)]}{\sum_{j=0}^{\infty}[G^*(\theta)]^j \int_0^T [F^{(j)}(t) - F^{(j+1)}(t)]\mathrm{d}t}, \end{aligned} \tag{2.59}$$

where $G^*(\theta) \equiv \int_0^\infty \mathrm{e}^{-\theta x}\mathrm{d}G(x)$, which represents the Laplace-Stieltjes transform of $G(x)$.

Furthermore, when shocks occur at a nonhomogeneous Poisson process with a cumulative hazard rate $H(t) \equiv \int_0^t h(u)\mathrm{d}u$, i.e., $h(t) \equiv \mathrm{d}H(t)/\mathrm{d}t$ and $F^{(j)}(t) = \sum_{i=j}^{\infty}[H(t)^i/i!]\mathrm{e}^{-H(t)}$ $(j = 0, 1, 2, \cdots)$ from (1.6), the expected cost rate in (2.59) is

$$C_R(T; \theta) = \frac{c_K - (c_K - c_T)\exp\{-[1 - G^*(\theta)]H(T)\}}{\int_0^T \exp\{-[1 - G^*(\theta)]H(t)\}\mathrm{d}t}. \tag{2.60}$$

If $h(t)$ increases strictly with t to $h(\infty)$, and

$$[1 - G^*(\theta)]h(\infty) \int_0^\infty \exp\left\{-[1 - G^*(\theta)]H(t)\right\}\mathrm{d}t > \frac{c_K}{c_K - c_T},$$

then there exists a finite and unique T_R^* $(0 < T_R^* < \infty)$ which satisfies (Problem 2.11)

$$\begin{aligned} &[1 - G^*(\theta)]h(T) \int_0^T \exp\left\{-[1 - G^*(\theta)]H(t)\right\}\mathrm{d}t \\ &+ \exp\left\{-[1 - G^*(\theta)]H(T)\right\} = \frac{c_K}{c_K - c_T}, \end{aligned} \tag{2.61}$$

and the resulting cost rate is

$$C_R(T_R^*; \theta) = (c_K - c_T)[1 - G^*(\theta)]h(T_R^*). \tag{2.62}$$

When the unit is replaced preventively only at shock N, the expected cost rate is

$$C_R(N; L) \equiv \lim_{T\to\infty} C_R(T, N; L)$$
$$= \frac{c_K - (c_K - c_N)\int_0^\infty G^{(N)}(x)\mathrm{d}L(x)}{\mu \sum_{j=0}^{N-1} \int_0^\infty G^{(j)}(x)\mathrm{d}L(x)} \quad (N = 1, 2, \cdots). \tag{2.63}$$

Forming the inequality $C_R(N + 1; L) - C_R(N; L) \geq 0$,

$$Q_7(N) \sum_{j=0}^{N-1} \int_0^\infty G^{(j)}(x)\mathrm{d}L(x) + \int_0^\infty G^{(N)}(x)\mathrm{d}L(x) \geq \frac{c_K}{c_K - c_N}, \tag{2.64}$$

where

$$Q_7(N) \equiv \frac{\int_0^\infty [G^{(N)}(x) - G^{(N+1)}(x)]\mathrm{d}L(x)}{\int_0^\infty G^{(N)}(x)\mathrm{d}L(x)} \leq 1.$$

If $Q_7(N)$ increases strictly with N to 1, then the left-hand side of (2.64) increases strictly with N to $\int_0^\infty [1 + M_G(x)]\mathrm{d}L(x)$, where $M_G(x)$ is given in (2.7). Therefore, if

$$\int_0^\infty M_G(x)\mathrm{d}L(x) > \frac{c_N}{c_K - c_N},$$

then there exists a finite and unique minimum N_R^* $(1 \leq N_R^* < \infty)$ which satisfies (2.64), and the resulting cost rate is (Problem 2.12)

$$(c_K - c_N)Q_7(N_R^* - 1) < \mu C_R(N_R^*; L) \leq (c_K - c_N)Q_7(N_R^*). \tag{2.65}$$

In particular, when $L(x) = 1 - \mathrm{e}^{-\theta x}$, $Q_7(N) = 1 - G^*(\theta)$, and $N_R^* = \infty$. Table 2.7 presents optimum T_R^* and its cost rate $C_R(T_R^*; \theta)/c_T$ for $G^*(\theta)$ and c_K/c_T when $H(t) = t^{2.0}$. Obviously, optimum T_R^* increases with $G^*(\theta)$ and decreases with c_K/c_T, and its cost rate decreases with $G^*(\theta)$ and increases with c_K/c_T.

Table 2.7 Optimum T_R^* and its cost rate $C_R(T_R^*; \theta)/c_T$ when $H(t) = t^2$

c_K/c_T	$G^*(\theta) = 0.1$		$G^*(\theta) = 0.5$		$G^*(\theta) = 0.9$	
	T_R^*	$C_R(T_R^*; \theta)/c_T$	T_R^*	$C_R(T_R^*; \theta)/c_T$	T_R^*	$C_R(T_R^*; \theta)/c_T$
5	0.538	3.875	0.772	2.888	1.615	1.292
10	0.355	5.746	0.476	4.283	1.064	1.915
15	0.283	7.142	0.380	5.322	0.850	2.380
20	0.243	8.305	0.326	6.191	0.729	2.769
30	0.196	10.247	0.263	7.640	0.589	3.416
50	0.151	13.303	0.202	9.918	0.452	4.434

2.3 Double Failure Modes

In crack growth models [27, 28] for aircrafts, it has been well-known that the unit would be failure when the size of one crack exceeds a threshold level, or when the total sizes of all cracks attain to a certain level, e.g., multi-site damage, which is defined as the simultaneous occurrence of many tiny fatigue cracks at multiple locations in the same structural element, and has become recently a major issue of aging aircrafts since the Aloha Airlines affair in 1988 [29].

This section takes up one extended model where the unit fails when the total damage has exceeded a failure threshold K $(0 < K < \infty)$ in Sect. 2.1, and also fails when the total number of shocks reaches to a certain value of N $(N = 1, 2, \cdots)$, whichever occurs first, and corrective replacement is done immediately when the failure is detected. As preventive replacement policies, the unit is replaced preventively at time T $(0 < T \leq \infty)$, or at damage Z $(0 < Z \leq K)$, whichever occurs first.

The probability that the unit is replaced at time T, at shock N, at damage Z, and at failure K are respectively given in (2.1)–(2.4), and the mean time to replacement is given in (2.5). Then, the expected replacement cost rate is

$$C_N(T, Z) = \frac{\begin{array}{c} c_K - (c_K - c_T)\sum_{j=0}^{N-1}[F^{(j)}(T) - F^{(j+1)}(T)]G^{(j)}(Z) \\ -(c_K - c_Z)\sum_{j=0}^{N-1} F^{(j+1)}(T)\int_0^Z [G(K - x) - G(Z - x)]\mathrm{d}G^{(j)}(x) \end{array}}{\sum_{j=0}^{N-1} G^{(j)}(Z)\int_0^T [F^{(j)}(t) - F^{(j+1)}(t)]\mathrm{d}t}, \tag{2.66}$$

where c_T and c_Z are defined in (2.6), and c_K is replacement cost at shock N and at damage K where $c_K > c_T$ and $c_K > c_Z$.

When the unit is replaced preventively only at time T $(0 < T \leq \infty)$,

$$C_N(T) \equiv \lim_{Z \to K} C_N(T, Z) = \frac{c_K - (c_K - c_T)\sum_{j=0}^{N-1}[F^{(j)}(T) - F^{(j+1)}(T)]G^{(j)}(K)}{\sum_{j=0}^{N-1} G^{(j)}(K)\int_0^T [F^{(j)}(t) - F^{(j+1)}(t)]\mathrm{d}t}, \tag{2.67}$$

and when the unit is replaced preventively only at damage Z $(0 < Z \leq K)$,

$$C_N(Z) \equiv \lim_{T \to \infty} C_N(T, Z) = \frac{c_K - (c_K - c_Z)\sum_{j=0}^{N-1}\int_0^Z [G(K - x) - G(Z - x)]\mathrm{d}G^{(j)}(x)}{\mu\sum_{j=0}^{N-1} G^{(j)}(Z)}. \tag{2.68}$$

(1) Optimum T_N^*

We find optimum T_N^* to minimize $C_N(T)$ in (2.67) for given N and K. In particular, when $N = 1$,

$$C_1(T) = \frac{c_K - (c_K - c_T)\overline{F}(T)}{\int_0^T \overline{F}(t)\mathrm{d}t}, \tag{2.69}$$

which corresponds to the expected cost rate of the standard age replacement policy [1]. When shocks occur at a nonhomogeneous Poisson process with a cumulative hazard rate $H(t)$, i.e., $F(t) = 1 - \mathrm{e}^{-H(t)}$, and $h(t) \equiv \mathrm{d}H(t)/\mathrm{d}t$ increases strictly with t to $h(\infty) \equiv \lim_{t\to\infty} h(t)$ and $\mu h(\infty) > c_K/(c_K - c_T)$, optimum T_1^* $(0 < T_1^* < \infty)$ satisfies

$$h(T)\int_0^T \overline{F}(t)\mathrm{d}t + \overline{F}(T) = \frac{c_K}{c_K - c_T}, \tag{2.70}$$

and the resulting cost rate is

$$C_1(T_1^*) = (c_K - c_T)h(T_1^*). \tag{2.71}$$

From (2.67),

$$C_N(0) \equiv \lim_{T\to 0} C_N(T) = \infty,$$

$$C_N(\infty) \equiv \lim_{T\to\infty} C_N(T) = \frac{c_K}{\mu\sum_{j=0}^{N-1} G^{(j)}(K)}. \tag{2.72}$$

Thus, there exist a positive T_N^* $(0 < T_N^* \le \infty)$ to minimize $C_N(T)$.

Differentiating $C_N(T)$ with respect to T and setting it equal to zero,

$$Q_8(T,N)\sum_{j=0}^{N-1} G^{(j)}(K)\int_0^T [F^{(j)}(t) - F^{(j+1)}(t)]\mathrm{d}t$$

$$+\sum_{j=0}^{N-1}[F^{(j)}(T) - F^{(j+1)}(T)]G^{(j)}(K) = \frac{c_K}{c_K - c_T}, \tag{2.73}$$

where

$$Q_8(T,N) \equiv \frac{-\sum_{j=0}^{N-1}[f^{(j)}(T) - f^{(j+1)}(T)]G^{(j)}(K)}{\sum_{j=0}^{N-1}[F^{(j)}(T) - F^{(j+1)}(T)]G^{(j)}(K)},$$

and $f^{(j)}(t) \equiv \mathrm{d}F^{(j)}(t)/\mathrm{d}t$. If $Q_8(T,N)$ increases strictly with T to $Q_8(\infty, N)$, then left-hand side of (2.73) also increases strictly from 1 to $\mu Q_8(\infty, N)\sum_{j=0}^{N-1} G^{(j)}(K)$. Therefore, if

$$\mu Q_8(\infty, N)\sum_{j=0}^{N-1} G^{(j)}(K) > \frac{c_K}{c_K - c_T},$$

then there exists a finite and unique T_N^* $(0 < T_N^* < \infty)$ which satisfies (2.73), and the resulting cost rate is

$$C_N(T_N^*) = (c_K - c_T)Q_8(T_N^*, N). \tag{2.74}$$

In particular, when $F(t) = 1 - \mathrm{e}^{-\lambda t}$, the expected cost rate in (2.67) is

$$\frac{C_N(T)}{\lambda} = \frac{c_K - (c_K - c_T)\sum_{j=0}^{N-1} F^{(j)}(T)G^{(j)}(K)}{\sum_{j=0}^{N-1} F^{(j+1)}(T)G^{(j)}(K)} + (c_K - c_T), \tag{2.75}$$

which decreases strictly with T. Thus, $T_N^* = \infty$, and the resulting cost rate is given in (2.72).

(2) Optimum Z_N^*

We find optimum Z_N^* to minimize $C_N(Z)$ in (2.68) for given N and K. In particular, when $N = 1$, $C_1(Z)$ increases with Z, and hence, $Z_1^* = 0$, and when $N \to \infty$, $C_N(Z)$ is given in (2.16).

Differentiating $C_N(Z)$ with respect to Z and setting it equal to zero for $N \geq 2$,

$$Q_9(Z)\sum_{j=0}^{N-1} G^{(j)}(Z) + \sum_{j=0}^{N-1}\int_0^Z [\overline{G}(K - Z) + \overline{G}(Z - x) - \overline{G}(K - x)]\mathrm{d}G^{(j)}(x)$$
$$= \frac{c_K}{c_K - c_Z}, \tag{2.76}$$

where

$$Q_9(Z) \equiv \frac{g^{(N)}(Z)}{\sum_{j=1}^{N-1} g^{(j)}(Z)},$$

and $g^{(N)}(x) \equiv \mathrm{d}G^{(N)}(x)/\mathrm{d}x$.

If $Q_9(Z)$ increases strictly with Z from 0 to $Q_9(K)$, then the left-hand side of (2.76) also increases strictly with Z from 1 to $[Q_9(K) + 1]\sum_{j=0}^{N-1} G^{(j)}(K)$. Therefore, if $[Q_9(K) + 1]\sum_{j=0}^{N-1} G^{(j)}(K) > c_K/(c_K - c_Z)$, then there exists a finite and unique Z_N^* $(0 < Z_N^* < K)$ which satisfies (2.76), and the resulting cost rate is

$$\mu C_N(Z_N^*) = (c_K - c_Z)[Q_9(Z_N^*) + \mathrm{e}^{-\omega(K - Z_N^*)}]. \tag{2.77}$$

Conversely, if $[Q_9(K) + 1]\sum_{j=0}^{N-1} G^{(j)}(K) \leq c_K/(c_K - c_Z)$, then $Z_N^* = K$, and the resulting cost rate is given in (2.72).

When $G(x) = 1 - \mathrm{e}^{-\omega x}$,

$$Q_9(Z) = \frac{(\omega Z)^{N-1}/(N-1)!}{\sum_{j=0}^{N-2}[(\omega Z)^j/j!]} \quad (N = 2, 3, \cdots),$$

which agrees with $\widetilde{r}_N(Z)$ in (2.55), and increases strictly with Z from 0 and decreases strictly with N to 0 (Problem 2.4). Thus, (2.76) becomes

$$[\widetilde{r}_N(Z) + 1]\sum_{j=0}^{N-1} G^{(j)}(Z) - [1 - \mathrm{e}^{-\omega(K-Z)}]\sum_{j=1}^{N} G^{(j)}(Z) = \frac{c_K}{c_K - c_Z}, \quad (2.78)$$

whose left-hand side increases strictly with Z from 1. Therefore, if

$$[\widetilde{r}_N(K) + 1]\sum_{j=0}^{N-1} G^{(j)}(K) > \frac{c_K}{c_K - c_Z},$$

then there exists a finite and unique Z_N^* $(0 < Z_N^* < K)$ which satisfies (2.78). If $\omega K > c_Z/(c_K - c_Z)$, then a finite Z_N^* might exist.

2.4 Problem 2

2.1 Prove that (2.1)+(2.2)+(2.3)+(2.4)=1 and derive (2.5).

2.2 Prove that if $Q_1(T)$ increases strictly with T to $Q_1(\infty)$, then the left-hand side of (2.10) increases strictly to $Q_1(\infty)\mu[1 + M_G(K)] - 1$. Furthermore, prove in (2.23) that when $F(t) = 1 - \mathrm{e}^{-\lambda t}$ and $r_{j+1}(x)$ increases strictly with j,

$$\widetilde{Q}_1(T, N) = \frac{\sum_{j=0}^{N-1}[(\lambda T)^j/j!][G^{(j)}(K) - G^{(j+1)}(K)]}{\sum_{j=0}^{N-1}[(\lambda T)^j/j!]G^{(j)}(K)}$$

increases strictly with T from $\overline{G}(K)$ to $r_N(K)$ for $N \geq 2$, and increases strictly with N from $\overline{G}(K)$ to $Q_1(T)/\lambda$.

2.3 Show why the inequality $C(N+1) - C(N) \geq 0$ should be formulated.

2.4 Prove that if $r_{N+1}(K)$ increases strictly with N, then the left-hand side of (2.14) increases with N to $r_\infty(K)[1 + M_G(K)] - 1$. Furthermore, prove that when $G(x) = 1 - \mathrm{e}^{-\omega x}$ for $0 < x \leq K$,

$$r_{N+1}(x) = \frac{(\omega x)^N/N!}{\sum_{j=N}^{\infty}(\omega x)^j/j!} \quad (N = 0, 1, 2, \cdots)$$

increases strictly with N from $\mathrm{e}^{-\omega K}$ to 1, and decreases strictly with x from 1 to $r_{N+1}(K)$. Prove that

$$\widetilde{r}_{N+1}(x) = \frac{(\omega x)^N/N!}{\sum_{j=0}^{N-1}(\omega x)^j/j!} \quad (N = 1, 2, \cdots)$$

decreases strictly with N from ωx to 0 and increases strictly with x from 0 to $\widetilde{r}_{N+1}(K)$.

2.5 Derive (2.17) and prove that its left-hand side increases strictly with Z from 0 to $M_G(K)$.

2.6 Prove that

$$\frac{\sum_{j=0}^{\infty}[(\lambda T)^j/j!]\int_0^Z \overline{G}(K-x)\mathrm{d}G^{(j)}(x)}{\sum_{j=0}^{\infty}[(\lambda T)^j/j!]G^{(j)}(Z)}$$

increases with T from $\overline{G}(K)$ and is less than $\overline{G}(K-Z)$.

2.7 Prove that (2.33) does not hold for any Z, and show that there does not exist any finite N_F^* which satisfies (2.32).

2.8 Prove that $Q_4(T, N)$ increases strictly with T from 0 to $G^{(N)}(K)/G^{(N-1)}(K)$.

2.9 Challenge to derive optimum N_F^* for given T for $c_N > c_T$.

2.10 Prove that $Q_6(T, Z)$ increases strictly with Z from $F(T)/F^{(2)}(T)$ to $Q_6(T, K)$, and decreases strictly with T from ∞ to 1.

2.11 Derive (2.61) and prove its result.

2.12 Derive (2.64) and prove its result.

Chapter 3
Replacement Last Policies

In theory, a new counter approach of *whichever triggering event occurs last* for the replacement policy with two PR scenarios should be considered rather than only the approach of *whichever triggering event occurs first* in Chap. 2 has been applied. Replacement models with two PR scenarios, such as age and usage number, age and failure number, etc., have been observed to make the operating system or unit more safety in theory [30]. The alternative replacement plans can also be found in real situations, e.g., appropriate replacement actions for parts of an aircraft are usually scheduled at a total hours of operation and at a specified number of flights since the last major overhaul [31].

The approach of replacement first has the primary advantage to prevent failures due to late scheduled PRs, and also, it has the absolute priority between CR and PR (or PRs) to balance failure losses and replacement costs, which has been indicated in Chap. 2. However, the only classical mode of whichever occurs first is not competent to plan two or more alternative PR scenarios: If any PR action is scheduled too early prior to other PRs, a waste of replacement is incurred because the system or unit might run for an additional period of time to complete more operations. That is, replacement plans can be scheduled: The unit undergoes CR immediately after failure, and is replaced preventively before failure at some thresholds or planned measurements, e.g., at time T, at shock N or at damage Z, *whichever occurs last*, which is named as *replacement last*.

The notion of *whichever triggering event occurs last* in replacement modeling can be originally found in [30], where replacement is scheduled before failure at a planned time T or at the Nth working time (or at the Nth failure), whichever occurs last. The approach of whichever occurs last was supposed to be modeled for the cases when replacement costs suffered for failures are estimated to be not so high [4, 5]. Replacement policies acting on the approaches of first and last was named as *replacement first* and *replacement last* and comparative method for their optimum times to determine which policy could save more expected cost rates have been originally found in [7]. From the point of performance, the comparisons of discrete

X. Zhao and T. Nakagawa, *Advanced Maintenance Policies for Shock and Damage Models*, Springer Series in Reliability Engineering,
https://doi.org/10.1007/978-3-319-70456-2_3

number N of working times for replacement first and last were next discussed [9]. To make effective use of the system, the approach of whichever occurs last was modeled with minimal repairs [13]. When the operating system or unit degrades with shock and damage process, several interesting comparative results of maintenance time T, shock number N and damage level Z for maintenance first and last were obtained [19].

In this chapter, we take up replacement last policies under the same assumptions as those in Chap. 2: Shocks occur at a renewal process according to an interarrival distribution $F(t)$ with finite mean time μ. Each damage due to shocks is additive and has an identical distribution $G(x)$ with finite mean $1/\omega$. The unit fails when the total damage has exceeded a failure threshold K and CR is done immediately. In Sect. 3.1, we obtain the expected cost rate of replacement last that are planned at time T, at shock N or at damage Z, whichever occurs last. Optimum polices bounded with two variables are discussed in Sect. 3.2 and respective optimum polices of T, N and Z are compared with those of replacement first in Sect. 3.3 and computed in Sect. 3.4.

3.1 Three Replacement Policies

The unit is replaced preventively at planned time T $(0 \le T \le \infty)$, at shock number N $(N = 0, 1, 2 \cdots)$, or at damage level Z $(0 \le Z \le K)$, whichever occurs last, which is called *replacement last* (RL). Compared with RF in Chap. 2, it can be understood for RL that the unit could operate as long as possible before failure until any PR is last triggered.

The probability that the unit is replaced at time T is

$$\sum_{j=N}^{\infty}[F^{(j)}(T) - F^{(j+1)}(T)][G^{(j)}(K) - G^{(j)}(Z)], \tag{3.1}$$

the probability that it is replaced at shock N is

$$[1 - F^{(N)}(T)][G^{(N)}(K) - G^{(N)}(Z)], \tag{3.2}$$

and the probability that it is replaced at damage Z is

$$\sum_{j=N}^{\infty}[1 - F^{(j+1)}(T)] \int_0^Z [G(K - x) - G(Z - x)]\mathrm{d}G^{(j)}(x). \tag{3.3}$$

The probability that the unit is replaced at failure is divided into the following three cases: The probability that the total damage exceeds K at some shock before time T is

$$\sum_{j=0}^{\infty} F^{(j+1)}(T)[G^{(j)}(K) - G^{(j+1)}(K)],$$

the probability that the total damage exceeds K at some shock after T and N is

$$\sum_{j=N}^{\infty}[1 - F^{(j+1)}(T)] \int_0^Z \overline{G}(K - x)\mathrm{d}G^{(j)}(x),$$

and the probability that the total damage exceeds K at some shock when the shock number is less than or equal to N is

$$\sum_{j=0}^{N-1}[1 - F^{(j+1)}(T)][G^{(j)}(K) - G^{(j+1)}(K)].$$

By summing up the above three failure probabilities (Problem 3.1),

$$1 - \sum_{j=N}^{\infty}[1 - F^{(j+1)}(T)] \int_Z^K \overline{G}(K - x)\mathrm{d}G^{(j)}(x), \tag{3.4}$$

where note that (3.1)+(3.2)+(3.3)+(3.4)=1. The mean time to replacement is

$$\begin{aligned}
&T \sum_{j=N}^{\infty}[F^{(j)}(T) - F^{(j+1)}(T)][G^{(j)}(K) - G^{(j)}(Z)] \\
&+ [G^{(N)}(K) - G^{(N)}(Z)] \int_T^{\infty} t\mathrm{d}F^{(N)}(t) \\
&+ \sum_{j=N}^{\infty} \int_T^{\infty} t\mathrm{d}F^{(j+1)}(t) \int_0^Z [G(K - x) - G(Z - x)]\mathrm{d}G^{(j)}(x) \\
&+ \sum_{j=0}^{\infty}[G^{(j)}(K) - G^{(j+1)}(K)] \int_0^T t\mathrm{d}F^{(j+1)}(t) \\
&+ \sum_{j=N}^{\infty} \int_T^{\infty} t\mathrm{d}F^{(j+1)}(t) \int_0^Z \overline{G}(K - x)\mathrm{d}G^{(j)}(x) \\
&+ \sum_{j=0}^{N-1}[G^{(j)}(K) - G^{(j+1)}(K)] \int_T^{\infty} t\mathrm{d}F^{(j+1)}(t) \\
&= \sum_{j=N}^{\infty} G^{(j)}(Z) \int_T^{\infty} [F^{(j)}(t) - F^{(j+1)}(t)]\mathrm{d}t
\end{aligned}$$

$$+\sum_{j=0}^{N-1} G^{(j)}(K)\int_T^{\infty}[F^{(j)}(t)-F^{(j+1)}(t)]\mathrm{d}t$$
$$+\sum_{j=0}^{\infty} G^{(j)}(K)\int_0^{T}[F^{(j)}(t)-F^{(j+1)}(t)]\mathrm{d}t. \tag{3.5}$$

Therefore, the expected cost rate is

$$C_L(T,N,Z)=$$
$$\frac{\begin{array}{l}c_K-(c_K-c_T)\sum_{j=N}^{\infty}[F^{(j)}(T)-F^{(j+1)}(T)][G^{(j)}(K)-G^{(j)}(Z)]\\ -(c_K-c_N)[1-F^{(N)}(T)][G^{(N)}(K)-G^{(N)}(Z)]\\ -(c_K-c_Z)\sum_{j=N}^{\infty}[1-F^{(j+1)}(T)]\int_0^Z[G(K-x)-G(Z-x)]\mathrm{d}G^{(j)}(x)\end{array}}{\begin{array}{l}\sum_{j=N}^{\infty}G^{(j)}(Z)\int_T^{\infty}[F^{(j)}(t)-F^{(j+1)}(t)]\mathrm{d}t\\ +\sum_{j=0}^{N-1}G^{(j)}(K)\int_T^{\infty}[F^{(j)}(t)-F^{(j+1)}(t)]\mathrm{d}t\\ +\sum_{j=0}^{\infty}G^{(j)}(K)\int_0^{T}[F^{(j)}(t)-F^{(j+1)}(t)]\mathrm{d}t\end{array}}, \tag{3.6}$$

where c_T, c_N, c_Z and c_K are given in (2.6).

We have known that the model of replacement last is a counter example to that of replacement first, and both of which are extensions for the models with one variable in Sect. 2.1.1. So that, it can be shown that

$$\lim_{\substack{T\to 0\\ N\to 0\\ Z\to 0}} C_L(T,N,Z)=\lim_{\substack{T\to\infty\\ N\to\infty\\ Z\to K}} C_F(T,N,Z)=C,$$

which is given in (2.7),

$$\lim_{\substack{N\to 0\\ Z\to 0}} C_L(T,N,Z)=\lim_{\substack{N\to\infty\\ Z\to K}} C_F(T,N,Z)=C(T),$$

which is given in (2.9),

$$\lim_{\substack{T\to 0\\ Z\to 0}} C_L(T,N,Z)=\lim_{\substack{T\to\infty\\ Z\to K}} C_F(T,N,Z)=C(N),$$

which is given in (2.13), and

$$\lim_{\substack{T\to 0\\ N\to 0}} C_L(T,N,Z)=\lim_{\substack{T\to\infty\\ N\to\infty}} C_F(T,N,Z)=C(Z),$$

which is given in (2.16).

In other words, both replacement first and last would degrade into the same replacement models with respective T, N and Z, when the required conditions do not meet for optimization results, which will be shown in the following discussions.

3.2 Optimum Policies

We have stated in Chap. 2 that the model of replacement first is absolutely reasonable when a single preventive replacement scenario is planned, and optimum policies with one variable such as T^*, N^* and Z^* have been obtained in Sect. 2.1.1. However, when we optimize models of replacement last, optimum polices should be bounded with another variables, which will be the focuses in Sects. 3.2 and 3.3.

(1) Optimum T_L^* and N_L^*

Suppose that the unit is replaced preventively at time T $(0 \le T \le \infty)$ or at shock N $(N = 0, 1, 2 \cdots)$, whichever occurs last. When $c_T = c_N$, putting that $Z \to 0$ in (3.6), the expected cost rate is

$$C_L(T, N) = \frac{c_T + (c_K - c_T)\{1 - \sum_{j=N}^{\infty}[1 - F^{(j+1)}(T)][G^{(j)}(K) - G^{(j+1)}(K)]\}}{\sum_{j=0}^{N-1} G^{(j)}(K)\int_T^{\infty}[F^{(j)}(t) - F^{(j+1)}(t)]\mathrm{d}t + \sum_{j=0}^{\infty} G^{(j)}(K)\int_0^T[F^{(j)}(t) - F^{(j+1)}(t)]\mathrm{d}t}. \tag{3.7}$$

We find optimum (T_L^*, N_L^*) to minimize $C_L(T, N)$. Forming the inequality $C_L(T, N+1) - C_L(T, N) \ge 0$,

$$\frac{r_{N+1}(K)[1 - F^{(N+1)}(T)]}{\int_T^{\infty}[F^{(N)}(t) - F^{(N+1)}(t)]\mathrm{d}t}\left\{\sum_{j=0}^{N-1} G^{(j)}(K)\int_T^{\infty}[F^{(j)}(t) - F^{(j+1)}(t)]\mathrm{d}t + \sum_{j=0}^{\infty} G^{(j)}(K)\int_0^T[F^{(j)}(t) - F^{(j+1)}(t)]\mathrm{d}t\right\}$$
$$+ \sum_{j=N}^{\infty}[1 - F^{(j+1)}(T)][G^{(j)}(K) - G^{(j+1)}(K)] - 1 \ge \frac{c_T}{c_K - c_T}, \tag{3.8}$$

where $r_{N+1}(x)$ is given in (2.14).

Differentiating $C_L(T, N)$ with respect to T and setting it equal to zero,

$$Q_1(T,N)\left\{\sum_{j=0}^{N-1} G^{(j)}(K)\int_T^{\infty}[F^{(j)}(t)-F^{(j+1)}(t)]\mathrm{d}t\right.$$
$$\left.+\sum_{j=0}^{\infty} G^{(j)}(K)\int_0^{T}[F^{(j)}(t)-F^{(j+1)}(t)]\mathrm{d}t\right\}$$
$$+\sum_{j=N}^{\infty}[1-F^{(j+1)}(T)][G^{(j)}(K)-G^{(j+1)}(K)]-1=\frac{c_T}{c_K-c_T}, \tag{3.9}$$

where

$$Q_1(T,N)\equiv\frac{\sum_{j=N}^{\infty} f^{(j+1)}(T)[G^{(j)}(K)-G^{(j+1)}(K)]}{\sum_{j=N}^{\infty}[F^{(j)}(T)-F^{(j+1)}(T)]G^{(j)}(K)},$$

and $f^{(j)}(t)\equiv \mathrm{d}F^{(j)}(t)/\mathrm{d}t$ $(j=1,2,\cdots)$.

Substituting (3.8) for (3.9),

$$Q_1(T,N)\le\frac{r_{N+1}(K)[1-F^{(N+1)}(T)]}{\int_T^{\infty}[F^{(N)}(t)-F^{(N+1)}(t)]\mathrm{d}t}. \tag{3.10}$$

Thus, if the inequality (3.10) does not hold for any N, there does not exist any positive T_L^* to satisfy (3.10), i.e., the optimum policy is $(T_L^*=0,\ N_L^*=N^*)$, where N^* is given in (2.14).

For example, when $F(t)=1-\mathrm{e}^{-\lambda t}$ and $r_{j+1}(x)$ increases strictly with j to 1, (3.10) is

$$\widetilde{Q}_1(T,N)\equiv\frac{\sum_{j=N}^{\infty}[(\lambda T)^j/j!][G^{(j)}(K)-G^{(j+1)}(K)]}{\sum_{j=N}^{\infty}[(\lambda T)^j/j!]G^{(j)}(K)}\le r_{N+1}(K),$$

whose left-hand side increases with T from $r_{N+1}(K)$ to 1. That is, the above inequality does not hold for $T>0$ (Problem 3.2). When optimum N_L^* satisfies (3.8), the left-hand side of (3.9) is greater than or equal to $c_T/(c_K-c_T)$, which means $C_L(T,N)$ increases with T, and hence, $T_L^*=0$. This concludes that if the inequality (3.10) does not hold and $c_T\ge c_N$, then the optimum policy which minimizes $C_L(T,N)$ is $(T_L^*=0,\ N_L^*=N^*)$.

(2) Optimum T_L^* and Z_L^*

Suppose that the unit is replaced preventively at time T $(0\le T\le\infty)$ or at damage Z $(0\le Z\le K)$, whichever occurs last. When $c_T=c_Z$, putting that $N\to 0$ in (3.6), the expected cost rate is

$$C_L(T,Z) = \frac{\begin{array}{l} c_T + (c_K - c_T)\{1 - \sum_{j=0}^{\infty}[F^{(j)}(T) - F^{(j+1)}(T)][G^{(j)}(K) - G^{(j)}(Z)] \\ - \sum_{j=0}^{\infty}[1 - F^{(j+1)}(T)]\int_0^Z [G(K-x) - G(Z-x)]\mathrm{d}G^{(j)}(x)\} \end{array}}{\begin{array}{l} \sum_{j=0}^{\infty} G^{(j)}(Z)\int_T^{\infty}[F^{(j)}(t) - F^{(j+1)}(t)]\mathrm{d}t \\ + \sum_{j=0}^{\infty} G^{(j)}(K)\int_0^T [F^{(j)}(t) - F^{(j+1)}(t)]\mathrm{d}t \end{array}}. \tag{3.11}$$

We find optimum (T_L^*, Z_L^*) to minimize $C_L(T,Z)$. Differentiating $C_L(T,Z)$ with respect to Z and setting it equal to zero,

$$\begin{aligned} &Q_2(T,Z)\left\{\sum_{j=0}^{\infty} G^{(j)}(Z)\int_T^{\infty}[F^{(j)}(t) - F^{(j+1)}(t)]\mathrm{d}t \right. \\ &\left. + \sum_{j=0}^{\infty} G^{(j)}(K)\int_0^T [F^{(j)}(t) - F^{(j+1)}(t)]\mathrm{d}t\right\} \\ &+ \sum_{j=0}^{\infty}[F^{(j)}(T) - F^{(j+1)}(T)][G^{(j)}(K) - G^{(j)}(Z)] \\ &+ \sum_{j=0}^{\infty}[1 - F^{(j+1)}(T)]\int_0^Z [G(K-x) - G(Z-x)]\mathrm{d}G^{(j)}(x) - 1 = \frac{c_T}{c_K - c_T}, \end{aligned} \tag{3.12}$$

where

$$Q_2(T,Z) \equiv \frac{\overline{G}(K-Z)\sum_{j=1}^{\infty} g^{(j)}(Z)[1 - F^{(j+1)}(T)]}{\sum_{j=1}^{\infty} g^{(j)}(Z)\int_T^{\infty}[F^{(j)}(t) - F^{(j+1)}(t)]\mathrm{d}t},$$

and $g^{(j)}(x) \equiv \mathrm{d}G^{(j)}(x)/\mathrm{d}x$ $(j = 1, 2, \cdots)$.

Differentiating $C_L(T,Z)$ with respect to T and setting it equal to 0,

$$\begin{aligned} &Q_3(T,Z)\left\{\sum_{j=0}^{\infty} G^{(j)}(Z)\int_T^{\infty}[F^{(j)}(t) - F^{(j+1)}(t)]\mathrm{d}t \right. \\ &\left. + \sum_{j=0}^{\infty} G^{(j)}(K)\int_0^T [F^{(j)}(t) - F^{(j+1)}(t)]\mathrm{d}t\right\} \\ &+ \sum_{j=0}^{\infty}[F^{(j)}(T) - F^{(j+1)}(T)][G^{(j)}(K) - G^{(j)}(Z)] \end{aligned}$$

$$+\sum_{j=0}^{\infty}[1-F^{(j+1)}(T)]\int_0^Z [G(K-x)-G(Z-x)]\mathrm{d}G^{(j)}(x)-1=\frac{c_T}{c_K-c_T}, \tag{3.13}$$

where

$$Q_3(T,Z)\equiv\frac{\sum_{j=0}^{\infty} f^{(j+1)}(T)\int_Z^K \overline{G}(K-x)\mathrm{d}G^{(j)}(x)}{\sum_{j=0}^{\infty}[F^{(j)}(T)-F^{(j+1)}(T)][G^{(j)}(K)-G^{(j)}(Z)]}.$$

Substituting (3.12) for (3.13),

$$Q_3(T,Z)=Q_2(T,Z). \tag{3.14}$$

Thus, if $Q_3(T,Z)>Q_2(T,Z)$ for any Z, then there does not exist any positive T_L^* to satisfy (3.14), i.e., the optimum policy is $(T_L^*=0, Z_L^*=Z^*)$, where Z^* is given in (2.17).

For example, when $F(t)=1-\mathrm{e}^{-\lambda t}$, it is proved that (Problem 3.3)

$$\frac{\sum_{j=0}^{\infty}[(\lambda T)^j/j!]\int_Z^K \overline{G}(K-x)\mathrm{d}G^{(j)}(x)}{\sum_{j=0}^{\infty}[(\lambda T)^j/j!][G^{(j)}(K)-G^{(j)}(Z)]}>\overline{G}(K-Z),$$

which always holds for any $T>0$, i.e., the inequality (3.14) does not hold. When optimum optimum Z_L^* satisfies (3.12), the left-hand side of (3.13) is greater than $c_T/(c_K-c_T)$, which means $C_L(T,Z)$ increases with T, and hence, $T_L^*=0$. This concludes that if $Q_3(T,Z)>Q_2(T,Z)$ and $c_T\geq c_Z$, then the optimum policy which minimizes $C_L(T,Z)$ is $(T_L^*=0, Z_L^*=Z^*)$.

(3) Optimum N_L^* and Z_L^*

Suppose that the unit is replaced preventively at shock N $(N=0,1,2\cdots)$ or at damage Z $(0\leq Z\leq K)$, whichever occurs last. When $c_N=c_Z$, putting that $T\to 0$ in (3.6), the expected cost rate is

$$C_L(N,Z)=\frac{c_N+(c_K-c_N)[1-G^{(N)}(K)+\sum_{j=N}^{\infty}\int_0^Z \overline{G}(K-x)\mathrm{d}G^{(j)}(x)]}{\mu[\sum_{j=N}^{\infty}G^{(j)}(Z)+\sum_{j=0}^{N-1}G^{(j)}(K)]}, \tag{3.15}$$

where $\mu\equiv\int_0^{\infty}\overline{F}(t)\mathrm{d}t$.

We find optimum (N_L^*, Z_L^*) to minimize $C_L(N,Z)$. Differentiating $C_L(N,Z)$ with respect to Z and setting it equal to zero,

$$
\overline{G}(K-Z)\left[\sum_{j=N}^{\infty}G^{(j)}(Z)+\sum_{j=0}^{N-1}G^{(j)}(K)\right]
$$
$$
-\left[1-G^{(N)}(K)+\sum_{j=N}^{\infty}\int_0^Z\overline{G}(K-x)\mathrm{d}G^{(j)}(x)\right]=\frac{c_N}{c_K-c_N}. \qquad (3.16)
$$

Forming the inequality $C_L(N+1,Z)-C_L(N,Z)\ge 0$,

$$
\frac{\int_Z^K\overline{G}(K-x)\mathrm{d}G^{(N)}(x)}{G^{(N)}(K)-G^{(N)}(Z)}\left[\sum_{j=N}^{\infty}G^{(j)}(Z)+\sum_{j=0}^{N-1}G^{(j)}(K)\right]
$$
$$
-\left[1-G^{(N)}(K)+\sum_{j=N}^{\infty}\int_0^Z\overline{G}(K-x)\mathrm{d}G^{(j)}(x)\right]\ge\frac{c_N}{c_K-c_N}. \qquad (3.17)
$$

Substituting (3.16) for (3.17),

$$
\frac{\int_Z^K\overline{G}(K-x)\mathrm{d}G^{(N)}(x)}{G^{(N)}(K)-G^{(N)}(Z)}\ge\overline{G}(K-Z), \qquad (3.18)
$$

which always holds for any Z (Problem 3.3). When optimum Z_L^* satisfies (3.16), the left-hand side of (3.17) is equal to or greater than $c_N/(c_K-c_N)$, which means $C_L(N,Z)$ increases with N, and hence, $N_L^*=0$. This concludes that if $c_N\ge c_Z$, then the optimum policy which minimizes $C_L(N,Z)$ is $(N_L^*=0,\ Z_L^*=Z^*)$, where Z^* is given in (2.17).

3.3 Comparisons of Replacement First and Last

When $c_T=c_N=c_Z$, shocks occur at a Poisson process with rate λ, i.e., $F(t)=1-\mathrm{e}^{-\lambda t}$ and $F^{(j)}(t)=\sum_{i=j}^{\infty}[(\lambda t)^i/i!]\mathrm{e}^{-\lambda t}$ $(j=0,1,2,\cdots)$, and $r_{j+1}(x)$ increases strictly with j to 1, we show there exist optimum T_L^* for given N, N_L^* for given T, and Z_L^* for given T to minimize their cost rates $C_L(T,N)$ in (3.7) and $C_L(T,Z)$ in (3.11). The above optimum policies are compared with optimum T_F^* for given N, N_F^* for given T, and Z_F^* for given T that minimize their cost rates $C_F(T,N)$ in (2.45) and $C_F(T,Z)$ in (2.49), and the cases in which replacement first or last should be adopted are found.

(1) Time T for Shock N

From (3.9), optimum T_L^* for given N satisfies (Problem 3.4)

$$\widetilde{Q}_1(T,N)\left\{\sum_{j=0}^{N-1}[1-F^{(j+1)}(T)]G^{(j)}(K)+\sum_{j=0}^{\infty}F^{(j+1)}(T)G^{(j)}(K)\right\}$$
$$+\sum_{j=N}^{\infty}[1-F^{(j+1)}(T)][G^{(j)}(K)-G^{(j+1)}(K)]-1=\frac{c_T}{c_K-c_T}, \tag{3.19}$$

where

$$\widetilde{Q}_1(T,N)\equiv\frac{\sum_{j=N}^{\infty}[(\lambda T)^j/j!][G^{(j)}(K)-G^{(j+1)}(K)]}{\sum_{j=N}^{\infty}[(\lambda T)^j/j!]G^{(j)}(K)}.$$

Note that when $r_{j+1}(K)$ increases strictly with j to 1, $\widetilde{Q}_1(T,N)$ increases strictly with T from $r_{N+1}(K)$ to 1 and increases strictly with N from $\widetilde{Q}_1(T)$ to 1 (Problem 3.2). Denoting the left-hand side of (3.19) by $V_L(T,N)$,

$$\frac{\mathrm{d}V_L(T,N)}{\mathrm{d}T}=$$
$$\frac{\mathrm{d}\widetilde{Q}_1(T,N)}{\mathrm{d}T}\left\{\sum_{j=0}^{N-1}[1-F^{(j+1)}(T)]G^{(j)}(K)+\sum_{j=0}^{\infty}F^{(j+1)}(T)G^{(j)}(K)\right\}>0,$$

which follows that $V_L(T,N)$ increases strictly with T from the left-hand side of (2.14) to $M_G(K)\equiv\sum_{j=1}^{\infty}G^{(j)}(K)$. Therefore, if

$$r_{N+1}(K)\sum_{j=0}^{N-1}G^{(j)}(K)-[1-G^{(N)}(K)]<\frac{c_T}{c_K-c_T}<M_G(K),$$

then there exists a finite and unique T_L^* $(0<T_L^*<\infty)$ which satisfies (3.19), and the resulting cost rate is

$$\frac{C_L(T_L^*,N)}{\lambda}=(c_K-c_T)\widetilde{Q}_1(T_L^*,N). \tag{3.20}$$

If $M_G(K)\le c_T/(c_K-c_T)$, then $T_L^*=\infty$, and the resulting cost rate is given in (2.7). If

$$r_{N+1}(K)\sum_{j=0}^{N-1}G^{(j)}(K)-[1-G^{(N)}(K)]\ge\frac{c_T}{c_K-c_T},$$

then $T_L^*=0$, and the resulting cost rate is given in (2.13).

Furthermore, because $\widetilde{Q}_1(T,N)$ increases strictly with N and $\widetilde{Q}_1(T,N+1)>r_{N+1}(K)$ (Problem 3.2),

$$V_L(T, N+1) - V_L(T, N) =$$

$$[\widetilde{Q}_1(T, N+1) - \widetilde{Q}_1(T, N)]\left\{\sum_{j=0}^{N-1}[1 - F^{(j+1)}(T)]G^{(j)}(K) + \sum_{j=0}^{\infty} F^{(j+1)}(T)G^{(j)}(K)\right\} + [\widetilde{Q}_1(T, N+1) - r_{N+1}(K)] \times [1 - F^{(N+1)}(T)]G^{(N)}(K) > 0.$$

Thus, noting that $V_L(T, N)$ increases strictly with N from the left hand side of (2.10), T_L^* in (3.19) decreases with N from T^* given in (2.10).

For replacement first in **(1)** of Sect. 2.1.2, it has been obtained that the left-hand side of (2.23) increases strictly with T from 0 to that of (2.14), and there exists a finite T_F^* for given N when $N > N^*$, and when $N \leq N^*$, $T_F^* = \infty$, where N^* is given in (2.14).

Therefore, when a finite N^* in (2.14) exists, we obtain the following comparative results for RF and RL:

1. If given N in (3.19) and (2.23) is less than N^*, then replacement first with time T becomes meaningless as $T_F^* = \infty$; however, a finite T_L^* exists for replacement last. That is, we need to adopt the policy acting on the approach of whichever occurs last.
2. If given N is greater than N^*, then $T_L^* = 0$ for replacement last, which becomes the same case when $T_F^* = \infty$; however, a finite T_F^* exists for replacement first. That is, we need to adopt the policy acting on the approach of whichever occurs first.
3. If given N is equal to N^*, $C_F(\infty, N^*) = C_L(0, N^*) = C(N^*)$ in (2.13).

(2) Shock N for Time T

I. Step 1

From (3.8), optimum N_L^* for given T satisfies

$$r_{N+1}(K)\left\{\sum_{j=0}^{N-1}[1 - F^{(j+1)}(T)]G^{(j)}(K) + \sum_{j=0}^{\infty} F^{(j+1)}(T)G^{(j)}(K)\right\}$$

$$+\sum_{j=N}^{\infty}[1 - F^{(j+1)}(T)][G^{(j)}(K) - G^{(j+1)}(K)] - 1 \geq \frac{c_N}{c_K - c_N}. \tag{3.21}$$

Denoting the left-hand side of (3.21) by $V_L(T, N)$,

$$V_L(T,N+1)-V_L(T,N)=$$
$$[r_{N+2}(K)-r_{N+1}(K)]\left[\sum_{j=N}^{\infty}F^{(j+1)}(T)G^{(j)}(K)+\sum_{j=0}^{N-1}G^{(j)}(K)\right]>0,$$

which follows that $V_L(T,N)$ increases strictly with N from

$$\sum_{j=0}^{\infty}F^{(j+1)}(T)G^{(j)}(K)[r_1(K)-r_{j+1}(K)]<0$$

to $M_G(K)\equiv\sum_{j=1}^{\infty}G^{(j)}(K)$. Therefore, if $M_G(K)>c_N/(c_K-c_N)$, then there exists a unique and minimum N_L^* $(1\le N_L^*<\infty)$ which satisfies (3.21), and the resulting cost rate is

$$(c_K-c_N)r_{N_L^*}(K)<\frac{C_L(T,N_L^*)}{\lambda}\le(c_K-c_N)r_{N_L^*+1}(K). \tag{3.22}$$

If $M_G(K)\le c_N/(c_K-c_N)$, then $N_L^*=\infty$ and the resulting cost rate is given in (2.7). Furthermore,

$$\frac{\mathrm{d}V_L(T,N)}{\mathrm{d}T}=\sum_{j=N}^{\infty}\frac{\lambda(\lambda T)^j}{j!}\mathrm{e}^{-\lambda T}G^{(j)}(K)[r_{N+1}(K)-r_{j+1}(K)]<0,$$

which indicates that N_L^* in (3.21) increases strictly with T from N^* given in (2.14).

II. Step 2

From (2.20), optimum N_F^* for given T satisfies

$$\sum_{j=0}^{N-1}F^{(j+1)}(T)G^{(j)}(K)[r_{N+1}(K)-r_{j+1}(K)]\ge\frac{c_N}{c_K-c_N}. \tag{3.23}$$

Denoting the left-hand side of (3.23) by $V_F(T,N)$,

$$V_F(T,N+1)-V_F(T,N)=$$
$$[r_{N+2}(K)-r_{N+1}(K)]\sum_{j=0}^{N}F^{(j+1)}(T)G^{(j)}(K)>0,$$

which follows that $V_F(N)$ increases strictly with N to $\sum_{j=1}^{\infty}F^{(j)}(T)G^{(j)}(K)$. Therefore, if $\sum_{j=1}^{\infty}F^{(j)}(T)G^{(j)}(K)>c_N/(c_K-c_N)$, then there exists a unique and minimum N_F^* $(1\le N_F^*<\infty)$ which satisfies (3.23), and the resulting cost rate is

$$(c_K - c_N) r_{N_F^*}(K) < \frac{C_F(T, N_F^*)}{\lambda} \leq (c_K - c_N) r_{N_F^*+1}(K). \tag{3.24}$$

If $\sum_{j=1}^{\infty} F^{(j)}(T) G^{(j)}(K) \leq c_N/(c_K - c_N)$, then $N_F^* = \infty$ and the resulting cost rate is given in (2.7). Furthermore, because $V_F(T, N)$ increases strictly with T to the left-hand side of (2.14), N_F^* in (3.23) decreases strictly with T to N^* given in (2.14).

III. Step 3

There exist both unique N_L^* $(1 \leq N_L^* < \infty)$ and N_F^* $(1 \leq N_F^* < \infty)$ which satisfy (3.21) and (3.23) for given T when $\sum_{j=1}^{\infty} F^{(j)}(T) G^{(j)}(K) > c_N/(c_K - c_N)$. Compare the left-hand side of (3.21) and (3.23) by denoting

$$A(N) \equiv V_L(T, N) - V_F(T, N).$$

Then,

$$A(N+1) - A(N) = [r_{N+2}(K) - r_{N+1}(K)] \left\{ \sum_{j=0}^{N} [1 - F^{(j+1)}(T)] G^{(j)}(K) + \sum_{j=N+1}^{\infty} F^{(j+1)}(T) G^{(j)}(K) \right\} > 0,$$

which follows that $A(N)$ increases strictly with N to (Problem 3.5)

$$\lim_{N \to \infty} A(N) = \sum_{j=0}^{\infty} [1 - F^{(j)}(T)] G^{(j)}(K) > 0.$$

Thus, there exists a unique and minimum N_A^* $(1 \leq N_A^* < \infty)$ which satisfies $A(N) \geq 0$.

From (3.23), denote that

$$L(N_A^*) \equiv \sum_{j=0}^{N_A^*-1} F^{(j+1)}(T) G^{(j)}(K) [r_{N_A^*+1}(K) - r_{j+1}(K)]. \tag{3.25}$$

Then, the following comparative results can be given:

1. If $L(N_A^*) < c_N/(c_K - c_N)$, then $N_L^* \leq N_F^*$, and $C_L(T, N_L^*) \leq C_F(T, N_F^*)$, i.e., replacement last should be adopted.
2. If $L(N_A^* - 1) > c_N/(c_K - c_N)$, then $N_F^* \leq N_L^*$, and $C_F(T, N_F^*) \leq C_L(T, N_L^*)$, i.e., replacement first should be adopted.
3. If $L(N_A^* - 1) \leq c_N/(c_K - c_N) \leq L(N_A^*)$, then either replacement last or first might be better than the other, or the same with each other.

IV. Step 4

Denoting the left-hand side of (2.14) be $V(N)$ and $B_L(N) \equiv V(N) - V_L(T, N)$, $B_L(\infty) = 0$, and

$$B_L(N+1) - B_L(N) = \\ -[r_{N+2}(K) - r_{N+1}(K)] \sum_{j=N+1}^{\infty} F^{(j+1)}(T)G^{(j)}(K) < 0. \tag{3.26}$$

Similarly, denoting $B_F(N) \equiv V(N) - V_F(T, N)$, $B_F(1) = \overline{F}(T)[r_2(K) - r_1(K)]$, and

$$B_F(N+1) - B_F(N) = \\ [r_{N+2}(K) - r_{N+1}(K)] \sum_{j=0}^{N} [1 - F^{(j+1)}(T)]G^{(j)}(K) > 0, \tag{3.27}$$

which follows that $N^* \leq N_L^*$ and $N^* \leq N_F^*$, i.e., $C(N^*) \leq C_L(T, N_L^*)$ from (2.15) and (3.22), and $C(N^*) \leq C_F(T, N_F^*)$ from (2.15) and (3.24). That is, when the same PR cost c_N is provided for $C(N)$ in (2.13), $C_F(T, N)$ in (2.19) and $C_L(T, N)$ in (3.7), we can obviously know both replacement first and last policies cost more than the policy with only one variable N.

(3) Damage Z for Time T

I. Step 1

From (3.12), optimum Z_L^* for given T satisfies

$$\sum_{j=0}^{\infty} [1 - F^{(j+1)}(T)] \int_Z^K [\overline{G}(K - x) - \overline{G}(K - Z)] \mathrm{d}G^{(j)}(x) \\ + \overline{G}(K - Z) \sum_{j=0}^{\infty} G^{(j)}(K) - 1 = \frac{c_Z}{c_K - c_Z}. \tag{3.28}$$

Denoting the left-hand side of (3.28) by $V_L(T, Z)$,

$$\frac{\mathrm{d}V_L(T, Z)}{\mathrm{d}Z} = \\ \frac{\mathrm{d}\overline{G}(K - Z)}{\mathrm{d}Z} \left[\sum_{j=0}^{\infty} [1 - F^{(j+1)}(T)]G^{(j)}(Z) + \sum_{j=0}^{\infty} F^{(j+1)}(T)G^{(j)}(K) \right] > 0,$$

which follows that $V_L(T, Z)$ increases strictly with Z from

$$\sum_{j=0}^{\infty} F^{(j+1)}(T) \int_0^K [\overline{G}(K) - \overline{G}(K-x)]\mathrm{d}G^{(j)}(x) < 0$$

to $M_G(K)$. Therefore, if $M_G(K) > c_Z/(c_K - c_Z)$, then there exists a unique Z_L^* $(0 < Z_L^* < K)$ that satisfies (3.28), and the resulting cost rate is

$$\frac{C_L(T, Z_L^*)}{\lambda} = (c_K - c_Z)\overline{G}(K - Z_L^*). \tag{3.29}$$

If $M_G(K) \leq c_Z/(c_K - c_Z)$, then $Z_L^* = K$, and the resulting cost rate is given in (2.7). Furthermore, noting that $V_L(T, Z)$ decreases strictly with T from the left-hand side of (2.17), Z_L^* in (3.28) increases strictly with T from Z^* given in (2.17).

II. Step 2

From (2.26), optimum Z_F^* for given T satisfies

$$\sum_{j=0}^{\infty} F^{(j+1)}(T) \int_0^Z [\overline{G}(K-Z) - \overline{G}(K-x)]\mathrm{d}G^{(j)}(x) = \frac{c_Z}{c_K - c_Z}. \tag{3.30}$$

Denoting the left-hand side of (3.30) by $V_F(T, Z)$,

$$\frac{\mathrm{d}V_F(T, Z)}{\mathrm{d}Z} = \frac{\mathrm{d}\overline{G}(K-Z)}{\mathrm{d}Z} \sum_{j=0}^{\infty} F^{(j+1)}(T) G^{(j)}(Z) > 0,$$

which follows that $V_F(T, Z)$ increases strictly with Z from 0 to

$$\sum_{j=1}^{\infty} F^{(j)}(T) G^{(j)}(K).$$

Therefore, if $\sum_{j=1}^{\infty} F^{(j)}(T) G^{(j)}(K) > c_Z/(c_K - c_Z)$, then there exists a unique Z_F^* $(0 < Z_F^* < K)$ that satisfies (3.30), and the resulting cost rate is

$$\frac{C_F(T, Z_F^*)}{\lambda} = (c_K - c_T)\overline{G}(K - Z_F^*). \tag{3.31}$$

If $\sum_{j=1}^{\infty} F^{(j)}(T) G^{(j)}(K) \leq c_Z/(c_K - c_Z)$, then $Z_F^* = K$, and the resulting cost rate is given in (2.9). Furthermore, noting that $V_F(T, Z)$ increases strictly with T to the left-hand side of (2.17), Z_F^* in (3.30) decreases strictly with T to Z^* given in (2.17).

III. Step 3

Because $M_G(K) \geq \sum_{j=1}^{\infty} F^{(j)}(T) G^{(j)}(K)$, there exist both unique Z_L^* $(0 < Z_L^* < K)$ and Z_F^* $(0 < Z_F^* < K)$ which satisfy (3.28) and (3.30) when $\sum_{j=1}^{\infty} F^{(j)}(T) G^{(j)}(K) > c_Z/(c_F - c_Z)$. Compare the left-hand side of (3.28) and (3.30) by denoting

$$A(Z) \equiv V_L(T, Z) - V_F(T, Z).$$

Then,

$$\lim_{Z\to 0} A(Z) = \sum_{j=0}^{\infty} F^{(j+1)}(T)\int_0^K [\overline{G}(K) - \overline{G}(K-x)]\mathrm{d}G^{(j)}(x) < 0,$$

$$\lim_{Z\to K} A(Z) = \sum_{j=1}^{\infty}[1 - F^{(j)}(T)]G^{(j)}(K) > 0,$$

$$\frac{\mathrm{d}A(Z)}{\mathrm{d}Z} = \frac{\mathrm{d}\overline{G}(K-Z)}{\mathrm{d}Z}\left\{\sum_{j=0}^{\infty}[1 - F^{(j+1)}(T)]G^{(j)}(Z) + \sum_{j=0}^{\infty} F^{(j+1)}(T)[G^{(j)}(K) - G^{(j)}(Z)]\right\} > 0.$$

Thus, there exists a unique Z_A^* $(0 < Z_A^* < K)$ which satisfies $A(Z) = 0$.

From (3.30), denoting that

$$L(Z_A^*) \equiv \sum_{j=0}^{\infty} F^{(j+1)}(T)\int_0^{Z_A^*} [\overline{G}(K - Z_A^*) - \overline{G}(K-x)]\mathrm{d}G^{(j)}(x), \tag{3.32}$$

the following comparative results can be given:

1. If $L(Z_A^*) < c_Z/(c_K - c_Z)$, then $Z_L^* < Z_F^*$, and hence, from (3.29) and (3.31), $C_L(T, Z_L^*) < C_F(T, Z_F^*)$, i.e., replacement last should be adopted.
2. If $L(Z_A^*) > c_Z/(c_K - c_Z)$, then $Z_F^* < Z_L^*$ and $C_F(T, Z_F^*) < C_L(T, Z_L^*)$, i.e., replacement first should be adopted.
3. If $L(Z_A^*) = c_Z/(c_K - c_Z)$, then replacement first is the same with replacement last.

In addition, $V_L(T, Z)$ in (3.28) decreases with T and $V_F(T, Z)$ in (3.30) increases with T, that is, optimum Z_L^* increases with T while Z_F^* decreases with T. It also can be easily found that Z_A^* increases with T, in other words, RF would show more superior cases than RL when given T becomes smaller.

IV. Step 4

Denoting the left-hand side of (2.17) be $V(Z)$ and $B_L(Z) \equiv V(Z) - V_L(T, Z)$, $B_L(0) > 0$, $B_L(K) = 0$, and

$$\frac{\mathrm{d}B_L(Z)}{\mathrm{d}Z} = \frac{\mathrm{d}\overline{G}(K-Z)}{\mathrm{d}Z}\sum_{j=0}^{\infty} F^{(j+1)}(T)[G^{(j)}(Z) - G^{(j)}(K)] < 0. \tag{3.33}$$

Similarly, denoting $B_F(Z) \equiv V(Z) - V_F(T, Z)$, $B_F(0) = 0$, $B_F(K) > 0$, and

$$\frac{\mathrm{d}B_F(Z)}{\mathrm{d}Z} = \frac{\mathrm{d}\overline{G}(K-Z)}{\mathrm{d}Z} \sum_{j=0}^{\infty} [1 - F^{(j+1)}(T)] G^{(j)}(Z) > 0, \tag{3.34}$$

which follows that $V(Z) > V_L(T, Z)$ and $V(Z) > V_F(T, Z)$, i.e., $Z^* < Z_L^*$ and $Z^* < Z_F^*$, and moreover, from (2.18), (3.29), and (3.31), $C(Z^*) < C_L(T, Z_L^*)$ and $C(Z^*) < C_F(T, Z_F^*)$.

3.4 Numerical Examples

When $F(t) = 1 - \mathrm{e}^{-\lambda t}$, $G(x) = 1 - \mathrm{e}^{-\omega x}$ and the same values $c_i/(c_K - c_i)$ $(i = T, N, Z)$ are assigned, we compute optimum times λT_i^* for N, N_i^* for T and ωZ_i^* for T, their cost rates $C_i(T_i^*, N)/(\lambda c_T)$, $C_i(T, N_i^*)/(\lambda c_N)$ and $C_i(T, Z_i^*)/(\lambda c_Z)$ where $i = L, F$, and critical points of comparisons.

From Table 3.1, the comparative point of replacement last and first is to compare the relative size of the given N in (3.19) and optimum N^* in (2.14).

1. When $N = 2$, i.e., when N for PR is too early scheduled, then $\lambda T_F^* = \infty$, i.e., replacement first degrades into an action where PR should be done only at $N = 2$; however, finite optimum λT_L^* of replacement last can be found as PR is delayed at suitable times by λT_L^*.
2. When $N = 10$, i.e., when N for PR is too late scheduled, replacement first is more effective than replacement last to prevent failure losses, in which case, $\lambda T_L^* = 0.0$ without unreasonable delay and finite optimum λT_F^* can be found.
3. When N is scheduled at suitable size, e.g., $N = 5$, replacement last is dominant in cases when failure cost c_K is small. For examples, when $c_T/(c_K - c_T) = 0.01$, $\lambda T_L^* = 0.0$ and $\lambda T_F^* = 2.809$, and when $c_T/(c_K - c_T) = 0.2$, $\lambda T_L^* = 1.276$ and $\lambda T_F^* = \infty$.

It also can be found in Table 3.1 that when the failure cost c_K is relatively large, replacement last is still effective to obtain finite optimum λT_L^*, e.g., when $N = 2$ and $c_T/(c_K - c_T) = 0.01$, $\lambda T_L^* = 1.040$ and $\lambda T_F^* = \infty$.

In Table 3.2, $L(N_A^* - 1)$ and $L(N_A^*)$ are computed to compare with given $c_N/(c_K - c_N)$. Obviously, when given λT is too early scheduled, e.g., $\lambda T = 2.0$, $N_L^* \le N_F^*$ and $C_L(T, N_L^*)/(\lambda c_N) < C_F(T, N_F^*)/(\lambda c_N)$, i.e., replacement last saves more cost rate than replacement first does, and vice versa, e.g., $\lambda T = 10.0$, whose reasons are the same as those in Table 3.1. We next take $\lambda T = 5.0$ as an example:

1. When $c_N/(c_K - c_N) = 0.01$ 0.1 which is less than $L(N_A^* - 1) = 0.142$, $C_F(T, N_F^*)/(\lambda c_N) < C_L(T, N_L^*)/(\lambda c_N)$, i.e., replacement first is better than replacement last, though $N_L^* = N_F^* = 4$ for $c_N/(c_K - c_N) = 0.02$ and $N_L^* = N_F^* = 5$ for $c_N/(c_K - c_N) = 0.1$.

Table 3.1 Optimum λT_L^* and λT_F^*, and their cost rates $C_L(T_L^*, N)/(\lambda c_T)$ and $C_F(T_F^*, N)/(\lambda c_T)$ when $\omega K = 10.0$

$\frac{c_T}{c_K - c_T}$	$N = 2$				
	λT_L^*	$C_L(T_L^*, N)/(\lambda c_T)$	λT_F^*	$C_F(T_F^*, N)/(\lambda c_T)$	N^*
0.01	1.040	0.519	∞	0.525	3
0.02	1.856	0.479	∞	0.512	4
0.05	3.007	0.401	∞	0.505	4
0.1	4.001	0.337	∞	0.503	5
0.2	5.204	0.276	∞	0.501	6
0.5	7.422	0.209	∞	0.501	7
1.0	10.116	0.168	∞	0.500	9
$\frac{c_T}{c_K - c_T}$	$N = 5$				
	λT_L^*	$C_L(T_L^*, N)/(\lambda c_T)$	λT_F^*	$C_F(T_F^*, N)/(\lambda c_T)$	N^*
0.01	0.0	0.787	2.809	0.623	3
0.02	0.0	0.494	4.272	0.467	4
0.05	0.0	0.318	14.518	0.318	4
0.1	0.0	0.259	∞	0.259	5
0.2	1.276	0.230	∞	0.230	6
0.5	6.093	0.201	∞	0.212	7
1.0	9.677	0.167	∞	0.206	9
$\frac{c_T}{c_K - c_T}$	$N = 10$				
	λT_L^*	$C_L(T_L^*, N)/(\lambda c_T)$	λT_F^*	$C_F(T_F^*, N)/(\lambda c_T)$	N^*
0.01	0.0	5.082	2.092	0.707	3
0.02	0.0	2.596	2.582	0.567	4
0.05	0.0	1.103	3.426	0.426	4
0.1	0.0	0.606	4.300	0.343	5
0.2	0.0	0.357	5.549	0.275	6
0.5	0.0	0.208	8.779	0.202	7
1.0	0.0	0.158	21.013	0.158	9

2. When $c_N/(c_K - c_N) = 0.5, 1.0$ which is larger than $L(N_A^*) = 0.270$, $C_L(T, N_L^*)/(\lambda c_N) < C_F(T, N_F^*)/(\lambda c_N)$, i.e., replacement last is better than replacement first.
3. When $L(N_A^* - 1) = 0.142 < c_N/(c_K - c_N) = 0.2 < L(N_A^*) = 0.270$, $N_L^* = N_F^* = 6$, and we can not give analytical comparison of their cost rates; however, $C_L(T, N_L^*)/(\lambda c_N) < C_F(T, N_F^*)/(\lambda c_N)$ in the numerical computation, in which case, replacement last is better than replacement first.

The reason behind the case when either replacement last or first might be better than the other for $L(N_A^* - 1) < c_N/(c_K - c_N) < L(N_A^*)$ is that N_L^* and N_F^* are discrete decision variables, and their optimum cost rates are also discrete.

Table 3.2 Optimum N_L^* and N_F^*, and their cost rates $C_L(T, N_L^*)/(\lambda c_N)$ and $C_F(T, N_F^*)/(\lambda c_N)$ when $\omega K = 10.0$

$\frac{c_N}{c_K - c_N}$	$\lambda T = 2.0$			
	N_L^*	$C_L(T, N_L^*)/(\lambda c_N)$	N_F^*	$C_F(T, N_F^*)/(\lambda c_N)$
0.01	3	0.494	3	0.620
0.02	4	0.398	4	0.575
0.05	4	0.307	5	0.530
0.1	5	0.260	6	0.521
0.2	6	0.224	7	0.511
0.5	7	0.183	11	0.505
1.0	9	0.156	19	0.503
$N_A^* = 3$	$L(N_A^*) = 0.012$		$L(N_A^* - 1) = 0.003$	
$\frac{c_N}{c_K - c_N}$	$\lambda T = 5.0$			
	N_L^*	$C_L(T, N_L^*)/(\lambda c_N)$	N_F^*	$C_F(T, N_F^*)/(\lambda c_N)$
0.01	4	1.662	3	0.441
0.02	4	0.904	4	0.394
0.05	5	0.467	4	0.324
0.1	5	0.320	5	0.288
0.2	6	0.242	6	0.260
0.5	7	0.185	8	0.232
1.0	9	0.156	10	0.219
$N_A^* = 6$	$L(N_A^*) = 0.270$		$L(N_A^* - 1) = 0.142$	
$\frac{c_N}{c_K - c_N}$	$\lambda T = 10.0$			
	N_L^*	$C_L(T, N_L^*)/(\lambda c_N)$	N_F^*	$C_F(T, N_F^*)/(\lambda c_N)$
0.01	6	5.328	3	0.426
0.02	6	2.721	4	0.380
0.05	6	1.157	4	0.303
0.1	6	0.635	5	0.260
0.2	7	0.374	6	0.226
0.5	8	0.217	7	0.186
1.0	9	0.163	9	0.161
$N_A^* = 9$	$L(N_A^*) = 1.210$		$L(N_A^* - 1) = 0.871$	

Table 3.3 shows similar comparative results with Table 3.2. As the decision variables Z_L^* and Z_F^*, and their cost rates are continuous ones, we can compare the relative size of Z_L^* and Z_F^* to determine whether $C_L(T, Z_L^*)/(\lambda c_Z)$ is less than $C_F(T, Z_F^*)/(\lambda c_Z)$ or not by comparing $L(Z_A^*)$ with $c_Z/(c_K - c_Z)$.

Table 3.3 Optimum ωZ_L^* and ωZ_F^*, and their cost rates $C_L(T, Z_L^*)/(\lambda c_Z)$ and $C_F(T, Z_F^*)/(\lambda c_Z)$ when $\omega K = 10.0$

$\frac{c_Z}{c_K - c_Z}$	$\lambda T = 2.0$			
	ωZ_L^*	$C_L(T; Z_L^*)/(\lambda c_Z)$	ωZ_F^*	$C_F(T, Z_F^*)/(\lambda c_Z)$
0.01	4.245	0.317	4.788	0.545
0.02	4.694	0.248	5.452	0.530
0.05	5.380	0.197	6.344	0.517
0.1	5.942	0.173	7.026	0.511
0.2	6.527	0.155	7.711	0.507
0.5	7.320	0.137	8.621	0.504
1.0	7.931	0.126	9.312	0.502
	$\omega Z_A^* = 3.682$		$L(Z_A^*) = 0.003$	
$\frac{c_Z}{c_K - c_Z}$	$\lambda T = 5.0$			
	ωZ_L^*	$C_L(T, Z_L^*)/(\lambda c_Z)$	ωZ_F^*	$C_F(T, Z_F^*)/(\lambda c_Z)$
0.01	5.649	1.289	4.184	0.298
0.02	5.751	0.714	4.790	0.273
0.05	6.002	0.367	5.618	0.250
0.1	6.309	0.249	6.259	0.237
0.2	6.721	0.188	6.911	0.228
0.5	7.391	0.147	7.787	0.219
1.0	7.959	0.130	8.458	0.214
	$\omega Z_A^* = 6.376$		$L(Z_A^*) = 0.113$	
$\frac{c_Z}{c_K - c_Z}$	$\lambda T = 10.0$			
	ωZ_L^*	$C_L(T, Z_L^*)/(\lambda c_Z)$	ωZ_F^*	$C_F(T, Z_F^*)/(\lambda c_Z)$
0.01	7.012	5.038	4.022	0.253
0.02	7.032	2.571	4.591	0.224
0.05	7.092	1.092	5.365	0.194
0.1	7.183	0.598	5.965	0.177
0.2	7.344	0.351	6.576	0.163
0.5	7.709	0.202	7.399	0.148
1.0	8.115	0.152	8.032	0.140
	$\omega Z_A^* = 8.371$		$L(Z_A^*) = 1.444$	

3.5 Problem 3

3.1 Derive (3.4) by summing up the three cases of the probabilities that the unit is replaced at failure.

3.2 Prove that the inequality (3.10) does not hold for $T > 0$ and $N \geq 0$, when $r_{j+1}(K)$ increases strictly with j to 1, and prove that

$$\widetilde{Q}_1(T, N) \equiv \frac{\sum_{j=N}^{\infty}[(\lambda T)^j/j!][G^{(j)}(K) - G^{(j+1)}(K)]}{\sum_{j=N}^{\infty}[(\lambda T)^j/j!]G^{(j)}(K)}$$

increases strictly with T from $r_{N+1}(K)$ to 1, and increases strictly with N to 1.

3.3 Prove that $Q_3(T, Z) > Q_2(T, Z)$ always holds for any $T > 0$ and $Z \geq 0$, and prove that the inequality (3.18) always holds for any $N > 0$ and $Z \geq 0$.

3.4 Derive (3.19) from (3.9) when $F(t) = 1 - \mathrm{e}^{-\lambda t}$.

3.5 When $r_{N+1}(K)$ increases strictly with N to 1, formulate $A(N)$ and prove that $A(N)$ increases strictly with N to $\sum_{j=0}^{\infty}[1 - F^{(j)}(T)]G^{(j)}(K)$.

Chapter 4
Replacement Overtime and Middle Policies

Originally, replacement last was discussed to replace an operating unit without stopping successive working cycles [7]; however, the approach of *replacing over a planned measure*, i.e., delaying replacement actions over the planned time T until the running works have been completed [30], was provided as another choice to make replacement easier. Obviously, this kind of replacement increases failure possibility due to the action delay but saves replacement cost at the end of working cycles [8, 10]. More preventive replacement policies with overtime approaches have been summarized in [16]. In this chapter, the approach of overtime is observed for an operating unit in shock and damage models, and its replacement policy is named as *replacement overtime*.

In addition, both replacement first and replacement last become special cases for a trivariate PR policy or the policy including more than three PR scenarios, and we normally plan PR policies in practical applications at suitable middles times [13, 19]. In theory, a theoretical added approach of *whichever triggering event occurs middle* for the compound preventive replacement scenarios should be considered, rather than only the approaches of *whichever triggering event occurs first and last* have been applied. Here, the compound PRs include at least three PR scenarios to be planned, and the approach of first is still used between PRs and CR. That is, replacements can be planned: The unit undergoes CR immediately after failure, and is replaced preventively before failure at some thresholds or planned measurements, e.g., at time T, at shock N or at damage Z, *whichever occurs middle*, which is named as *replacement middle*.

In this chapter, we discuss replacement overtime and replacement middle policies under the same assumptions as those in Chap. 2: Shocks occur at a renewal process according to an interarrival distribution $F(t)$ with finite mean time μ. Each damage due to shocks is additive and has an identical distribution $G(x)$ with finite mean $1/\omega$. The unit fails when the total damage has exceeded a failure threshold K and CR is done immediately. In Sect. 4.1, two models of replacement overtime policies,

X. Zhao and T. Nakagawa, *Advanced Maintenance Policies for Shock and Damage Models*, Springer Series in Reliability Engineering,
https://doi.org/10.1007/978-3-319-70456-2_4

i.e., over time T policy and over damage Z policy, are optimized. The replacement overtime policies are also compared with replacement first from the point of cost rate and their modified replacement costs and times are computed in Sect. 4.1.2. Replacement middle policies will be formulated and computed in Sect. 4.2.

4.1 Replacement Overtime Policies

The unit is supposed to be operating successive works without stops except when it is damaged by shocks, that is, maintenance actions, e.g., inspection, repair, or replacement, are done only at shock times. We have known from the above models that the policy acting on time T could not be competent for replacing such an operating unit; however, it can be modified to be done at discrete shock times. That is, we can replace the operating unit at discrete shock times over a continuous measure.

Replacement policy acting on the approach of *replacing over a planned measure* is named as *replacement overtime*, and the following two models are given: (i) Replacement is done at the forthcoming shock over time T, and (ii) replacement is done at the next shock over damage Z.

4.1.1 Optimum Policies

(1) Replacement with Time T

Suppose that the unit is replaced correctively before time T ($0 \leq T \leq \infty$) when the total damage has exceeded failure K, and after T, it is replaced preventively at the forthcoming shock time, i.e., preventive replacement is done at the forthcoming shock over time T. Then, the probability that the unit is replaced over time T is

$$\sum_{j=0}^{\infty}[F^{(j)}(T) - F^{(j+1)}(T)]G^{(j+1)}(K), \tag{4.1}$$

and the probability that it is replaced at failure K is

$$\sum_{j=0}^{\infty} F^{(j)}(T)[G^{(j)}(K) - G^{(j+1)}(K)], \tag{4.2}$$

of which the failure possibility increases due to the delayed replacement acting over time T. Note that (4.1) + (4.2) = 1. Thus, the mean time to replacement is (Problem 4.1)

$$\sum_{j=0}^{\infty}[G^{(j)}(K)-G^{(j+1)}(K)]\left\{\int_0^T\left[\int_{T-t}^{\infty}(t+u)\mathrm{d}F(u)\right]\mathrm{d}F^{(j)}(t)\right.$$
$$\left.+\int_0^T t\mathrm{d}F^{(j+1)}(t)\right\}+\sum_{j=0}^{\infty}G^{(j+1)}(K)\int_0^T\left[\int_{T-t}^{\infty}(t+u)\mathrm{d}F(u)\right]\mathrm{d}F^{(j)}(t)$$
$$=\mu\sum_{j=0}^{\infty}F^{(j)}(T)G^{(j)}(K), \tag{4.3}$$

where $\mu \equiv \int_0^{\infty}\overline{F}(t)\mathrm{d}t < \infty$.

Therefore, the expected replacement cost rate is

$$C_O(T)=\frac{c_K-(c_K-c_O)\sum_{j=0}^{\infty}[F^{(j)}(T)-F^{(j+1)}(T)]G^{(j+1)}(K)}{\mu\sum_{j=0}^{\infty}F^{(j)}(T)G^{(j)}(K)}, \tag{4.4}$$

where c_O = replacement cost over time T, and c_K is given in (2.6) and $c_K > c_O$. Clearly, $\lim_{T\to 0}C_O(T)=C(1)$ in (2.13), and $\lim_{T\to\infty}C_O(T)=C$ in (2.7).

When $F(t)=1-\mathrm{e}^{-\lambda t}$, i.e., $F^{(j)}(t)=\sum_{i=j}^{\infty}[(\lambda t)^i/i!]\mathrm{e}^{-\lambda t}$ $(j=0,1,2,\ldots)$, the expected cost rate in (4.4) is

$$\frac{C_O(T)}{\lambda}=\frac{c_K-(c_K-c_O)\sum_{j=0}^{\infty}[(\lambda T)^j/j!]\mathrm{e}^{-\lambda T}G^{(j+1)}(K)}{\sum_{j=0}^{\infty}F^{(j)}(T)G^{(j)}(K)}. \tag{4.5}$$

Differentiating $C_O(T)$ with respect to T and setting it equal to zero,

$$Q_1(T)\sum_{j=0}^{\infty}F^{(j)}(T)G^{(j)}(K)-\sum_{j=0}^{\infty}F^{(j)}(T)[G^{(j)}(K)-G^{(j+1)}(K)]=\frac{c_O}{c_K-c_O},$$

or

$$\sum_{j=0}^{\infty}F^{(j)}(T)G^{(j)}(K)[Q_1(T)-r_{j+1}(K)]=\frac{c_O}{c_K-c_O}, \tag{4.6}$$

where for $N\geq 2$,

$$Q_1(T,N)\equiv\frac{\sum_{j=0}^{N-2}[(\lambda T)^j/j!][G^{(j+1)}(K)-G^{(j+2)}(K)]}{\sum_{j=0}^{N-2}[(\lambda T)^j/j!]G^{(j+1)}(K)}, \tag{4.7}$$

and

$$Q_1(T)\equiv\lim_{N\to\infty}Q_1(T,N)=\frac{\sum_{j=0}^{\infty}[(\lambda T)^j/j!][G^{(j+1)}(K)-G^{(j+2)}(K)]}{\sum_{j=0}^{\infty}[(\lambda T)^j/j!]G^{(j+1)}(K)}.$$

If $r_{j+1}(x) \equiv [G^{(j)}(x) - G^{(j+1)}(x)]/G^{(j)}(x)$ increases strictly with j to 1, then $Q_1(T)$ increases strictly with T from $r_2(K)$ to 1 (Problem 4.2). Thus, the left-hand side of (4.6) increases strictly with T from $[G(K)^2 - G^{(2)}(K)]/G(K)$ to $M_G(K) \equiv \sum_{j=1}^{\infty} G^{(j)}(K)$.

Therefore, we have the following optimum policies:

1. If $[G(K)^2 - G^{(2)}(K)]/G(K) \geq c_O/(c_K - c_O)$, then $T_O^* = 0$, i.e., the unit is replaced at the first shock, and the expected cost rate is given in $C(1)$ in (2.13).
2. If $[G(K)^2 - G^{(2)}(K)]/G(K) < c_O/(c_K - c_O) < M_G(K)$, then there exists a finite and unique T_O^* $(0 < T_O^* < \infty)$ which satisfies (4.6), and the resulting cost rate is

$$\frac{C_O(T_O^*)}{\lambda} = (c_K - c_O)Q_1(T_O^*). \tag{4.8}$$

3. If $M_G(K) \leq c_O/(c_K - c_O)$, the $T_O^* = \infty$, i.e., the unit is replaced only at failure, and the expected cost rate is given in C in (2.7).

In particular, when $G(x) = 1 - \mathrm{e}^{-\omega x}$, then $G^{(j)}(x) = \sum_{i=j}^{\infty}[(\omega x)^i/i!]\mathrm{e}^{-\omega x}$ and $M_G(x) = \omega x$, and $r_{j+1}(x)$ increases strictly with j from $\mathrm{e}^{-\omega K}$ to 1, and $Q_1(T)$ increases strictly with T from $\omega K/(\mathrm{e}^{\omega K} - 1)$ to 1 (Problem 4.2). Therefore, if

$$\frac{\omega K + \mathrm{e}^{-\omega K} - 1}{\mathrm{e}^{\omega K} - 1} < \frac{c_O}{c_K - c_O} < \omega K,$$

then there exists a finite and unique T_O^* $(0 < T_O^* < \infty)$ which satisfies (4.6). Because $(\omega K + \mathrm{e}^{-\omega K} - 1)/(\mathrm{e}^{\omega K} - 1) < \omega K/2$, if $\omega K/2 < c_O/(c_K - c_O) < \omega K$, then a finite T_O^* exists (Problem 4.3).

(2) Replacement with Damage Z

Suppose that the unit is replaced correctively at failure when the total damage has exceeded K, and preventive replacement is done at the next shock over damage Z $(0 \leq Z \leq K)$. Then, the probability that the unit is replaced over damage Z is

$$\sum_{j=0}^{\infty} \int_0^Z \left[\int_{Z-x}^{K-x} G(K - x - y)\mathrm{d}G(y)\right] \mathrm{d}G^{(j)}(x), \tag{4.9}$$

and the probability that it is replaced at failure K is

$$\sum_{j=0}^{\infty} \int_0^Z \overline{G}(K - x)\mathrm{d}G^{(j)}(x) + \sum_{j=0}^{\infty} \int_0^Z \left[\int_{Z-x}^{K-x} \overline{G}(K - x - y)\mathrm{d}G(y)\right] \mathrm{d}G^{(j)}(x), \tag{4.10}$$

where note that (4.9) + (4.10) = 1. Thus, the mean time to replacement is

$$\mu \left\{ \sum_{j=0}^{\infty}(j+2)\int_0^Z \left[\int_{Z-x}^{K-x} G(K-x-y)\mathrm{d}G(y)\right]\mathrm{d}G^{(j)}(x) \right.$$
$$+\sum_{j=0}^{\infty}(j+1)\int_0^Z \overline{G}(K-x)\mathrm{d}G^{(j)}(x)$$
$$\left. +\sum_{j=0}^{\infty}(j+2)\int_0^Z \left[\int_{Z-x}^{K-x} \overline{G}(K-x-y)\mathrm{d}G(y)\right]\mathrm{d}G^{(j)}(x) \right\}$$
$$= \mu\left[1+G(K)+\int_0^Z G(K-x)\mathrm{d}M_G(x)\right]. \qquad (4.11)$$

Therefore, the expected replacement cost rate is

$$C_O(Z) = \frac{c_K-(c_K-c_O)\{\int_Z^K G(K-x)\mathrm{d}G(x) + \int_0^Z[\int_{Z-x}^{K-x} G(K-x-y)\mathrm{d}G(y)]\mathrm{d}M_G(x)\}}{\mu[1+G(K)+\int_0^Z G(K-x)\mathrm{d}M_G(x)]}. \qquad (4.12)$$

Differentiating $C_O(Z)$ with respect to Z and setting it equal to zero,

$$r_2(K-Z)\left[1+G(K)+\int_0^Z G(K-x)\mathrm{d}M_G(x)\right]+\int_Z^K G(K-x)\mathrm{d}G(x)$$
$$+\int_0^Z\left[\int_{Z-x}^{K-x} G(K-x-y)\mathrm{d}G(y)\right]\mathrm{d}M_G(x)-1=\frac{c_O}{c_K-c_O}. \qquad (4.13)$$

When $r_2(K-Z)$ increases strictly with Z from $r_2(K)$ to 1, the left-hand side of (4.13) increases strictly with Z from $[G(K)^2-G^{(2)}(K)]/G(K)$ to $M_G(K)$, which corresponds to that of (4.6) (Problem 4.4).

Therefore, we have the following optimum policies:

1. If $[G(K)^2-G^{(2)}(K)]/G(K) \geq c_O/(c_K-c_O)$, then $Z_O^*=0$, and the resulting cost rate is

$$\mu C_O(0)=\frac{c_K-(c_K-c_O)G^{(2)}(K)}{1+G(K)}. \qquad (4.14)$$

2. If $[G(K)^2-G^{(2)}(K)]/G(K) < c_O/(c_K-c_O) < M_G(K)$, then there exists a finite and unique Z_O^* $(0 < Z_O^* < K)$ which satisfies (4.13), and the resulting cost rate is

$$\mu C_O(Z_O^*)=(c_K-c_O)r_2(K-Z_O^*). \qquad (4.15)$$

3. If $M_G(K) \le c_O/(c_K - c_O)$, then $Z_O^* = K$, and the expected cost rate is C given in (2.7).

It is of great interest that if

$$\frac{G(K)^2 - G^{(2)}(K)}{G(K)} < \frac{c_O}{c_K - c_O} < M_G(K),$$

then both finite T_O^* and Z_O^* exist.

When $F(t) = 1 - e^{-\lambda t}$ and $G(x) = 1 - e^{-\omega x}$, the expected cost rate in (4.12) is

$$\frac{C_O(Z)}{\lambda} = \frac{c_K - (c_K - c_O)\{1 - [1 + \omega(K - Z)]e^{-\omega(K-Z)}\}}{2 + \omega Z - e^{-\omega(K-Z)}}. \tag{4.16}$$

Optimum Z_O^* to minimize $C_O(Z)$ satisfies, from (4.13),

$$e^{-\omega(K-Z)}\left[\frac{\omega(K - Z)(1 + \omega Z)}{1 - e^{-\omega(K-Z)}} - 1\right] = \frac{c_O}{c_K - c_O}, \tag{4.17}$$

whose left-hand increases strictly with Z from $(\omega K + e^{-\omega K} - 1)/(e^{\omega K} - 1)$ to ωK. Therefore, if

$$\frac{\omega K + e^{-\omega K} - 1}{e^{\omega K} - 1} < \frac{c_O}{c_K - c_O} < \omega K,$$

then there exists a unique Z_O^* $(0 < Z_O^* < K)$ which satisfies (4.17), and the resulting cost rate is

$$\frac{C_O(Z_O^*)}{\lambda} = (c_K - c_O)\frac{\omega(K - Z_O^*)}{e^{\omega(K-Z_O^*)} - 1} = \frac{c_O + (c_K - c_O)e^{-\omega(K-Z_O^*)}}{1 + \omega Z_O^*}, \tag{4.18}$$

which agrees with (2.42) when $c_O = c_Z$ and $Z_O^* = Z$.

4.1.2 Comparisons of Replacement First and Overtime

When $F(t) = 1 - e^{-\lambda t}$, $G(x) = 1 - e^{-\omega x}$, and $c_O = c_T = c_N = c_Z$, we compare the above replacement overtime policies with those done at time T, at shock N, and at damage Z obtained in Sect. 2.1.1. The comparative case when $c_O \ne c_T, c_O \ne c_N$, and $c_O \ne c_Z$ will be addressed with numerical analyses.

(1) Replacement with Time T

We compare $C(T)$ in (2.36) and $C_O(T)$ in (4.5), of which optimum T^* satisfies (2.37) and its cost rate is given in (2.38), and optimum T_O^* satisfies (4.6) and its cost rate is given in (4.8).

Note that $\widetilde{Q}_1(T)$ in (2.37) and $Q_1(T)$ in (4.6) mean physically the probabilities of failures at respective $(j+1)$th and $(j+2)$th shocks, given that the unit has not failed at the jth and $(j+1)$th shocks, then $\widetilde{Q}_1(T)$ should be less than $Q_1(T)$ for any T $(0 \le T < \infty)$. It can be shown (Problem 4.5) that when $G^{(j+1)}(K)/G^{(j)}(K)$ decreases strictly with j, $\widetilde{Q}_1(T) < Q_1(T)$.

Next, comparing the left-hand side of (4.6) with that of (2.37) when $G(x) = 1 - \mathrm{e}^{-\omega x}$,

$$Q_1(T)\sum_{j=0}^{\infty} F^{(j)}(T)G^{(j)}(K) - \widetilde{Q}_1(T)\sum_{j=0}^{\infty} F^{(j+1)}(T)G^{(j)}(K)$$

$$-\sum_{j=0}^{\infty}\frac{(\lambda T)^j}{j!}\mathrm{e}^{-\lambda T}\frac{(\omega K)^j}{j!}\mathrm{e}^{-\omega K}$$

$$= [Q_1(T) - \widetilde{Q}_1(T)]\sum_{j=0}^{\infty} F^{(j)}(T)G^{(j)}(K) > 0,$$

which follows that $T_O^* < T^*$.

Furthermore, because

$$\widetilde{Q}_1(T)\sum_{j=0}^{\infty} F^{(j+1)}(T)G^{(j)}(K) - \sum_{j=0}^{\infty} F^{(j+1)}(T)\frac{(\omega K)^j}{j!}\mathrm{e}^{-\omega K}$$

$$= \widetilde{Q}_1(T)\sum_{j=0}^{\infty} F^{(j)}(T)G^{(j)}(K) - \sum_{j=0}^{\infty} F^{(j)}(T)\frac{(\omega K)^j}{j!}\mathrm{e}^{-\omega K},$$

in (2.37), (2.38) is

$$\frac{C(T^*)}{\lambda} = \frac{c_K - (c_K - c_O)\sum_{j=0}^{\infty}[(\lambda T^*)^j/j!]\mathrm{e}^{-\lambda T^*}G^{(j+1)}(K)}{\sum_{j=0}^{\infty} F^{(j)}(T^*)G^{(j)}(K)},$$

which agrees with (4.5) when $c_O = c_T$ and $T^* = T$. Therefore, from $T_O^* < T^*$, replacement overtime saves more cost than replacement first dose.

(2) Replacement with Shock N

We compare $C(N)$ in (2.39) with $C_O(T)$ in (4.5). For this purpose, we propose an extended replacement policy named as *replacement overtime first*: The unit is replaced preventively at the forthcoming shock over time T $(0 \le T \le \infty)$ or at shock N $(N = 1, 2, \ldots)$, whichever occurs first.

The probability that the unit is replaced at shock N before time T is

$$F^{(N)}(T)G^{(N)}(K), \tag{4.19}$$

the probability that it is replaced over time T is

$$\sum_{j=0}^{N-1}[F^{(j)}(T) - F^{(j+1)}(T)]G^{(j+1)}(K), \tag{4.20}$$

and the probability that it is replaced at failure K is

$$\sum_{j=0}^{N-1} F^{(j)}(T)[G^{(j)}(K) - G^{(j+1)}(K)], \tag{4.21}$$

where note that (4.19) + (4.20) + (4.21) = 1. The mean time to replacement is

$$G^{(N)}(K)\int_0^T t\mathrm{d}F^{(N)}(t) + \sum_{j=0}^{N-1} G^{(j+1)}(K)\int_0^T \left[\int_{T-u}^{\infty}(t+u)\mathrm{d}F(t)\right]\mathrm{d}F^{(j)}(u)$$
$$+\sum_{j=0}^{N-1}[G^{(j)}(K) - G^{(j+1)}(K)]\left\{\int_0^T \left[\int_{T-u}^{\infty}(t+u)\mathrm{d}F(t)\right]\mathrm{d}F^{(j)}(t)\right.$$
$$\left.+\int_0^T t\mathrm{d}F^{(j+1)}(t)\right\} = \frac{1}{\lambda}\sum_{j=0}^{N-1} F^{(j)}(T)G^{(j)}(K). \tag{4.22}$$

Therefore, the expected cost rate is

$$\frac{C_{OF}(T,N)}{\lambda} = \frac{c_O + (c_K - c_O)\sum_{j=0}^{N-1} F^{(j)}(T)[G^{(j)}(K) - G^{(j+1)}(K)]}{\sum_{j=0}^{N-1} F^{(j)}(T)G^{(j)}(K)}. \tag{4.23}$$

Clearly, when $c_O = c_N$, $\lim_{T\to\infty} C_{OF}(T,N) = C(N)$ in (2.39), $\lim_{N\to\infty} C_{OF}(T,N) = C_O(T)$ in (4.5), and $C_{OF}(0,N) = C_{OF}(T,1) = C_{OF}(0,1) = C(1)$ in (2.39).

We find optimum T_{OF}^* and N_{OF}^* to minimize $C_{OF}(T,N)$ for $0 < T < \infty$ and $N \geq 2$, using the same method in **(1)** of Sect. 2.1.2.

Forming the inequality $C_{OF}(T,N-1) - C_{OF}(T,N) > 0$,

$$\sum_{j=0}^{N-1} F^{(j)}(T)G^{(j)}(K)[r_N(K) - r_{j+1}(K)] < \frac{c_O}{c_K - c_O}. \tag{4.24}$$

Differentiating $C_{OF}(T,N)$ with respect to T and setting it equal to zero,

$$\sum_{j=0}^{N-1} F^{(j)}(T)G^{(j)}(K)[Q_1(T,N) - r_{j+1}(K)] = \frac{c_O}{c_K - c_O}, \tag{4.25}$$

where $Q_1(T,N)$ is given in (4.7).

Substituting (4.24) for (4.25),

$$Q_1(T, N) > r_N(K),$$

which does not hold for any N, because $Q_1(T, N) \le r_N(K)$ for $N \ge 2$, i.e., $C_{OF}(T, N)$ decreases with T and $T_{OF}^* = \infty$. In other words, the optimum policy to minimize $C_{OF}(T, N)$ degrades into the case when only finite N^* could be found in (2.40), i.e., replacement with shock N in (2.39) is better than replacement overtime in (4.5).

(3) Replacement with Damage *Z*

We compare $C(Z)$ in (2.42) and $C_O(Z)$ in (4.16), of which optimum Z^* satisfies (2.43) and its cost rate is given in (2.44), and optimum Z_O^* satisfies (4.17) and its cost rate is given in (4.18).

Next, comparing the left-hand side of (4.17) with that of (2.43),

$$\mathrm{e}^{-\omega(K-Z)}\left[\frac{\omega(K-Z)(1+\omega Z)}{1-\mathrm{e}^{-\omega(K-Z)}} - 1\right] - \omega Z \mathrm{e}^{-\omega(K-Z)}$$
$$= \frac{(1+\omega Z)\mathrm{e}^{-\omega(K-Z)}}{1-\mathrm{e}^{-\omega(K-Z)}}\left[\omega(K-Z) - 1 + \mathrm{e}^{-\omega(K-Z)}\right] > 0,$$

which follows that $Z_O^* < Z^*$. Thus, from (4.18), replacement with Z in (2.42) saves more cost than replacement overtime does.

4.1.3 Numerical Examples

(a) Case of $c_O = c_T = c_N = c_Z$

When $F(t) = 1 - \mathrm{e}^{-\lambda t}$ and $G(x) = 1 - \mathrm{e}^{-\omega x}$, Tables 4.1 and 4.2 present optimum λT_O^* and ωZ_O^*, and their cost rates $C_O(T_O^*)/(\lambda c_O)$ and $C_O(Z_O^*)/(\lambda c_O)$ for c_K/c_O and ωK. Optimum λT_O^* and ωZ_O^* increase with ωK and decrease with c_K/c_O, and their cost rates decrease with ωK and increase with c_K/c_O, all of which correspond to Tables 2.1 and 2.3 so that similar explanations can be given.

Our concern is to observe optimum λT_O^* and $C_O(T_O^*)/(\lambda c_O)$ in Table 4.1 comparing with λT^* $C(T^*)/(\lambda c_T)$ in Table 2.1, and to observe optimum ωZ_O^* and $C_O(Z_O^*)/(\lambda c_O)$ in Table 4.2 comparing with ωZ^* $C(Z^*)/(\lambda c_Z)$ in Table 2.3:

(a) $\lambda T_O^* < \lambda T^*$, which is explained as even though replacement overtime has higher failure possibility than that of replacement first due to delayed action, but the optimum T_O^* can be determined earlier. It can also be found that when $\omega K = 5.0$ for all of c_K/c_O, $C_O(T_O^*)/(\lambda c_O) < C(T^*)/(\lambda c_T)$, i.e., replacement overtime saves more cost rate than replacement first does. We will show when replacement overtime (or replacement first) would cost less by considering $c_O \ne c_T$ in the following discussion.

(b) $\omega Z_O^* < \omega Z^*$ and $C_O(Z_O^*)/(\lambda c_O) > C(Z^*)/(\lambda c_Z)$. Obviously, replacement first is better than replacement overtime from the viewpoint of cost saving. We have

Table 4.1 Optimum λT_O^* and its cost rate $C_O(T_O^*)/(\lambda c_O)$

c_K/c_O	$\omega K = 5.0$		$\omega K = 10.0$		$\omega K = 20.0$	
	λT_O^*	$C_O(T_O^*)/(\lambda c_O)$	λT_O^*	$C_O(T_O^*)/(\lambda c_O)$	λT_O^*	$C_O(T_O^*)/(\lambda c_O)$
5	1.265	0.613	3.683	0.262	9.459	0.108
10	0.611	0.833	2.727	0.327	8.010	0.124
15	0.349	0.969	2.311	0.368	7.341	0.134
20	0.194	1.066	2.056	0.398	6.918	0.141
30	0.009	1.194	1.741	0.444	6.378	0.152
50	0.000	1.981	1.402	0.507	5.772	0.165

Table 4.2 Optimum λZ_O^* and its cost rate $C_O(Z_O^*)/(\lambda c_O)$

c_K/c_O	$\omega K = 5.0$		$\omega K = 10.0$		$\omega K = 20.0$	
	ωZ_O^*	$C_O(Z_O^*)/(\lambda c_O)$	ωZ_O^*	$C_O(Z_O^*)/(\lambda c_O)$	ωZ_O^*	$C_O(Z_O^*)/(\lambda c_O)$
5	1.527	0.445	5.248	0.166	14.136	0.067
10	0.876	0.611	4.423	0.191	13.245	0.071
15	0.565	0.745	3.991	0.207	12.770	0.073
20	0.369	0.866	3.700	0.220	12.444	0.075
30	0.124	1.087	3.306	0.241	11.997	0.078
50	0.000	1.662	2.831	0.271	11.450	0.081

known that replacement acting on damage level Z ($Z < K$) provides a precise action to prevent failure threshold K, so that any delay for such a replacement would increase failure possibility much and increase the whole replacement cost rate. Even though, we will find the case when replacement overtime would cost less than replacement first dose by considering $c_O \neq c_Z$ in the following discussion.

(b) Case of $c_O \neq c_T, c_O \neq c_N, c_O \neq c_Z$

In general, the cost c_T for replacement first in (2.36) would be greater than c_O for replacement overtime in (4.5) because of the penalty of operational interruption. We next obtain a modified cost $\widehat{c}_O$ for c_O and its optimum replacement time $\widehat{T}_O$ analytically for the overtime policy, when the modified cost rate in (4.5) equals to that in (2.44) for given c_K and c_T. That is, we solve a modified cost $\widehat{c}_O$ to find a critical point at which replacement overtime saves more cost than replacement first does.

Compute a finite T^* in (2.37) and $C(T^*)$ in (2.38) for given c_K/c_T. Using T^* and $C(T^*)$, we compute a modified cost $\widehat{c}_O$ and its optimum replacement time $\widehat{T}_O$ which satisfies, from (2.38), (4.6), and (4.8),

$$(c_K - \widehat{c}_O)Q_4(\widehat{T}_O) = (c_K - c_T)\widetilde{Q}_1(T^*), \tag{4.26}$$

$$Q_4(\widehat{T}_O)\sum_{j=0}^{\infty} F^{(j)}(\widehat{T}_O)G^{(j)}(K) + \sum_{j=0}^{\infty}\frac{(\lambda\widehat{T}_O)^j}{j!}\mathrm{e}^{-\lambda\widehat{T}_O}G^{(j+1)}(K) = \frac{c_K}{c_K - \widehat{c}_O}. \tag{4.27}$$

That is, compute $\widehat{T}_O$ which satisfies

$$\sum_{j=0}^{\infty} F^{(j)}(\widehat{T}_O)G^{(j)}(K) + \frac{1}{Q_4(\widehat{T}_O)}\sum_{j=0}^{\infty}\frac{(\lambda\widehat{T}_O)^j}{j!}\mathrm{e}^{-\lambda\widehat{T}_O}G^{(j+1)}(K)$$
$$= \frac{c_K}{c_K - c_T}\frac{1}{\widetilde{Q}_1(T^*)},$$

and then, compute $\widehat{c}_O$ from (4.26) or (4.27).

It has been obtained analytically that replacement acting on shock N is better than the policy with overtime T when $c_O = c_N$. However, if the modified cost $\widehat{c}_O$ is less than c_N, replacement overtime would be at an advantage in cost saving (Problem 4.5).

We next observe the modified cost $\widehat{c}_O$ and its optimum damage level $\widehat{Z}_O$ in (4.16). Compute a finite Z^* in (2.43) and $C(Z^*)$ in (2.44). Using Z^* and $C(Z^*)$, we compute a modified cost $\widehat{c}_O$ for c_O and its optimum damage level $\widehat{Z}_O$ which satisfies, from (2.44), from (4.17), and (4.18),

$$(c_K - \widehat{c}_O)\frac{\omega(K - \widehat{Z}_O)\mathrm{e}^{-\omega(K-\widehat{Z}_O)}}{1 - \mathrm{e}^{-\omega(K-\widehat{Z}_O)}} = (c_K - c_Z)\mathrm{e}^{-\omega(K-Z^*)}, \tag{4.28}$$

$$\mathrm{e}^{-\omega(K-\widehat{Z}_O)}\frac{\omega(K - \widehat{Z}_O)(1 + \omega\widehat{Z}_O)}{1 - \mathrm{e}^{-\omega(K-\widehat{Z}_O)}} + [1 - \mathrm{e}^{-\omega(K-\widehat{Z}_O)}] = \frac{c_K}{c_K - \widehat{c}_O}. \tag{4.29}$$

That is, compute $\widehat{Z}_O$ which satisfies

$$(1 + \omega\widehat{Z}_O) + \frac{[1 - \mathrm{e}^{-\omega(K-\widehat{Z}_O)}]^2}{\omega(K - \widehat{Z}_O)\mathrm{e}^{-\omega(K-\widehat{Z}_O)}} = \frac{c_K}{c_K - c_Z}\frac{1}{\mathrm{e}^{-\omega(K-Z^*)}},$$

and then, compute $\widehat{c}_O$ from (4.28) or (4.29).

Table 4.3 presents modified cost $\widehat{c}_O/c_T$ and its optimum $\lambda\widehat{T}_O$ for c_K/c_T and ωK. Comparing with Table 4.1, it is shown that $\lambda\widehat{T}_O > \lambda T_O^*$ when a modified cost $\widehat{c}_O$ is greater than c_T, and $\lambda\widehat{T}_O < \lambda T_O^*$ when $\widehat{c}_O$ is less than c_T.

The modified computations also correspond to the comparisons between Tables 2.1 and 4.1, we take $\omega K = 10.0$ and $c_K/c_O = 5$ as an example, $C(T^*)/(\lambda c_T) = 0.260 < C_O(T_O^*)/(\lambda c_O) = 0.262$ when $c_T = c_O$, and from Table 4.6, modified cost $\widehat{c}_O$ for c_O could be computed as $\widehat{c}_O/c_T = 0.987$ to determine $C(T^*)/(\lambda c_T) = C_O(T_O^*)/(\lambda\widehat{c}_O)$, and its optimum replacement time $\lambda\widehat{T}_O = 3.660 < \lambda T_O^* = 3.683$. That is, if we estimate previously that the cost for replacement overtime is less than or equal to $0.987c_T$, then overtime policy should be adopted.

Table 4.3 Modified cost $\widehat{c}_O/c_T$ and its optimum $\lambda\widehat{T}_O$

c_K/c_T	$\omega K = 5.0$		$\omega K = 10.0$		$\omega K = 20.0$	
	$\widehat{c}_O/c_T$	$\lambda\widehat{T}_O$	$\widehat{c}_O/c_T$	$\lambda\widehat{T}_O$	$\widehat{c}_O/c_T$	$\lambda\widehat{T}_O$
5	1.013	1.282	0.987	3.660	0.436	7.773
10	1.160	0.725	1.029	2.759	0.994	7.999
15	1.275	0.498	1.056	2.361	1.001	7.342
20	1.375	0.366	1.075	2.117	1.005	6.926
30	1.553	0.212	1.108	1.816	1.013	6.393
50	1.858	0.057	1.149	1.491	1.021	5.795

Table 4.4 Modified cost $\widehat{c}_O/c_Z$ and its optimum $\omega\widehat{Z}_O$

c_K/c_Z	$\omega K = 5.0$		$\omega K = 10.0$		$\omega K = 20.0$	
	$\widehat{c}_O/c_Z$	$\omega\widehat{Z}_O$	$\widehat{c}_O/c_Z$	$\omega\widehat{Z}_O$	$\widehat{c}_O/c_Z$	$\omega\widehat{Z}_O$
5	0.746	1.230	0.876	5.079	0.939	14.049
10	0.627	0.523	0.843	4.238	0.929	13.157
15	0.539	0.167	0.822	3.792	0.923	12.678
20	–	–	0.808	3.489	0.919	12.350
30	–	–	0.785	3.076	0.913	11.899
50	–	–	0.753	2.575	0.908	11.345

In Table 4.3, we have also obtained the cases when $\widehat{c}_O/c_T > 1.0$, e.g., when $\omega K = 5.0$, that is, replacement overtime will be absolutely better than replacement first, which is also shown from Tables 2.1 and 4.1. Actually, the real c_O is much less than c_T, so that this newly proposed replacement overtime has an obvious advantage from the viewpoint of cost savings.

Table 4.4 presents modified cost $\widehat{c}_O/c_Z$ and its optimum $\omega\widehat{Z}_O$ for c_K/c_Z and ωK. Comparing with Table 4.2, it is shown that $\omega\widehat{Z}_O < \omega Z_O^*$ as all of modified cost $\widehat{c}_O$ are less than c_Z. We have known from Tables 2.3 and 4.2 that, $C(Z^*)/(\lambda c_Z) < C_O(Z_O^*)/(\lambda c_O)$ when $c_O = c_Z$, so that if we need to find then cases when replacement first could be cost saving, the modified $\widehat{c}_O/c_Z$ should be less than 1, as shown in Table 4.4. However, there are several cases when replacement first would be never better, which are marked with "-" in Table 4.4.

4.2 Replacement Middle Policies

We have obtained the comparative results of two approaches of whichever triggering event occurs first and last in Sect. 3.3 from the points of cost and performability. However, both approaches of first and last will become special cases for a trivariate PR policy or the policy including more than three PR scenarios. We cannot accept the frequent and unnecessary PRs acting on the approach of first, but we also cannot stand the PR delay with high failure possibility due to the approach of last.

In this section, we model replacement policies with three PR scenarios, i.e., planned time T, shock number N, and damage level Z, which could be done at suitable middle times, which are named as *replacement middle*, and comparisons with *replacement first* in Chap. 2 and *replacement last* in Chap. 3 are computed numerically.

We denote t_N and t_Z be the respective PR times at shock N and at damage Z, and t_K be the CR time at failure K. The PR policies acting on the approaches of whichever triggering event occurs first and last can be respectively described as: The unit is replaced preventively at $\min\{T, t_N, t_Z\}$ for replacement first and at $\max\{T, t_N, t_Z\}$ for replacement last.

From (2.6), the expected cost rate of replacement first is

$$C_F(T,N,Z) = \frac{c_P + (c_K - c_P)\sum_{j=0}^{N-1} F^{(j+1)}(T)\int_0^Z \overline{G}(K-x)\mathrm{d}G^{(j)}(x)}{\sum_{j=0}^{N-1} G^{(j)}(Z)\int_0^T [F^{(j)}(t) - F^{(j+1)}(t)]\mathrm{d}t}, \tag{4.30}$$

where c_P = replacement cost at T, N, or Z, and c_K = replacement cost at failure given in (2.6), where $c_K > c_P$. Optimum policies T_F^*, N_F^*, and Z_F^* for one PR scenario such that

$$\lim_{\substack{N\to\infty\\ Z\to K}} C_F(T_F^*, N, Z), \quad \lim_{\substack{T\to\infty\\ Z\to K}} C_F(T, N_F^*, Z), \quad \lim_{\substack{T\to\infty\\ N\to\infty}} C_F(T, N, Z_F^*)$$

and optimum policies T_F^*, N_F^*, and Z_F^* for the two PR scenarios such that

$$\lim_{Z\to K} C_F(T_F^*, N_F^*, Z), \quad \lim_{T\to\infty} C_F(T, N_F^*, Z_F^*), \quad \lim_{N\to\infty} C_F(T_F^*, N, Z_F^*)$$

have been discussed in Chap. 2.

From (3.6), the expected cost rate of replacement last is

$$C_L(T,N,Z) = \frac{c_P + (c_K - c_P)\{1 - \sum_{j=N}^{\infty}[1 - F^{(j+1)}(T)]\int_Z^K \overline{G}(K-x)\mathrm{d}G^{(j)}(x)\}}{\begin{array}{l}\sum_{j=N}^{\infty} G^{(j)}(Z)\int_T^{\infty}[F^{(j)}(t) - F^{(j+1)}(t)]\mathrm{d}t\\ +\sum_{j=0}^{N-1} G^{(j)}(K)\int_T^{\infty}[F^{(j)}(t) - F^{(j+1)}(t)]\mathrm{d}t\\ +\sum_{j=0}^{\infty} G^{(j)}(K)\int_0^T [F^{(j)}(t) - F^{(j+1)}(t)]\mathrm{d}t\end{array}}, \tag{4.31}$$

whose optimum policies T_L^*, N_L^*, and Z_L^* for the two PR scenarios such that

$$\lim_{Z\to 0} C_L(T_L^*, N_L^*, Z), \quad \lim_{T\to 0} C_L(T, N_L^*, Z_L^*), \quad \lim_{N\to 0} C_L(T_L^*, N, Z_L^*)$$

have been discussed in Chap. 3.

When $F(t) = 1 - \mathrm{e}^{-\lambda t}$ and $G(x) = 1 - \mathrm{e}^{-\omega x}$, we give three discussions in the following numerical examples, when $\lambda = 1.0$, $\omega = 1.0$ and $K = 10.0$:

1. Both given PR actions are lately planned before K arrives, e.g., $T, N, Z = 8.0$.
2. One of given PR actions is early planned while the other is lately done, e.g., $T, N, Z = 2.0$ and 8.0.
3. Both given PR actions are early planned before K arrives, e.g., $T, N, Z = 2.0$.

Tables 4.5 and 4.6 presents optimum T_F^*, N_F^* and Z_F^* in (4.30), and optimum T_L^*, N_L^* and Z_L^* minimizing $C_L(T, N, Z)$ in (4.31) for given T, N and Z for above three cases.

Obviously, finite T_F^* and N_F^* can be only found for case (a), and finite T_L^* and N_L^* only exist for case (c). This just fits the respective features of the approaches of first and last, that is, replacement first makes the unit safe while replacement last lets the unit operate longer. However, neither of finite T_i^*, N_i^* $(i = F, L)$ could be found for case (b). In other words, both approaches of first and last are not working for

Table 4.5 Optimum T_F^*, N_F^* and Z_F^* when $\lambda = 1.0$, $\mu = 1.0$ and $K = 10.0$

$\frac{c_P}{c_K - c_P}$	$N = 8,\ Z = 8.0$	$N = 2,\ Z = 8.0$	$N = 2,\ Z = 2.0$
	T_F^*	T_F^*	T_F^*
0.01	3.0092	∞	∞
0.02	3.9956	∞	∞
0.05	6.5001	∞	∞
0.10	13.8952	∞	∞
0.20	∞	∞	∞
0.50	∞	∞	∞
$\frac{c_P}{c_K - c_P}$	$T = 8.0,\ Z = 8.0$	$T = 2.0,\ Z = 8.0$	$T = 2.0,\ Z = 2.0$
	N_F^*	N_F^*	N_F^*
0.01	4	∞	∞
0.02	4	∞	∞
0.05	5	∞	∞
0.10	7	∞	∞
0.20	9	∞	∞
0.50	∞	∞	∞
$\frac{c_P}{c_K - c_P}$	$T = 8.0,\ N = 8$	$T = 2.0,\ N = 8$	$T = 2.0,\ N = 2$
	Z_F^*	Z_F^*	Z_F^*
0.01	4.0507	4.7880	5.0373
0.02	4.6306	5.4523	5.7215
0.05	5.4235	6.3443	6.6316
0.10	6.0410	7.0259	7.3225
0.20	6.6717	7.7114	8.0144
0.50	7.5235	8.6216	8.9299

Table 4.6 Optimum T_L^*, N_L^* and Z_L^* when $\lambda = 1.0, \omega = 1.0$ and $K = 10.0$

$\frac{c_P}{c_K - c_P}$	$N = 8,\ Z = 8.0$	$N = 2,\ Z = 8.0$	$N = 2,\ Z = 2.0$
	T_L^*	T_L^*	T_L^*
0.01	0.0	0.0	0.0
0.02	0.0	0.0	0.3401
0.05	0.0	0.0	1.8259
0.10	0.0	0.0	3.0883
0.20	0.0	0.0	4.5559
0.50	0.0	0.0	7.0848
$\frac{c_P}{c_K - c_P}$	$T = 8.0,\ Z = 8.0$	$T = 2.0,\ Z = 8.0$	$T = 2.0,\ Z = 2.0$
	N_L^*	N_L^*	N_L^*
0.01	0	0	2
0.02	0	0	3
0.05	0	0	4
0.10	0	0	5
0.20	0	0	6
0.50	0	0	7
$\frac{c_P}{c_K - c_P}$	$T = 8.0,\ N = 8$	$T = 2.0,\ N = 8$	$T = 2.0,\ N = 2$
	Z_L^*	Z_L^*	Z_L^*
0.01	6.7911	6.4278	4.2507
0.02	6.8179	6.4686	4.6973
0.05	6.8940	6.5817	5.3815
0.10	7.0088	6.7447	5.9430
0.20	7.2038	7.0045	6.5269
0.50	7.6246	7.5167	7.3200

the optimizations in case (b). In addition, optimal solutions of Z_F^* and Z_L^* always exist for all cases, and $Z_F^* < Z_L^*$ for case (a) and $Z_F^* > Z_L^*$ for case (c), which also indicates the features of the approaches of first and last.

4.2.1 *Model I*

We next give replacement models to show how the above three PR actions could be done at suitable middle times. For this purpose, the combined approach of first and last is employed, and the new approach of whichever occurs middle is then proposed.

For replacement middle I (RM-I), we consider an operating unit that is replaced preventively at time T or at $\max\{t_N, t_Z\}$, whichever occurs first. Here, the approach of whichever occurs last is used for variables t_N and t_Z, from the operation's perspective, to let the unit run as long as possible while it is limited by damage level Z. Meanwhile, the engineer uses the approach of whichever occurs first for T and $\max\{t_N, t_Z\}$ to keep the unit running in safety even though replacement is delayed by t_N or t_Z.

The following PR probabilities are considered: When PR is done at time T for $T < t_N < t_Z, T < t_Z < t_N, t_Z < T < t_N$ or $t_N < T < t_Z$,

$$\sum_{j=0}^{N-1} G^{(j)}(K)[F^{(j)}(T) - F^{(j+1)}(T)] + \sum_{j=N}^{\infty} G^{(j)}(Z)[F^{(j)}(T) - F^{(j+1)}(T)], \tag{4.32}$$

where we can find that replacement planned at T will be executed in the middle of t_N or t_Z to balance the profit of operation and loss of failure except t_N and t_Z are largely planned.

When PR is done at shock N for $t_Z \le t_N \le T$,

$$F^{(N)}(T)[G^{(N)}(K) - G^{(N)}(Z)], \tag{4.33}$$

and when PR is done at damage Z for $t_N < t_Z \le T$,

$$\sum_{j=N}^{\infty} F^{(j+1)}(T) \int_0^Z [G(K-x) - G(Z-x)]\mathrm{d}G^{(j)}(x), \tag{4.34}$$

where both cases are replaced in the middle times.

However, the failure probability increases in some degree due to replacement delay caused by the approach of last, but this will be improved by the approach of first, i.e., CR done at failure is divided into the following two cases of $t_K \le t_N \le T$ or $t_N < t_K \le T$:

$$\sum_{j=0}^{N-1} F^{(j+1)}(T)[G^{(j)}(K) - G^{(j+1)}(K)] + \sum_{j=N}^{\infty} F^{(j+1)}(T) \int_0^Z \overline{G}(K-x)\mathrm{d}G^{(j)}(x). \tag{4.35}$$

The mean time to replacement is (Problem 4.6)

$$\sum_{j=N}^{\infty} G^{(j)}(Z) \int_0^T [F^{(j)}(t) - F^{(j+1)}(t)]\mathrm{d}t + \sum_{j=0}^{N-1} G^{(j)}(K) \int_0^T [F^{(j)}(t) - F^{(j+1)}(t)]\mathrm{d}t. \tag{4.36}$$

Therefore, the expected cost rate is

$$C_{MI}(T,N,Z)=\frac{\begin{array}{c}c_P+(c_K-c_P)\{\sum_{j=0}^{N-1}F^{(j+1)}(T)[G^{(j)}(K)-G^{(j+1)}(K)]\\+\sum_{j=N}^{\infty}F^{(j+1)}(T)\int_0^Z\overline{G}(K-x)\mathrm{d}G^{(j)}(x)\}\end{array}}{\begin{array}{c}\sum_{j=N}^{\infty}G^{(j)}(Z)\int_0^T[F^{(j)}(t)-F^{(j+1)}(t)]\mathrm{d}t\\+\sum_{j=0}^{N-1}G^{(j)}(K)\int_0^T[F^{(j)}(t)-F^{(j+1)}(t)]\mathrm{d}t\end{array}}. \tag{4.37}$$

Next we suppose that the unit should be replaced at the forthcoming shock over time T for the above replacement. It can be shown that this replacement is modified to be done at discrete shock times and has the same replacement probabilities in (4.32)–(4.35). Letting t_T be the replacement time over T, the following probabilities for replacement actions are considered: The probability that PR is done for $t_T \le t_N$ and $t_T \le t_Z$ is

$$\sum_{j=0}^{N-1}G^{(j+1)}(Z)[F^{(j)}(T)-F^{(j+1)}(T)], \tag{4.38}$$

the probability that PR is done for $t_T > t_N$ and $t_T > t_Z$ is

$$\sum_{j=N}^{\infty}[F^{(j)}(T)-F^{(j+1)}(T)]\int_0^Z[G(K-x)-G(Z-x)]\mathrm{d}G^{(j)}(x), \tag{4.39}$$

the probability that PR is done for $t_N < t_T \le t_Z$ is

$$\sum_{j=N}^{\infty}G^{(j+1)}(Z)[F^{(j)}(T)-F^{(j+1)}(T)], \tag{4.40}$$

the probability that PR is done for $t_Z < t_T \le t_N$ is

$$\sum_{j=0}^{N-1}[F^{(j)}(T)-F^{(j+1)}(T)]\left\{\int_0^Z[G(K-x)-G(Z-x)]\mathrm{d}G^{(j)}(x)\right.$$
$$\left.+\int_Z^K G(K-x)\mathrm{d}G^{(j)}(x)\right\}, \tag{4.41}$$

and the failure probability due to overtime delay is

$$\sum_{j=0}^{N-1}[F^{(j)}(T)-F^{(j+1)}(T)][G^{(j)}(K)-G^{(j+1)}(K)]$$
$$+\sum_{j=N}^{\infty}[F^{(j)}(T)-F^{(j+1)}(T)]\int_0^Z\overline{G}(K-x)\mathrm{d}G^{(j)}(x). \tag{4.42}$$

The mean time to replacement is

$$\begin{aligned}
&\sum_{j=N}^{\infty}[G^{(j)}(Z)-G^{(j+1)}(Z)]\int_0^T t\mathrm{d}F^{(j+1)}(t)\\
&+\sum_{j=0}^{N-1}G^{(j)}(K)\int_0^T\left[\int_{T-t}^{\infty}(u+t)\mathrm{d}F(u)\right]\mathrm{d}F^{(j)}(t)\\
&+\sum_{j=N}^{\infty}G^{(j)}(Z)\int_0^T\left[\int_{T-t}^{\infty}(u+t)\mathrm{d}F(u)\right]\mathrm{d}F^{(j)}(t)\\
&+[G^{(N)}(K)-G^{(N)}(Z)]\int_0^T t\mathrm{d}F^{(N)}(t)\\
&+\sum_{j=0}^{N-1}[G^{(j)}(K)-G^{(j+1)}(K)]\int_0^T t\mathrm{d}F^{(j+1)}(t)\\
&=\mu\left[\sum_{j=0}^{N-1}F^{(j)}(T)G^{(j)}(K)+\sum_{j=N}^{\infty}F^{(j)}(T)G^{(j)}(Z)\right].
\end{aligned}\tag{4.43}$$

Therefore, the expected cost rate is

$$\widetilde{C}_{MI}(T,N,Z)=\frac{\begin{array}{c}c_P+(c_K-c_P)\{\sum_{j=0}^{N-1}F^{(j)}(T)[G^{(j)}(K)-G^{(j+1)}(K)]\\+\sum_{j=N}^{\infty}F^{(j)}(T)\int_0^Z\overline{G}(K-x)\mathrm{d}G^{(j)}(x)\}\end{array}}{\mu[\sum_{j=0}^{N-1}F^{(j)}(T)G^{(j)}(K)+\sum_{j=N}^{\infty}F^{(j)}(T)G^{(j)}(Z)]}.\tag{4.44}$$

Clearly, $\lim_{T\to\infty}\widetilde{C}_{MI}(T,N,Z)=\lim_{T\to\infty}C_{MI}(T,N,Z)$.

Table 4.7 presents optimum λT^*_{MI} in (4.37) and $\lambda\widetilde{T}^*_{MI}$ in (4.44) when $F(t)=1-\mathrm{e}^{-\lambda t}$, $G(x)=1-\mathrm{e}^{-\omega x}$ and $\omega K=10.0$. Obviously, it is easy to understand that $\widetilde{T}^*_{MI}<T^*_{MI}$ because of replacement delay. Comparing Table 4.7 to Tables 4.5 and 4.6, replacement policies in RM-I for case (b) can be optimized and finite solutions exist. That is, replacement middle I should be adopted when either action N or Z is early or lately planned. Not only that, the policies here are also working for case (a) when both N and Z are largely planned, which may provide engineers with more choices to compare them with the policy of replacement first.

4.2.2 Model II

As a theoretical added for the approaches of whichever occurs first and last in replacement modelings with three PR scenarios, we next propose a new approach of

Table 4.7 Optimum λT_{MI}^* and $\lambda \widetilde{T}_{MI}^*$ when $F(t) = 1 - e^{-\lambda t}$, $G(x) = 1 - e^{-\omega x}$ and $\omega K = 10.0$

$\frac{c_P}{c_K - c_P}$	$N = 8,\ \omega Z = 8.0$	$N = 2,\ \omega Z = 8.0$	$N = 2,\ \omega Z = 2.0$
	λT_{MI}^*	λT_{MI}^*	λT_{MI}^*
0.01	2.1081	2.9609	∞
0.02	2.6243	3.8251	∞
0.05	3.5791	5.5298	∞
0.10	4.7247	7.6528	∞
0.20	7.2202	11.5769	∞
0.50	∞	∞	∞
$\frac{c_P}{c_K - c_P}$	$N = 8,\ \omega Z = 8.0$	$N = 2,\ \omega Z = 8.0$	$N = 2,\ \omega Z = 2.0$
	$\lambda \widetilde{T}_{MI}^*$	$\lambda \widetilde{T}_{MI}^*$	$\lambda \widetilde{T}_{MI}^*$
0.01	1.7331	2.4475	∞
0.02	2.2248	3.2371	∞
0.05	3.0939	4.7033	∞
0.10	4.0414	6.3481	∞
0.20	5.5439	8.8568	∞
0.50	14.5247	15.2879	∞

whichever triggering event occurs middle. That is, PRs are scheduled preventively at time T $(0 < T < \infty)$, at shock N $(N = 1, 2, \ldots)$, or at damage Z $(0 < Z < K)$, whichever occurs middle. This replacement middle II (RM-II) can be described as replacement is done preventively at middle$\{T, t_N, t_Z\}$.

The following replacement probabilities are considered: When PR is done at time T for $t_Z < T < t_N$ or $t_N < T < t_Z$,

$$\sum_{j=0}^{N-1}[F^{(j)}(T) - F^{(j+1)}(T)][G^{(j)}(K) - G^{(j)}(Z)]$$
$$+ \sum_{j=N}^{\infty} G^{(j)}(Z)[F^{(j)}(T) - F^{(j+1)}(T)], \tag{4.45}$$

when PR is done at shock N for $T < t_N < t_Z$ or $t_Z \le t_N < T$,

$$G^{(N)}(Z)[1 - F^{(N)}(T)] + F^{(N)}(T)[G^{(N)}(K) - G^{(N)}(Z)], \tag{4.46}$$

when PR is done at damage Z for $T < t_Z \le t_N$ or $t_N < t_Z \le T$,

$$\sum_{j=0}^{N-1}[1-F^{(j+1)}(T)]\int_0^Z[G(K-x)-G(Z-x)]\mathrm{d}G^{(j)}(x)$$
$$+\sum_{j=N}^{\infty}F^{(j+1)}(T)\int_0^Z[G(K-x)-G(Z-x)]\mathrm{d}G^{(j)}(x), \tag{4.47}$$

and when CR is done at failure K for $t_K \le t_N < T, t_N < t_K < T$ or $T < t_K \le t_N$,

$$\sum_{j=0}^{N-1}F^{(j+1)}(T)[G^{(j)}(K)-G^{(j+1)}(K)]+\sum_{j=N}^{\infty}F^{(j+1)}(T)\int_0^Z\overline{G}(K-x)\mathrm{d}G^{(j)}(x)$$
$$+\sum_{j=0}^{N-1}[1-F^{(j+1)}(T)]\int_0^Z\overline{G}(K-x)\mathrm{d}G^{(j)}(x). \tag{4.48}$$

The mean time to replacement is (Problem 4.7)

$$T\sum_{j=0}^{N-1}[F^{(j)}(T)-F^{(j+1)}(T)][G^{(j)}(K)-G^{(j)}(Z)]$$
$$+T\sum_{j=N}^{\infty}G^{(j)}(Z)[F^{(j)}(T)-F^{(j+1)}(T)]$$
$$+G^{(N)}(Z)\int_T^{\infty}t\mathrm{d}F^{(N)}(t)+[G^{(N)}(K)-G^{(N)}(Z)]\int_0^T t\mathrm{d}F^{(N)}(t)$$
$$+\sum_{j=N}^{\infty}[G^{(j)}(Z)-G^{(j+1)}(Z)]\int_0^T t\mathrm{d}F^{(j+1)}(t)$$
$$+\sum_{j=0}^{N-1}[G^{(j)}(Z)-G^{(j+1)}(Z)]\int_T^{\infty}t\mathrm{d}F^{(j+1)}(t)$$
$$+\sum_{j=0}^{N-1}[G^{(j)}(K)-G^{(j+1)}(K)]\int_0^T t\mathrm{d}F^{(j+1)}(t)$$
$$=\sum_{j=0}^{N-1}G^{(j)}(Z)\int_T^{\infty}[F^{(j)}(t)-F^{(j+1)}(t)]\mathrm{d}t$$
$$+\sum_{j=0}^{N-1}G^{(j)}(K)\int_0^T[F^{(j)}(t)-F^{(j+1)}(t)]\mathrm{d}t$$
$$+\sum_{j=N}^{\infty}G^{(j)}(Z)\int_0^T[F^{(j)}(t)-F^{(j+1)}(t)]\mathrm{d}t. \tag{4.49}$$

Therefore, the expected cost rate is

$$C_{MII}(T,N,Z)=\frac{\begin{array}{l}c_P+(c_K-c_P)\{\sum_{j=0}^{N-1}F^{(j+1)}(T)[G^{(j)}(K)-G^{(j+1)}(K)]\\ +\sum_{j=N}^{\infty}F^{(j+1)}(T)\int_0^Z\overline{G}(K-x)\mathrm{d}G^{(j)}(x)\\ +\sum_{j=0}^{N-1}[1-F^{(j+1)}(T)]\int_0^Z\overline{G}(K-x)\mathrm{d}G^{(j)}(x)\}\end{array}}{\begin{array}{l}\sum_{j=0}^{N-1}G^{(j)}(Z)\int_T^{\infty}[F^{(j)}(t)-F^{(j+1)}(t)]\mathrm{d}t\\ +\sum_{j=0}^{N-1}G^{(j)}(K)\int_0^T[F^{(j)}(t)-F^{(j+1)}(t)]\mathrm{d}t\\ +\sum_{j=N}^{\infty}G^{(j)}(Z)\int_0^T[F^{(j)}(t)-F^{(j+1)}(t)]\mathrm{d}t\end{array}}. \tag{4.50}$$

Clearly, $\lim_{T\to\infty}C_{MII}(T,N,Z)=\lim_{T\to\infty}C_{MI}(T,N,Z)$.

Table 4.8 presents optimum λT^*_{MII}, N^*_{MII} and ωZ^*_{MII} in (4.50) for given λT, N and ωZ when $F(t)=1-\mathrm{e}^{-\lambda t}$, $G(x)=1-\mathrm{e}^{-\omega x}$ and $\omega K=10.0$. Table 4.8 shows

Table 4.8 Optimum λT^*_{MII}, N^*_{MII} and ωZ^*_{MII} when $F(t)=1-\mathrm{e}^{-\lambda t}$, $G(x)=1-\mathrm{e}^{-\omega x}$ and $\omega K=10.0$

$\frac{c_P}{c_K-c_P}$	$N=8$, $\omega Z=8.0$	$N=2$, $\omega Z=8.0$	$N=2$, $\omega Z=2.0$
	λT^*_{MII}	λT^*_{MII}	λT^*_{MII}
0.01	0.0	2.1880	∞
0.02	0.0	3.3942	∞
0.05	0.0	5.3739	∞
0.10	0.0	7.6045	∞
0.20	0.0	11.5706	∞
0.50	0.0	32.0972	∞
$\frac{c_P}{c_K-c_P}$	$\lambda T=8.0$, $\omega Z=8.0$	$\lambda T=2.0$, $\omega Z=8.0$	$\lambda T=2.0$, $\omega Z=2.0$
	N^*_{MII}	N^*_{MII}	N^*_{MII}
0.01	1	3	4
0.02	1	4	5
0.05	1	5	7
0.10	1	7	∞
0.20	2	9	∞
0.50	4	15	∞
$\frac{c_P}{c_K-c_P}$	$\lambda T=8.0$, $N=8$	$\lambda T=2.0$, $N=8$	$\lambda T=2.0$, $N=2$
	ωZ^*_{MII}	ωZ^*_{MII}	ωZ^*_{MII}
0.01	6.1756	4.2528	4.5618
0.02	6.2343	4.7088	5.2182
0.05	6.3915	5.4093	6.1060
0.10	6.6072	5.9877	6.7865
0.20	6.9326	6.5935	7.4718
0.50	7.5314	7.4236	8.3817

that finite T_{MII}^* and N_{MII}^* exist for case (b), in which the optimization gab between approaches of first and last can be filled. Further, it also can be shown from Tables 4.5, 4.6, 4.7 and 4.8 that $T_F^* > T_{MII}^*(T_{MI}^*$ or $\widetilde{T}_{MI}^*) > T_L^*$ and $N_F^* \geq N_{MII}^* > N_L^*$. With respect to Z_{MII}^*, Z_{MII}^* exists for all of three cases, and $Z_F^* < Z_{MII}^* < Z_L^*$ for case (a) and $Z_F^* > Z_{MII}^* > Z_L^*$ for case (c). All this indicates that the approach of middle should be taken in considerations in replacement modelings with three PR scenarios.

4.2.3 Other Models

We next give another two extended models in which PRs may possibly be done at middle times: Suppose that an operating unit is replaced preventively at the forthcoming shock over time T, or at $\min\{t_N, t_Z\}$, whichever occurs first. Then, the probability that the unit is replaced over time T is

$$\sum_{j=0}^{N-1}[F^{(j)}(T) - F^{(j+1)}(T)]G^{(j)}(Z), \tag{4.51}$$

the probability that it is replaced at shock N is

$$F^{(N)}(T)G^{(N)}(Z), \tag{4.52}$$

the probability that it is replaced at damage Z is

$$\sum_{j=0}^{N-1} F^{(j+1)}(T) \int_0^Z [G(K-x) - G(Z-x)]\mathrm{d}G^{(j)}(x), \tag{4.53}$$

and the probability that it is replaced at failure K is

$$\sum_{j=0}^{N-1} F^{(j+1)}(T) \int_0^Z \overline{G}(K-x)\mathrm{d}G^{(j)}(x), \tag{4.54}$$

where note that (4.51) + (4.52) + (4.53) + (4.54) = 1. The mean time to replacement is

$$G^{(N)}(Z)\int_0^T t\mathrm{d}F^{(N)}(t) + \sum_{j=0}^{N-1}[G^{(j)}(Z) - G^{(j+1)}(Z)]\int_0^T t\mathrm{d}F^{(j+1)}(t)$$
$$+ \sum_{j=0}^{N-1} G^{(j)}(Z)\int_0^T \left[\int_{T-t}^{\infty}(t+u)\mathrm{d}F(u)\right]\mathrm{d}F^{(j)}(t)$$

$$= \mu \sum_{j=0}^{N-1} F^{(j)}(T) G^{(j)}(Z). \tag{4.55}$$

Therefore, the expected cost rate is

$$C_{MIII}(T,N,Z) = \frac{\begin{array}{l} c_K - (c_K - c_O)\sum_{j=0}^{N-1}[F^{(j)}(T) - F^{(j+1)}(T)]G^{(j)}(Z) \\ -(c_K - c_N)F^{(N)}(T)G^{(N)}(Z) \\ -(c_K - c_Z)\sum_{j=0}^{N-1} F^{(j+1)}(T)\int_0^Z [G(K-x) - G(Z-x)]\mathrm{d}G^{(j)}(x) \end{array}}{\mu \sum_{j=0}^{N-1} F^{(j)}(T)G^{(j)}(Z)}, \tag{4.56}$$

where c_O = replacement cost over time T, and c_N, c_Z and c_K are given in (2.6).

Furthermore, the probability in (4.51) is classified into three cases: The unit is replaced at shock N with probability

$$\sum_{j=0}^{N-1}[F^{(j)}(T) - F^{(j+1)}(T)]G^{(j+1)}(Z),$$

at damage Z with probability

$$\sum_{j=0}^{N-1}[F^{(j)}(T) - F^{(j+1)}(T)]\int_0^Z [G(K-x) - G(Z-x)]\mathrm{d}G^{(j)}(x),$$

and at failure with probability

$$\sum_{j=0}^{N-1}[F^{(j)}(T) - F^{(j+1)}(T)]\int_0^Z \overline{G}(K-x)\mathrm{d}G^{(j)}(x).$$

In this case, the expected cost rate is

$$\widetilde{C}_{MIII}(T,N,Z) = \frac{\begin{array}{l} c_N + (c_Z - c_N)\sum_{j=0}^{N-1} F^{(j)}(T)\int_0^Z [G(K-x) - G(Z-x)]\mathrm{d}G^{(j)}(x) \\ +(c_K - c_N)\sum_{j=0}^{N-1} F^{(j)}(T)\int_0^Z \overline{G}(K-x)\mathrm{d}G^{(j)}(x) \end{array}}{\mu \sum_{j=0}^{N-1} F^{(j)}(T)G^{(j)}(Z)}. \tag{4.57}$$

Suppose that the unit is replaced preventively at $\max\{t_N, t_Z\}$ before time T, at $\min\{t_N, t_Z\}$ after time T, or at the forthcoming shock time over time T, i.e., in cases of $\{t_N < T < t_Z\}$ and $\{t_Z < T < t_N\}$. Then, the probabilities of preventive and corrective replacements are the same as those in (4.45)–(4.48), and the mean time to replacement is

$$\begin{aligned}
&\sum_{j=N}^{\infty} G^{(j)}(Z) \int_0^T \left[\int_{T-t}^{\infty} (t+u)\mathrm{d}F(u)\right] \mathrm{d}F^{(j)}(t) \\
&+ \sum_{j=0}^{N-1} [G^{(j)}(K) - G^{(j)}(Z)] \int_0^T \left[\int_{T-t}^{\infty} (t+u)\mathrm{d}F(u)\right] \mathrm{d}F^{(j)}(t) \\
&+ [G^{(N)}(K) - G^{(N)}(Z)] \int_0^T t\mathrm{d}F^{(N)}(t) + G^{(N)}(Z) \int_T^{\infty} t\mathrm{d}F^{(N)}(t) \\
&+ \sum_{j=N}^{\infty} [G^{(j)}(Z) - G^{(j+1)}(Z)] \int_0^T t\mathrm{d}F^{(j+1)}(t) \\
&+ \sum_{j=0}^{N-1} [G^{(j)}(Z) - G^{(j+1)}(Z)] \int_T^{\infty} t\mathrm{d}F^{(j+1)}(t) \\
&+ \sum_{j=0}^{N-1} [G^{(j)}(K) - G^{(j+1)}(K)] \int_0^T t\mathrm{d}F^{(j+1)}(t) \\
&= \mu \left\{ \sum_{j=0}^{N-1} F^{(j)}(T)G^{(j)}(K) + \sum_{j=N}^{\infty} F^{(j)}(T)G^{(j)}(Z) \right. \\
&\quad \left. + \sum_{j=0}^{N-1} [1 - F^{(j)}(T)]G^{(j)}(Z) \right\}.
\end{aligned} \tag{4.58}$$

Therefore, the expected cost rate is

$$C_{MIV}(T, N, Z) = \frac{\begin{array}{l} c_K - (c_K - c_O)\{\sum_{j=N}^{\infty}[F^{(j)}(T) - F^{(j+1)}(T)]G^{(j)}(Z) \\ +\sum_{j=0}^{N-1}[F^{(j)}(T) - F^{(j+1)}(T)][G^{(j)}(K) - G^{(j)}(Z)]\} \\ -(c_K - c_N)\{F^{(N)}(T)[G^{(N)}(K) - G^{(N)}(Z)] + [1 - F^{(N)}(T)]G^{(N)}(Z)\} \\ -(c_K - c_Z)\{\sum_{j=N}^{\infty} F^{(j+1)}(T) \int_0^Z [G(K-x) - G(Z-x)]\mathrm{d}G^{(j)}(x) \\ +\sum_{j=0}^{N-1}[1 - F^{(j+1)}(T)] \int_0^Z [G(K-x) - G(Z-x)]\mathrm{d}G^{(j)}(x)\} \end{array}}{\begin{array}{l} \mu\{\sum_{j=0}^{N-1} F^{(j)}(T)G^{(j)}(K) + \sum_{j=N}^{\infty} F^{(j)}(T)G^{(j)}(Z) \\ +\sum_{j=0}^{N-1}[1 - F^{(j)}(T)]G^{(j)}(Z)\} \end{array}}. \tag{4.59}$$

Furthermore, the probability that the unit is replaced over time T is classified into three cases: The unit is replaced at shock N with probability

$$\sum_{j=N}^{\infty}[F^{(j)}(T)-F^{(j+1)}(T)]G^{(j+1)}(Z),$$

at damage Z with probability

$$\sum_{j=N}^{\infty}[F^{(j)}(T)-F^{(j+1)}(T)]\int_0^Z[G(K-x)-G(Z-x)]\mathrm{d}G^{(j)}(x)$$
$$+\sum_{j=0}^{N-1}[F^{(j)}(T)-F^{(j+1)}(T)]\int_Z^K G(K-x)\mathrm{d}G^{(j)}(x),$$

and at failure with probability

$$\sum_{j=N}^{\infty}[F^{(j)}(T)-F^{(j+1)}(T)]\int_0^Z \overline{G}(Z-x)\mathrm{d}G^{(j)}(x)$$
$$+\sum_{j=0}^{N-1}[F^{(j)}(T)-F^{(j+1)}(T)]\int_Z^K \overline{G}(K-x)\mathrm{d}G^{(j)}(x).$$

In this case, the expected cost rate is

$$\widetilde{C}_{MIV}(T,N,Z)=$$
$$\frac{\begin{array}{l}c_K-(c_K-c_N)\{F^{(N)}(T)[G^{(N)}(K)-G^{(N)}(Z)]+[1-F^N(T)]G^{(N)}(Z)\\ +\sum_{j=N}^{\infty}[F^{(j)}(T)-F^{(j+1)}(T)]G^{(j+1)}(Z)\}\\ -(c_K-c_Z)\{\sum_{j=0}^{N-1}[1-F^{(j+1)}(T)]\int_0^Z[G(K-x)-G(Z-x)]\mathrm{d}G^{(j)}(x)\\ +\sum_{j=N}^{\infty}F^{(j)}(T)\int_0^Z[G(K-x)-G(Z-x)]\mathrm{d}G^{(j)}(x)\\ +\sum_{j=0}^{N-1}[F^{(j)}(T)-F^{(j+1)}(T)]\int_Z^K G(K-x)\mathrm{d}G^{(j)}(x)\}\end{array}}{\begin{array}{l}\mu\{\sum_{j=0}^{N-1}F^{(j)}(T)G^{(j)}(K)+\sum_{j=N}^{\infty}F^{(j)}(T)G^{(j)}(Z)\\ +\sum_{j=0}^{N-1}[1-F^{(j)}(T)]G^{(j)}(Z)\}\end{array}}. \tag{4.60}$$

In general, it is very difficult to obtain optimum policies to minimize the above expected cost rates theoretically. However, adjusting and modifying these models to real situations would be possible with numerical analyses.

4.3 Problem 4

4.1 Derive the mean time to replacement in (4.3).

4.2 Prove that when $r_{j+1}(K)$ increases strictly with j to 1, $Q_1(T, N)$ increases strictly with T from $r_2(K)$ to $r_N(K)$ for $N \geq 2$ and increases strictly with N for $N \geq 2$ from $r_2(K)$ to $Q_1(T)$, and $Q_1(T)$ increases strictly from $r_2(K)$ to 1.

4.3 Prove that for $\omega K > 0$,

$$\frac{\omega K + \mathrm{e}^{-\omega K} - 1}{\mathrm{e}^{\omega K} - 1} < \frac{\omega K}{2}.$$

4.4 Prove that when $r_2(x)$ decreases strictly with x from 1, the left-hand side of (4.13) increases strictly with Z from $[G(K)^2 - G^{(2)}(K)]/G(K)$ to $M_G(K)$.

4.5 Prove that when $G^{(j+1)}(x)/G^{(j)}(x)$ decreases strictly with j, $\widetilde{Q}_1(T) < Q_1(T)$.

4.6 Show that (4.32)+(4.33)+(4.34)+(4.35) = 1 and derive (4.36).

4.7 Show that (4.45)+(4.46)+(4.47)+(4.48) = 1 and derive (4.49).

Chapter 5
Replacement Policies with Repairs

In real situation, the damaged unit has probabilities of minor failures (or malfunctions) at shock times, in which case, minimal repairs that cost less are always considered to resume quickly the operation of unit. Repair models have been studied especially for large and complex systems, which consist of many kinds of units [1, 22]. In recent works, models such as imperfect repair considering time-dependent repair effectiveness [32], repairable system subjected to minimal repair [33], inspection modeling for repairs [34], post-warranty maintenance with repair time threshold [35], random working models with replacement and minimal repair [14, 36], age-based replacement with repair for shocks and degradation [37, 38], etc., have been studied extensively.

In this chapter, a new unit with damage level 0 begins to operate at time 0 and degrades with damage produced by shocks, and the probability distributions $F(t)$ and $G(x)$ for shocks and damage have been supposed in Sect. 2.1. The unit fails with probability $p(x)$ when the total damage reaches x at some shock, and can be quickly resumed to operation after minimal repair at failure, where the function $p(x)$ increases strictly with damage x from $p(0) \equiv 0$. In addition, the definition of minimal repair, in periodic replacement modelings [1], has claimed that the instantaneous failure rate or simply the failure rate after repair has the same monotone property as it has been before failure. So that we suppose here for replacement policies that the total damage x at each failure remains undisturbed by minimal repair and the time for each repair is negligible.

Section 5.1 gives three preventive replacement policies that have been discussed in Chaps. 2 and 3, where $p(x)$ is considered and the failure threshold K is set to be infinity. Optimum policies with respective and combined time T, shock N and damage Z are derived analytically. Replacement last polices with minimal repairs are formulated and optimized in Sect. 5.2, in which comparisons of replacement first

X. Zhao and T. Nakagawa, *Advanced Maintenance Policies for Shock and Damage Models*, Springer Series in Reliability Engineering,
https://doi.org/10.1007/978-3-319-70456-2_5

and last are made. In Sects. 5.3 and 5.4, the approach of replacement overtime is used for modeling, and the policies of *replacement overtime first* and *replacement overtime last* are named and discussed. Finally, in Sect. 5.5, models of replacement middle policies with minimal repairs are given for further discussions.

5.1 Three Replacement Policies

We observe three replacement policies with minimal repairs, using the approach of *whichever triggering event occurs first* discussed in Chap. 2: Preventive replacement times are scheduled at planned time T $(0 < T \leq \infty)$, at shock number N $(N = 1, 2, \cdots)$, or at damage level Z $(0 < Z \leq \infty)$, whichever occurs first, which is named as *replacement first with minimal repair*. Putting that $K \to \infty$ in (2.1)–(2.5), the probability that the unit is replaced at time T is

$$\sum_{j=0}^{N-1}[F^{(j)}(T) - F^{(j+1)}(T)]G^{(j)}(Z), \tag{5.1}$$

the probability that it is replaced at shock N is

$$F^{(N)}(T)G^{(N)}(Z), \tag{5.2}$$

the probability that it is replaced at damage Z is

$$\sum_{j=0}^{N-1} F^{(j+1)}(T)[G^{(j)}(Z) - G^{(j+1)}(Z)], \tag{5.3}$$

where note that (5.1) + (5.2) + (5.3) = 1. The mean time to replacement is

$$\begin{aligned} &T\sum_{j=0}^{N-1}[F^{(j)}(T) - F^{(j+1)}(T)]G^{(j)}(Z) + G^{(N)}(Z)\int_0^T t\mathrm{d}F^{(N)}(t) \\ &+\sum_{j=0}^{N-1}[G^{(j)}(Z) - G^{(j+1)}(Z)]\int_0^T t\mathrm{d}F^{(j+1)}(t) \\ &=\sum_{j=0}^{N-1} G^{(j)}(Z)\int_0^T [F^{(j)}(t) - F^{(j+1)}(t)]\mathrm{d}t. \end{aligned} \tag{5.4}$$

The expected number of shocks until replacement is

$$\sum_{j=0}^{N-1} j[F^{(j)}(T) - F^{(j+1)}(T)]G^{(j)}(Z) + NF^{(N)}(T)G^{(N)}(Z)$$
$$+\sum_{j=0}^{N-1}(j+1)F^{(j+1)}(T)[G^{(j)}(Z) - G^{(j+1)}(Z)]$$
$$= \sum_{j=0}^{N-1} F^{(j+1)}(T)G^{(j)}(Z).$$

It is assumed that the number of failures is counted when the unit is replaced at shock N and at damage Z. Then, the expected number of failures, i.e., minimal repairs, until replacement is (Problem 5.1)

$$\sum_{j=0}^{N-1} F^{(j+1)}(T)\int_0^Z p(x)\mathrm{d}G^{(j)}(x), \tag{5.5}$$

where the failure is counted at shock N and damage Z.

Therefore, the expected replacement and repair cost rate is

$$C_F(T,N,Z) = \frac{\begin{array}{l} c_Z - (c_Z - c_T)\sum_{j=0}^{N-1}[F^{(j)}(T) - F^{(j+1)}(T)]G^{(j)}(Z) \\ -(c_Z - c_N)F^{(N)}(T)G^{(N)}(Z) \\ +c_M\sum_{j=0}^{N-1} F^{(j+1)}(T)\int_0^Z p(x)\mathrm{d}G^{(j)}(x) \end{array}}{\sum_{j=0}^{N-1} G^{(j)}(Z)\int_0^T [F^{(j)}(t) - F^{(j+1)}(t)]\mathrm{d}t}, \tag{5.6}$$

where c_T = replacement cost at time T, c_N = replacement cost at shock N, c_Z = replacement cost at damage Z, and c_M = cost for minimal repair at each failure.

When the failure should not be counted at shock N and damage Z, the expected cost rate is

$$\widetilde{C}_F(T,N,Z) = \frac{\begin{array}{l} c_Z - (c_Z - c_T)\sum_{j=0}^{N-1}[F^{(j)}(T) - F^{(j+1)}(T)]G^{(j)}(Z) \\ -(c_Z - c_N)F^{(N)}(T)G^{(N)}(Z) \\ +c_M\sum_{j=1}^{N-1} F^{(j)}(T)\int_0^Z p(x)\mathrm{d}G^{(j)}(x) \end{array}}{\sum_{j=0}^{N-1} G^{(j)}(Z)\int_0^T [F^{(j)}(t) - F^{(j+1)}(t)]\mathrm{d}t}. \tag{5.7}$$

In addition, when the unit fails with p_j at the jth ($j = 1, 2, \cdots$) shock, where p_j increases with j and $p_0 \equiv 0$, the expected cost rate is (Problem 5.2)

$$\widehat{C}_F(T,N,Z)=\frac{\begin{array}{l}c_Z-(c_Z-c_T)\sum_{j=0}^{N-1}[F^{(j)}(T)-F^{(j+1)}(T)]G^{(j)}(Z)\\ -(c_Z-c_N)F^{(N)}(T)G^{(N)}(Z)\\ +c_M\sum_{j=1}^{N-1}F^{(j+1)}(T)p_jG^{(j)}(Z)\end{array}}{\sum_{j=0}^{N-1}G^{(j)}(Z)\int_0^T[F^{(j)}(t)-F^{(j+1)}(t)]\mathrm{d}t}. \tag{5.8}$$

5.1.1 Optimum Policies with One Variable

We discuss analytically optimum time T^*, shock N^*, and damage Z^* to minimize the expected cost rates for T, N, and Z, respectively, when $p(x)=1-\mathrm{e}^{-\theta x}$ $(0<\theta<\infty)$. In this case, the probability that the unit fails at shock j is

$$\int_0^\infty p(x)\mathrm{d}G^{(j)}(x)=\int_0^\infty(1-\mathrm{e}^{-\theta x})\mathrm{d}G^{(j)}(x)=1-[G^*(\theta)]^j,$$

where $G^*(\theta)$ denotes the Laplace-Stieltjes transform of $G(x)$, i.e., $G^*(\theta)\equiv\int_0^\infty \mathrm{e}^{-\theta x}\mathrm{d}G(x)$ for $\theta>0$. In this case, letting $p_j\equiv 1-[G^*(\theta)]^j$, $C_F(T,N,\infty)$ in (5.6) is equal to $\widehat{C}_F(T,N,\infty)$ in (5.8).

(1) Optimum T^*

Suppose that the unit is replaced only at time T $(0<T\le\infty)$. Then, putting that $N\to\infty$ and $Z\to\infty$ in (5.6),

$$C(T)\equiv\lim_{\substack{N\to\infty\\ Z\to\infty}}C_F(T,N,Z)=\frac{1}{T}\left(c_T+c_M\sum_{j=0}^{\infty}F^{(j+1)}(T)\left\{1-[G^*(\theta)]^j\right\}\right). \tag{5.9}$$

We find optimum T^* to minimize $C(T)$. Differentiating $C(T)$ with respect to T and setting it equal to zero,

$$\sum_{j=0}^{\infty}[Tf^{(j+1)}(T)-F^{(j+1)}(T)]\left\{1-[G^*(\theta)]^j\right\}=\frac{c_T}{c_M}, \tag{5.10}$$

where $f(t)$ is a density function of $F(t)$, and $f^{(j)}(t)$ is the j-fold convolution of $f(t)$ with itself.

In particular, when shocks occur at a Poisson process with rate λ, i.e., $F(t)=1-\mathrm{e}^{-\lambda t}$, and $F^{(j)}(t)=\sum_{i=j}^{\infty}[(\lambda t)^i/i!]\mathrm{e}^{-\lambda t}$ $(j=0,1,2,\cdots)$, (5.10) becomes

$$\frac{1}{1-G^*(\theta)}\left(1-\{1+\lambda T[1-G^*(\theta)]\}\mathrm{e}^{-\lambda T[1-G^*(\theta)]}\right)=\frac{c_T}{c_M}, \tag{5.11}$$

whose left-hand increases strictly with T from 0 to $1/[1-G^*(\theta)]$. Thus, if $1/[1-G^*(\theta)] > c_T/c_M$, then there exists a finite and unique T^* $(0 < T^* < \infty)$ which satisfies (5.11), and the resulting cost rate is

$$\frac{C(T^*)}{\lambda} = c_M\left\{1 - G^*(\theta)\mathrm{e}^{-\lambda T^*[1-G^*(\theta)]}\right\}. \tag{5.12}$$

Note that

$$\sum_{j=0}^{\infty}\int_0^{\infty}\mathrm{e}^{-\theta x}\mathrm{d}G^{(j)}(x) = \frac{1}{1-G^*(\theta)},$$

which represents the expected number of shocks for non-failures without replacement, and in general, it is greater than the ratio c_T/c_M.

(2) Optimum N^*

Suppose that the unit is replaced only at shock N $(N = 1, 2, \cdots)$. Then, putting that $T \to \infty$ and $Z \to \infty$ in (5.6),

$$C(N) \equiv \lim_{\substack{T\to\infty \\ Z\to\infty}} C_F(T, N, Z) = \frac{1}{N\mu}\left(c_N + c_M\sum_{j=0}^{N-1}\{1-[G^*(\theta)]^j\}\right) \quad (N = 1, 2, \cdots), \tag{5.13}$$

where $\mu \equiv \int_0^\infty \overline{F}(t)\mathrm{d}t < \infty$.

We find optimum N^* to minimize $C(N)$. Forming the inequality $C(N+1) - C(N) \geq 0$,

$$\sum_{j=0}^{N-1}\left\{[G^*(\theta)]^j - [G^*(\theta)]^N\right\} \geq \frac{c_N}{c_M}, \tag{5.14}$$

whose left-hand side increases strictly with N from $1 - G^*(\theta)$ to $1/[1-G^*(\theta)]$.

Note that $1 - G^*(\theta)$ represents the probability that the unit fails at the first shock. Thus, if $1/[1-G^*(\theta)] > c_N/c_M$, then there exists a finite and unique minimum N^* $(1 \leq N^* < \infty)$ which satisfies (5.14), and the resulting cost rate is

$$c_M\{1-[G^*(\theta)]^{N^*-1}\} < \mu C(N^*) \leq c_M\{1-[G^*(\theta)]^{N^*}\}. \tag{5.15}$$

If $1 - G^*(\theta) \geq c_N/c_M$, then $N^* = 1$.

(3) Optimum Z^*

Suppose that the unit is replaced only at damage Z $(0 < Z \leq \infty)$. Then, putting that $T \to \infty$ and $N \to \infty$ in (5.6),

$$C(Z) \equiv \lim_{\substack{T\to\infty \\ N\to\infty}} C_F(T, N, Z) = \frac{c_Z + c_M \int_0^Z (1-\mathrm{e}^{-\theta x})\mathrm{d}M_G(x)}{\mu[1+M_G(Z)]}, \tag{5.16}$$

where $M_G(x) \equiv \sum_{j=1}^{\infty} G^{(j)}(x)$.

We find optimum Z^* to minimize $C(Z)$. Differentiating $C(Z)$ with respect to Z and setting it equal to zero,

$$\int_0^Z [1+M_G(x)]\theta\mathrm{e}^{-\theta x}\mathrm{d}x = \frac{c_Z}{c_M}, \tag{5.17}$$

whose left-hand side increases strictly with Z from 0 to $1/[1-G^*(\theta)]$. Thus, if $1/[1-G^*(\theta)] > c_Z/c_M$, then there exists a finite and unique Z^* $(0 < Z^* < \infty)$ which satisfies (5.17), and the resulting cost rate is

$$\mu C(Z^*) = c_M(1-\mathrm{e}^{-\theta Z^*}). \tag{5.18}$$

From the above results, we can obtain all of finite and unique T^*, N^* and Z^* to minimize their respective cost rates when $1/[1-G^*(\theta)] > c_i/c_M$ $(i = T, N, Z)$.

(4) Numerical Examples

When $F(t) = 1-\mathrm{e}^{-\lambda t}$, $G(x) = 1-\mathrm{e}^{-\omega x}$ and $p(x) = 1-\mathrm{e}^{-\theta x}$, Tables 5.1, 5.2 and 5.3 present respectively optimum λT^* in (5.11), N^* in (5.14) and ωZ^* in (5.17), and their cost rates $C(T^*)/(\lambda c_M)$, $C(N^*)/(\lambda c_M)$ and $C(Z^*)/(\lambda c_M)$ for c_i/c_M $(i = T, N, Z)$ and ω/θ. These indicate that λT^* are almost equal to N^* and ωZ^* are little smaller than N^*. It can be easily found from the point of resulting cost rates that $C(T^*) > C(N^*) > C(Z^*)$ for the same ω/θ and c_i/c_M, and their differences become smaller as c_i/c_M and ω/θ become smaller.

Table 5.1 Optimum λT^* and its cost rate $C(T^*)/(\lambda c_M)$

c_T/c_M	$\omega/\theta = 5.0$		$\omega/\theta = 10.0$		$\omega/\theta = 20.0$	
	λT^*	$C(T^*)/(\lambda c_M)$	λT^*	$C(T^*)/(\lambda c_M)$	λT^*	$C(T^*)/(\lambda c_M)$
0.1	1.168	0.314	1.554	0.211	2.119	0.139
0.2	1.701	0.372	2.244	0.259	3.041	0.176
0.5	2.863	0.483	3.705	0.351	4.954	0.248
1.0	4.386	0.599	5.526	0.450	7.261	0.326
2.0	7.133	0.746	8.512	0.581	10.847	0.432
5.0	19.410	0.967	16.909	0.805	19.507	0.624

Table 5.2 Optimum N^* and its cost rate $C(N^*)/(\lambda c_M)$

c_N/c_M	$\omega/\theta = 5.0$		$\omega/\theta = 10.0$		$\omega/\theta = 20.0$	
	N^*	$C(N^*)/(\lambda c_M)$	N^*	$C(N^*)/(\lambda c_M)$	N^*	$C(N^*)/(\lambda c_M)$
0.1	1	0.100	2	0.095	2	0.074
0.2	2	0.183	2	0.145	3	0.114
0.5	3	0.324	4	0.253	5	0.191
1.0	4	0.473	5	0.366	7	0.275
2.0	7	0.668	8	0.516	11	0.389
5.0	18	0.957	16	0.775	19	0.595

Table 5.3 Optimum ωZ^* and its cost rate $C(Z^*)/(\lambda c_M)$

c_Z/c_M	$\omega/\theta = 5.0$		$\omega/\theta = 10.0$		$\omega/\theta = 20.0$	
	ωZ^*	$C(Z^*)/(\lambda c_M)$	ωZ^*	$C(Z^*)/(\lambda c_M)$	ωZ^*	$C(Z^*)/(\lambda c_M)$
0.1	0.431	0.083	0.756	0.073	1.268	0.061
0.2	0.782	0.145	1.303	0.122	2.082	0.099
0.5	1.649	0.281	2.550	0.225	3.843	0.175
1.0	2.861	0.436	4.162	0.340	6.008	0.259
2.0	5.106	0.640	6.845	0.496	9.400	0.375
5.0	15.292	0.953	14.449	0.764	17.628	0.586

5.1.2 Optimum Policies with Two Variables

We next discuss analytically optimum policies with two variables to minimize the expected cost rates $\lim_{Z\to\infty} C_F(T, N, Z)$, $\lim_{N\to\infty} C_F(T, N, Z)$, $\lim_{T\to\infty} C_F(T, N, Z)$, respectively, when $c_T = c_N = c_Z$, $F(t) = 1 - e^{-\lambda t}$ and $p(x) = 1 - e^{-\theta x}$.

(1) Optimum T_F^* and N_F^*

Suppose that the unit is replaced at time T $(0 < T \le \infty)$ or at shock N $(N = 1, 2, \cdots)$, whichever occurs first. Putting that $Z \to \infty$ in (5.6), the expected cost rate is

$$\frac{C_F(T, N)}{\lambda} \equiv \lim_{Z\to\infty} \frac{C_F(T, N, Z)}{\lambda} = \frac{c_T + c_M \sum_{j=0}^{N-1} F^{(j+1)}(T)\{1 - [G^*(\theta)]^j\}}{\sum_{j=0}^{N-1} F^{(j+1)}(T)}. \tag{5.19}$$

We find optimum T_F^* and Z_F^* to minimize $C_F(T, N)$. When $N = 1$,

$$\frac{C_F(T,1)}{\lambda} = \frac{c_T}{F(T)},$$

and the optimum policy is $T_F^* = \infty$.

Forming the inequality $C_F(T, N+1) - C_F(T, N) \geq 0$,

$$\sum_{j=0}^{N-1} F^{(j+1)}(T)\{[G^*(\theta)]^j - [G^*(\theta)]^N\} \geq \frac{c_T}{c_M}, \tag{5.20}$$

whose left-hand side increases strictly with T to that of (5.17). Thus, if optimum N_F^* for given T exists, it would decrease with T to N^* given in (5.14).

Forming the inequality $C_F(T, N-1) - C_F(T, N) > 0$ (Problem 2.3),

$$\sum_{j=0}^{N-1} F^{(j+1)}(T)\{[G^*(\theta)]^j - [G^*(\theta)]^{N-1}\} < \frac{c_T}{c_M}. \tag{5.21}$$

Differentiating $C_F(T, N)$ with respect to T and setting it equal to zero for $N \geq 2$,

$$\frac{\sum_{j=0}^{N-1}[(\lambda T)^j/j!]\{1 - [G^*(\theta)]^j\}}{\sum_{j=0}^{N-1}[(\lambda T)^j/j!]} \sum_{j=0}^{N-1} F^{(j+1)}(T) - \sum_{j=0}^{N-1} F^{(j+1)}(T)\{1 - [G^*(\theta)]^j\} = \frac{c_T}{c_M}. \tag{5.22}$$

Substituting (5.21) for (5.22),

$$\frac{\sum_{j=0}^{N-1}[(\lambda T)^j/j!]\{1 - [G^*(\theta)]^j\}}{\sum_{j=0}^{N-1}[(\lambda T)^j/j!]} > 1 - [G^*(\theta)]^{N-1}. \tag{5.23}$$

However, the above inequality (5.23) does not hold for any N, in which case, $C_F(T, N)$ in (5.19) decreases with T (Problem 5.3). Thus, there does not exist any finite T_F^* which satisfies (5.22), i.e., the optimum policy is $(T_F^* = \infty, N_F^* = N^*)$, where N^* is given in (5.14).

(2) Optimum T_F^* and Z_F^*

Suppose that the unit is replaced at time T $(0 < T \leq \infty)$ or at damage Z $(0 < Z \leq \infty)$, whichever occurs first. Putting that $N \to \infty$ in (5.6), the expected cost rate is

$$\frac{C_F(T,Z)}{\lambda} \equiv \lim_{N\to\infty} \frac{C_F(T,N,Z)}{\lambda} = \frac{c_T + c_M \sum_{j=0}^{\infty} F^{(j+1)}(T) \int_0^Z p(x)\mathrm{d}G^{(j)}(x)}{\sum_{j=0}^{\infty} F^{(j+1)}(T) G^{(j)}(Z)}. \tag{5.24}$$

We find optimum T_F^* and Z_F^* to minimize $C_F(T,Z)$. Differentiating $C_F(T,Z)$ with respect to Z and setting it equal to zero,

$$\frac{\sum_{j=0}^{\infty}[(\lambda T)^j/j!]\int_0^Z p(x)\mathrm{d}G^{(j)}(x)}{\sum_{j=0}^{\infty}[(\lambda T)^j/j!]G^{(j)}(Z)}\sum_{j=0}^{\infty}F^{(j+1)}(T)G^{(j)}(Z)$$
$$-\sum_{j=0}^{\infty}F^{(j+1)}(T)\int_0^Z p(x)\mathrm{d}G^{(j)}(x)=\frac{c_T}{c_M}, \tag{5.25}$$

whose left-hand side increases strictly with T to that of (5.17) (Problem 5.4). Thus, if optimum Z_F^* exists, it decreases with T to Z^* given in (5.17).

Differentiating $C_F(T,Z)$ with respect to T and setting it equal to zero,

$$\sum_{j=0}^{\infty}F^{(j+1)}(T)\int_0^Z[p(Z)-p(x)]\mathrm{d}G^{(j)}(x)=\frac{c_T}{c_M}. \tag{5.26}$$

Substituting (5.25) for (5.26),

$$\frac{\sum_{j=0}^{\infty}[(\lambda T)^j/j!]\int_0^Z p(x)\mathrm{d}G^{(j)}(x)}{\sum_{j=0}^{\infty}[(\lambda T)^j/j!]G^{(j)}(Z)}=p(Z). \tag{5.27}$$

However, the above inequality (5.27) does not hold for any Z, in which case, $C_F(T,Z)$ in (5.24) decreases with T. Thus, there does not exist any finite T_F^* which satisfies (5.26), i.e., the optimum policy is $(T_F^*=\infty, Z_F^*=Z^*)$, where Z^* is given in (5.17).

(3) Optimum N_F^* and Z_F^*

Suppose that the unit is replaced at shock N $(N=1,2,\cdots)$ or at damage Z $(0<Z\le\infty)$, whichever occurs first. Putting that $T\to\infty$ in (5.6), the expected cost rate is

$$\frac{C_F(N,Z)}{\lambda}\equiv\lim_{T\to\infty}\frac{C_F(T,N,Z)}{\lambda}=\frac{c_N+c_M\sum_{j=0}^{N-1}\int_0^Z p(x)\mathrm{d}G^{(j)}(x)}{\sum_{j=0}^{N-1}G^{(j)}(Z)}. \tag{5.28}$$

We find optimum N_F^* and Z_F^* to minimize $C_F(N,Z)$. Differentiating $C_F(N,Z)$ with respect to Z and setting it equal to zero,

$$\sum_{j=0}^{N-1}\int_0^Z[p(Z)-p(x)]\mathrm{d}G^{(j)}(x)=\frac{c_N}{c_M}, \tag{5.29}$$

whose left-hand side increases strictly with N to that of (5.17). Thus, if optimum Z_F^* exists, it would decrease with N to Z^* given in (5.17).

Forming the inequality $C_F(N+1, Z) - C_F(N, Z) \geq 0$,

$$\frac{\int_0^Z p(x)\mathrm{d}G^{(N)}(x)}{G^{(N)}(Z)} \sum_{j=0}^{N-1} G^{(j)}(Z) - \sum_{j=0}^{N-1} \int_0^Z p(x)\mathrm{d}G^{(j)}(x) \geq \frac{c_N}{c_M}. \tag{5.30}$$

Substituting (5.29) for (5.30),

$$\frac{\int_0^Z p(x)\mathrm{d}G^{(N)}(x)}{G^{(N)}(Z)} \geq p(Z). \tag{5.31}$$

However, the above inequality (5.31) does not hold for any Z, in which case, $C_F(N, Z)$ in (5.28) decreases with N. Thus, there does not exist any finite N_F^* which satisfies (5.30), i.e., the optimum policy is $(N_F^* = \infty, Z_F^* = Z^*)$, where Z^* is given in (5.17).

5.2 Replacement Last Policies

The approach of *whichever triggering event occurs last* discussed in Chap. 3 is observed for the above three replacement polices: The unit undergoes minimal repairs at failures and is replaced at planned time T $(0 \leq T \leq \infty)$, at shock number N $(N = 0, 1, 2, \cdots)$, or at damage level Z $(0 \leq Z \leq \infty)$, whichever occurs last, which is named as *replacement last with minimal repair*. Putting that $K \to \infty$ in (3.1)–(3.5), the probability that the unit is replaced at time T is

$$\sum_{j=N}^{\infty} [F^{(j)}(T) - F^{(j+1)}(T)][1 - G^{(j)}(Z)], \tag{5.32}$$

the probability that it is replaced at shock N is

$$[1 - F^{(N)}(T)][1 - G^{(N)}(Z)], \tag{5.33}$$

and the probability that it is replaced at damage Z is

$$\sum_{j=N}^{\infty} [1 - F^{(j+1)}(T)][G^{(j)}(Z) - G^{(j+1)}(Z)]. \tag{5.34}$$

The mean time to replacement is

$$T\sum_{j=N}^{\infty}[F^{(j)}(T)-F^{(j+1)}(T)][1-G^{(j)}(Z)]+[1-G^{(N)}(Z)]\int_T^{\infty}t\mathrm{d}F^{(N)}(t)$$
$$+\sum_{j=N}^{\infty}[G^{(j)}(Z)-G^{(j+1)}(Z)]\int_T^{\infty}t\mathrm{d}F^{(j+1)}(t)$$
$$=T+\int_T^{\infty}[1-F^{(N)}(t)]\mathrm{d}t+\sum_{j=N}^{\infty}G^{(j)}(Z)\int_T^{\infty}[F^{(j)}(t)-F^{(j+1)}(t)]\mathrm{d}t. \quad (5.35)$$

The expected number of failures without replacement policies is

$$\sum_{j=0}^{\infty}\int_0^{\infty}p(x)\mathrm{d}G^{(j)}(x),$$

and the expected number of failures when replacement policies of T, N and Z have been done is

$$\sum_{j=N}^{\infty}[1-F^{(j+1)}(T)]\int_Z^{\infty}p(x)\mathrm{d}G^{(j)}(x).$$

Thus, the expected number of failures until replacement is (Problem 5.5)

$$\sum_{j=0}^{\infty}\int_0^{\infty}p(x)\mathrm{d}G^{(j)}(x)-\sum_{j=N}^{\infty}[1-F^{(j+1)}(T)]\int_Z^{\infty}p(x)\mathrm{d}G^{(j)}(x). \quad (5.36)$$

Therefore, the expected cost rate is

$$C_L(T,N,Z)=$$
$$\frac{\begin{array}{l}c_Z-(c_Z-c_T)\sum_{j=N}^{\infty}[F^{(j)}(T)-F^{(j+1)}(T)][1-G^{(j)}(Z)]\\-(c_Z-c_N)[1-F^{(N)}(T)][1-G^{(N)}(Z)]\\+c_M\{\sum_{j=0}^{\infty}\int_0^{\infty}p(x)\mathrm{d}G^{(j)}(x)-\sum_{j=N}^{\infty}[1-F^{(j+1)}(T)]\int_Z^{\infty}p(x)\mathrm{d}G^{(j)}(x)\}\end{array}}{T+\int_T^{\infty}[1-F^{(N)}(t)]\mathrm{d}t+\sum_{j=N}^{\infty}G^{(j)}(Z)\int_T^{\infty}[F^{(j)}(t)-F^{(j+1)}(t)]\mathrm{d}t}. \quad (5.37)$$

It can be shown that

$$\lim_{\substack{N\to 0\\ Z\to 0}}C_L(T,N,Z)=\lim_{\substack{N\to\infty\\ Z\to\infty}}C_F(T,N,Z)=C(T),$$

which is given in (5.9),

$$\lim_{\substack{T\to 0\\ Z\to 0}} C_L(T,N,Z) = \lim_{\substack{T\to\infty\\ Z\to\infty}} C_F(T,N,Z) = C(N),$$

which is given in (5.13), and

$$\lim_{\substack{T\to 0\\ N\to 0}} C_L(T,N,Z) = \lim_{\substack{T\to\infty\\ N\to\infty}} C_F(T,N,Z) = C(Z),$$

which is given in (5.16).

In addition, when the unit fails with p_j at the jth $(j = 1, 2, \cdots)$ shock, where p_j increases with j and $p_0 \equiv 0$, the expected cost rate is

$$\widehat{C}_L(T,N,Z) = \frac{\begin{array}{l} c_Z - (c_Z - c_T)\sum_{j=N}^{\infty}[F^{(j)}(T) - F^{(j+1)}(T)][1 - G^{(j)}(Z)] \\ -(c_Z - c_N)[1 - F^{(N)}(T)][1 - G^{(N)}(Z)] \\ +c_M\{\sum_{j=1}^{\infty} p_j - \sum_{j=N}^{\infty}[1 - F^{(j+1)}(T)]p_j[1 - G^{(j)}(Z)]\} \end{array}}{T + \int_T^{\infty}[1 - F^{(N)}(t)]\mathrm{d}t + \sum_{j=N}^{\infty} G^{(j)}(Z)\int_T^{\infty}[F^{(j)}(t) - F^{(j+1)}(t)]\mathrm{d}t}. \quad (5.38)$$

5.2.1 *Optimum Policies*

We have stated in Chap. 3 that optimum policies for replacement last should be bounded with two variables. In this section, we discuss analytically optimum (T_L^*, N_L^*), (T_L^*, Z_L^*), and (N_L^*, Z_L^*) to minimize $\lim_{Z\to 0} C_L(T,N,Z)$, $\lim_{N\to 0} C_L(T,N,Z)$, and $\lim_{T\to 0} C_L(T,N,Z)$ when $c_T = c_N = c_Z$, $F(t) = 1 - \mathrm{e}^{-\lambda t}$, and $p(x) = 1 - \mathrm{e}^{-\theta x}$, respectively.

(1) Optimum T_L^* and N_L^*

The unit is replaced at time T $(0 \le T \le \infty)$ or at shock N $(N = 0, 1, 2, \cdots)$, whichever occurs last. Putting that $Z \to 0$ in (5.37), the expected cost rate is

$$\frac{C_L(T,N)}{\lambda} \equiv \lim_{Z\to 0}\frac{C_L(T,N,Z)}{\lambda} = \frac{\begin{array}{l} c_T + c_M\big(\sum_{j=0}^{\infty} F^{(j+1)}(T)\{1 - [G^*(\theta)]^j\} \\ +\sum_{j=0}^{N-1}[1 - F^{(j+1)}(T)]\{1 - [G^*(\theta)]^j\}\big) \end{array}}{\lambda T + \sum_{j=0}^{N-1}[1 - F^{(j+1)}(T)]}. \quad (5.39)$$

Forming the inequality $C_L(T, N+1) - C_L(T,N) \ge 0$,

$$\{1-[G^*(\theta)]^N\}\left\{\lambda T+\sum_{j=0}^{N-1}[1-F^{(j+1)}(T)]\right\}$$
$$-\sum_{j=0}^{\infty}F^{(j+1)}(T)\{1-[G^*(\theta)]^j\}-\sum_{j=0}^{N-1}[1-F^{(j+1)}(T)]\{1-[G^*(\theta)]^j\}\geq\frac{c_T}{c_M},\tag{5.40}$$

whose left-hand side decreases strictly with T from that of (5.14) (Problem 5.6). If optimum N_L^* exists, it increases with T from N^* given in (5.14).

Differentiating $C_L(T,N)$ with respect to T and setting it equal to zero,

$$\frac{\sum_{j=N}^{\infty}[(\lambda T)^j/j!]\{1-[G^*(\theta)]^j\}}{\sum_{j=N}^{\infty}[(\lambda T)^j/j!]}\left\{\lambda T+\sum_{j=0}^{N-1}[1-F^{(j+1)}(T)]\right\}$$
$$-\sum_{j=0}^{\infty}F^{(j+1)}(T)\{1-[G^*(\theta)]^j\}-\sum_{j=0}^{N-1}[1-F^{(j+1)}(T)]\{1-[G^*(\theta)]^j\}=\frac{c_T}{c_M}.\tag{5.41}$$

Substituting (5.40) for (5.41),

$$\frac{\sum_{j=N}^{\infty}[(\lambda T)^j/j!]\{1-[G^*(\theta)]^j\}}{\sum_{j=N}^{\infty}[(\lambda T)^j/j!]}\leq 1-[G^*(\theta)]^N.\tag{5.42}$$

However, the above inequality (5.42) does not hold for any N, in which case, $C_L(T,N)$ in (5.39) increases with T (Problem 5.7). Thus, there does not exist any positive T_L^* which satisfies (5.41), i.e., the optimum policy is ($T_L^*=0$, $N_L^*=N^*$), where N^* is given in (5.14).

(2) Optimum T_L^* and Z_L^*

The unit is replaced at time T ($0\leq T\leq\infty$) or at damage Z ($0\leq Z\leq\infty$), whichever occurs last. Putting that $N\to 0$ in (5.37), the expected cost rate is

$$\frac{C_L(T,Z)}{\lambda}\equiv\lim_{N\to 0}\frac{C_L(T,N,Z)}{\lambda}$$
$$=\frac{c_T+c_M\left(\sum_{j=0}^{\infty}F^{(j+1)}(T)\{1-[G^*(\theta)]^j\}+\sum_{j=0}^{\infty}[1-F^{(j+1)}(T)]\int_0^Z(1-\mathrm{e}^{-\theta x})\mathrm{d}G^{(j)}(x)\right)}{\lambda T+\sum_{j=0}^{\infty}[1-F^{(j+1)}(T)]G^{(j)}(Z)}.\tag{5.43}$$

Differentiating $C_L(T,Z)$ with respect to Z and setting it equal to zero,

$$(1-\mathrm{e}^{-\theta Z})\left\{\lambda T+\sum_{j=0}^{\infty}[1-F^{(j+1)}(T)]G^{(j)}(Z)\right\}-\sum_{j=0}^{\infty}F^{(j+1)}(T)\{1-[G^*(\theta)]^j\}$$

$$-\sum_{j=0}^{\infty}[1-F^{(j+1)}(T)]\int_0^Z(1-\mathrm{e}^{-\theta x})\mathrm{d}G^{(j)}(x)=\frac{c_T}{c_M}, \tag{5.44}$$

whose left-hand side decreases strictly with T from that of (5.17) (Problem 5.8).Thus, if optimum Z_L^* exists, it increases with T from Z^* given in (5.17).

Differentiating $C_L(T,Z)$ with respect to T and setting it equal to zero,

$$\frac{\sum_{j=0}^{\infty}[(\lambda T)^j/j!]\int_Z^{\infty}(1-\mathrm{e}^{-\theta x})\mathrm{d}G^{(j)}(x)}{\sum_{j=0}^{\infty}[(\lambda T)^j/j!][1-G^{(j)}(Z)]}\left\{\lambda T+\sum_{j=0}^{\infty}[1-F^{(j+1)}(T)]G^{(j)}(Z)\right\}$$

$$-\sum_{j=0}^{\infty}F^{(j+1)}(T)\{1-[G^*(\theta)]^j\}$$

$$-\sum_{j=0}^{\infty}[1-F^{(j+1)}(T)]\int_0^Z(1-\mathrm{e}^{-\theta x})\mathrm{d}G^{(j)}(x)=\frac{c_T}{c_M}. \tag{5.45}$$

Substituting (5.44) for (5.45),

$$\frac{\sum_{j=0}^{\infty}[(\lambda T)^j/j!]\int_Z^{\infty}(1-\mathrm{e}^{-\theta x})\mathrm{d}G^{(j)}(x)}{\sum_{j=0}^{\infty}[(\lambda T)^j/j!][1-G^{(j)}(Z)]}=1-\mathrm{e}^{-\theta Z}. \tag{5.46}$$

However, the above inequality (5.46) does not hold for any Z, in which case, $C_L(T,Z)$ in (5.43) increases with T. Thus, there does not exist any positive T_L^* which satisfies (5.45), i.e., the optimum policy is ($T_L^*=0$, $Z_L^*=Z^*$), where Z^* is given in (5.17).

(3) Optimum N_L^* and Z_L^*

The unit is replaced at shock N ($N=0,1,2,\cdots$) or at damage Z ($0\le Z\le\infty$), whichever occurs last. Putting that $T\to 0$ in (5.37), the expected cost rate is

$$\frac{C_L(N,Z)}{\lambda}\equiv\lim_{T\to 0}\frac{C_L(T,N,Z)}{\lambda}$$
$$=\frac{c_N+c_M[\sum_{j=0}^{\infty}\int_0^Z p(x)\mathrm{d}G^{(j)}(x)+\sum_{j=0}^{N-1}\int_Z^{\infty}p(x)\mathrm{d}G^{(j)}(x)]}{N+\sum_{j=N}^{\infty}G^{(j)}(Z)}. \tag{5.47}$$

Differentiating $C_L(N,Z)$ with respect to Z and setting it equal to zero,

$$(1-\mathrm{e}^{-\theta Z})\left[N+\sum_{j=N}^{\infty}G^{(j)}(Z)\right]-\sum_{j=0}^{\infty}\int_0^Z(1-\mathrm{e}^{-\theta x})\mathrm{d}G^{(j)}(x)$$
$$-\sum_{j=0}^{N-1}\int_Z^{\infty}(1-\mathrm{e}^{-\theta x})\mathrm{d}G^{(j)}(x)=\frac{c_N}{c_M},\tag{5.48}$$

whose left-hand side decreases strictly with N from that of (5.17). Thus, if optimum Z_L^* exists, it increases with N from Z^* given in (5.17).

Forming the inequality $C_L(N-1,Z)-C_L(N,Z)>0$,

$$\frac{\int_Z^{\infty}(1-\mathrm{e}^{-\theta x})\mathrm{d}G^{(N-1)}(x)}{1-G^{(N-1)}(Z)}\left[N+\sum_{j=N}^{\infty}G^{(j)}(Z)\right]-\sum_{j=0}^{\infty}\int_0^Z(1-\mathrm{e}^{-\theta x})\mathrm{d}G^{(j)}(x)$$
$$-\sum_{j=0}^{N-1}\int_Z^{\infty}(1-\mathrm{e}^{-\theta x})\mathrm{d}G^{(j)}(x)<\frac{c_N}{c_M}.\tag{5.49}$$

Substituting (5.48) for (5.49),

$$\frac{\int_Z^{\infty}(1-\mathrm{e}^{-\theta x})\mathrm{d}G^{(N)}(x)}{1-G^{(N)}(Z)}<1-\mathrm{e}^{-\theta Z}.\tag{5.50}$$

However, the above inequality (5.50) does not hold for any Z, in which case, $C_L(N,Z)$ in (5.47) increases with N. Thus, there does not exist any positive N_L^* which satisfies (5.49), i.e., the optimum policy is ($N_L^*=0$, $Z_L^*=Z^*$), where Z^* is given in (5.17).

5.2.2 Comparisons of Replacement First and Last

When $F(t)=1-\mathrm{e}^{-\lambda t}$ and $p(x)=1-\mathrm{e}^{-\theta x}$, the above optimum results for replacement first and last have indicated that the policy with damage Z is still the best among three ones, the next one is the policy with shock N, and the third one is the policy with time T. We next compare optimum policies for $C_F(T,N)$ in (5.19) and $C_L(T,N)$ in (5.39) when $c_T=c_N$.

(1) Time T for Shock N

For given N, the left-hand side of (5.22) increases strictly with T from 0 to

$$\sum_{j=0}^{N-1}\left\{[G^*(\theta)]^j-[G^*(\theta)]^{N-1}\right\}<\sum_{j=0}^{N-1}\left\{[G^*(\theta)]^j-[G^*(\theta)]^{N}\right\},$$

which agrees with the left-hand side of (5.14). Thus, if $N > N^*$ given in (5.14), then

$$\sum_{j=0}^{N-1}\{[G^*(\theta)]^j - [G^*(\theta)]^{N-1}\} - \sum_{j=0}^{N^*-1}\{[G^*(\theta)]^j - [G^*(\theta)]^{N^*}\} > 0,$$

and hence, there exists a finite and unique T_F^* which satisfies (5.22). Conversely, if $N \le N^*$, then $T_F^* = \infty$.

The left-hand side of (5.41) increases with T from

$$\sum_{j=0}^{N-1}\{[G^*(\theta)]^j - [G^*(\theta)]^N\},$$

which agrees with the left-hand side of (5.14). Thus, if $N < N^*$, then there exists a finite and unique T_L^* which satisfies (5.41). Conversely, if $N \ge N^*$, then $T_L^* = 0$.

Therefore, when a finite N^* in (5.14) exists, we obtain the following comparative results for replacement first and last:

1. If given N in (5.22) and (5.41) is less than N^*, then $T_F^* = \infty$ and a finite T_L^* $(0 < T_L^* < \infty)$ exists. That is, we need to adopt replacement last.
2. If given N is greater than N^*, then $T_L^* = 0$ and a finite T_F^* $(0 < T_F^* < \infty)$ exists. That is, we need to adopt replacement first.
3. When given N is equal to N^*, $C_F(\infty, N^*) = C_L(0, N^*) = C(N^*)$ in (5.13).

(2) Shock N] for Time T

For given T, optimum N_F^* satisfies (5.20), and the resulting cost rate is

$$1 - [G^*(\theta)]^{N_F^*-1} < \frac{C_F(T, N_F^*)}{\lambda} \le 1 - [G^*(\theta)]^{N_F^*}. \tag{5.51}$$

Optimum N_L^* satisfies (5.40), and the resulting cost rate is

$$1 - [G^*(\theta)]^{N_L^*-1} < \frac{C_L(T, N_L^*)}{\lambda} \le 1 - [G^*(\theta)]^{N_L^*}. \tag{5.52}$$

Compare the left-hand side of (5.20) and (5.40). Denoting

$$A(T, N) \equiv \sum_{j=N}^{\infty} F^{(j+1)}(T)\{1 - [G^*(\theta)]^j\} + \sum_{j=0}^{N-1}[1 - F^{(j+1)}(T)]\{1 - [G^*(\theta)]^j\}$$
$$- \{1 - [G^*(\theta)]^N\}\left\{\sum_{j=N}^{\infty} F^{(j+1)}(T) + \sum_{j=0}^{N-1}[1 - F^{(j+1)}(T)]\right\}, \tag{5.53}$$

we obtain

$$\frac{\mathrm{d}A(T,N)}{\mathrm{d}T} = \lambda\left(\sum_{j=0}^{N-1}\frac{(\lambda T)^j}{j!}\mathrm{e}^{-\lambda T}\{[G^*(\theta)]^j - [G^*(\theta)]^N\}\right.$$
$$\left.-\sum_{j=N}^{\infty}\frac{(\lambda T)^j}{j!}\mathrm{e}^{-\lambda T}\{[G^*(\theta)]^N - [G^*(\theta)]^j\}\right) > 0,$$
$$\lim_{T\to 0} A(T,N) = \sum_{j=0}^{N-1}\{[G^*(\theta)]^N - [G^*(\theta)]^j\} < 0,$$
$$\lim_{T\to\infty} A(T,N) = \sum_{j=N}^{\infty}\{[G^*(\theta)]^N - [G^*(\theta)]^j\} > 0.$$

Thus, there exists a finite and unique T_A^* $(0 < T_A^* < \infty)$ which satisfies $A(T;N) = 0$ for any N.

Therefore, we obtain the following comparative results for replacement first and last:

1. If given T in (5.20) and (5.40) is less than T_A^*, then $N_L^* < N_F^*$, and hence, we need to adopt replacement last.
2. If given T is greater than or equal to T_A^*, then $N_F^* \le N_L^*$, and hence, we need to adopt replacement first.

(3) Numerical Examples

When $F(t) = 1 - \mathrm{e}^{-\lambda t}$, $G(x) = 1 - \mathrm{e}^{-\omega x}$ and $p(x) = 1 - \mathrm{e}^{-\theta x}$, Tables 5.4 and 5.5 present optimum λT_L^* and λT_F^* for given N, N_L^* and N_F^* for given λT, and their cost rates $C_L(T_L^*, N)/(\lambda c_M)$, $C_F(T_F^*, N)/(\lambda c_M)$, $C_L(T, N_L^*)/(\lambda c_M)$ and $C_F(T, N_F^*)/(\lambda c_M)$ for c_T/c_M when $\omega/\theta = 10.0$. The numerical results in two tables agree with that obtained analytically in above **(1)** and **(2)**, i.e., when $N < N^*$, $0 < T_L^* < \infty$ and $T_F^* = \infty$, when $N = N^*$, $T_F^* = \infty$ and $T_L^* = 0$, and when $N > N^*$, $0 < T_F^* < \infty$ and $T_L^* = 0$. Furthermore, when $T < T^*$, $N_L^* < N_F^*$, and when $T > T^*$, $N_L^* \ge N_F^*$.

5.3 Replacement Overtime First

We have discussed *replacement overtime first* in Sect. 4.1.2, where replacement planned at time T is modified to be done at the first shock after T has arrived. In this section, we obtain *replacement overtime first with minimal repair*: The unit undergoes minimal repairs at failures and is replaced at the forthcoming shock over time T $(0 \le T \le \infty)$, at shock N $(N = 1, 2, \cdots)$, or at damage Z $(0 < Z \le \infty)$, whichever occurs first.

The probabilities that the unit is replaced over time T, at shock N and at damage Z are obtained, respectively, in (5.1), (5.2) and (5.3). The mean time to replacement

Table 5.4 Optimum λT_L^* and λT_F^*, and their cost rates $C_L(T_L^*, N)/(\lambda c_M)$ and $C_F(T_F^*, N)/(\lambda c_M)$ when $\omega/\theta = 10.0$

c_T/c_M	λT_L^*	$C_L(T_L^*, N)/(\lambda c_M)$	λT_F^*	$C_F(T_F^*, N)/(\lambda c_M)$	N^*
	$N = 2$				
0.1	0.0	0.095	∞	0.095	2
0.2	0.0	0.145	∞	0.145	2
0.5	2.855	0.275	∞	0.295	4
1.0	5.315	0.393	∞	0.545	5
2.0	8.488	0.539	∞	1.045	8
5.0	16.909	0.785	∞	2.545	16
	$N = 5$				
0.1	0.0	0.186	1.624	0.131	2
0.2	0.0	0.206	2.543	0.180	2
0.5	0.0	0.266	6.696	0.265	4
1.0	0.0	0.366	∞	0.366	5
2.0	7.699	0.534	∞	0.566	8
5.0	16.904	0.785	∞	1.166	16
	$N = 10$				
0.1	0.0	0.334	1.554	0.132	2
0.2	0.0	0.344	2.245	0.185	2
0.5	0.0	0.374	3.729	0.286	4
1.0	0.0	0.424	5.816	0.393	5
2.0	0.0	0.524	13.364	0.524	8
5.0	16.440	0.784	∞	0.824	16

is

$$\sum_{j=0}^{N-1} G^{(j)}(Z) \int_0^T \left[\int_{T-t}^{\infty} (t+u)\mathrm{d}F(u)\right] \mathrm{d}F^{(j)}(t) + G^{(N)}(Z) \int_0^T t\mathrm{d}F^{(N)}(t)$$
$$+ \sum_{j=0}^{N-1} [G^{(j)}(Z) - G^{(j+1)}(Z)] \int_0^T t\mathrm{d}F^{(j+1)}(t)$$
$$= \mu \sum_{j=0}^{N-1} F^{(j)}(T) G^{(j)}(Z). \tag{5.54}$$

The expected number of shocks until replacement is

Table 5.5 Optimum N_L^* and N_F^*, and their cost rates $C_L(T, N_L^*)/(\lambda c_M)$ and $C_F(T, N_F^*)/(\lambda c_M)$ when $\omega/\theta = 10.0$

c_T/c_M	N_L^*	$C_L(T, N_L^*)/(\lambda c_M)$	N_F^*	$C_F(T, N_F^*)/(\lambda c_M)$	λT^*
	$\lambda T = 2.0$				
0.1	2	0.121	2	0.106	1.554
0.2	2	0.161	3	0.174	2.244
0.5	4	0.255	5	0.335	3.705
1.0	5	0.366	10	0.586	5.526
2.0	8	0.516	∞	1.086	8.512
5.0	16	0.775	∞	2.586	16.909
	$\lambda T = 5.0$				
0.1	3	0.214	2	0.096	1.554
0.2	3	0.233	2	0.147	2.244
0.5	4	0.289	4	0.259	3.705
1.0	5	0.377	6	0.387	5.526
2.0	8	0.517	10	0.596	8.512
5.0	16	0.775	∞	1.196	16.909
	$\lambda T = 10.0$				
0.1	5	0.353	2	0.095	1.554
0.2	5	0.363	2	0.145	2.244
0.5	6	0.393	4	0.253	3.705
1.0	7	0.442	5	0.367	5.526
2.0	9	0.538	8	0.521	8.512
5.0	16	0.775	20	0.843	16.909

$$\sum_{j=0}^{N-1}(j+1)[F^{(j)}(T) - F^{(j+1)}(T)]G^{(j)}(Z) + NF^{(N)}(T)G^{(N)}(Z)$$
$$+\sum_{j=0}^{N-1}(j+1)F^{(j+1)}(T)[G^{(j)}(Z) - G^{(j+1)}(Z)]$$
$$= \sum_{j=0}^{N-1} F^{(j)}(T)G^{(j)}(Z). \tag{5.55}$$

Thus, the expected number of failures until replacement is

$$\sum_{j=0}^{N-1} F^{(j)}(T)\int_0^Z p(x)\mathrm{d}G^{(j)}(x), \tag{5.56}$$

where note that when $\mu = p(x) = 1$, both (5.54) and (5.56) are equal to the expected number of shocks in (5.55).

Therefore, the expected cost rate is

$$C_{OF}(T, N, Z) = \frac{\begin{array}{l} c_Z - (c_Z - c_O) \sum_{j=0}^{N-1} [F^{(j)}(T) - F^{(j+1)}(T)] G^{(j)}(Z) \\ -(c_Z - c_N) F^{(N)}(T) G^{(N)}(Z) \\ +c_M \sum_{j=0}^{N-1} F^{(j)}(T) \int_0^Z p(x) \mathrm{d}G^{(j)}(x) \end{array}}{\mu \sum_{j=0}^{N-1} F^{(j)}(T) G^{(j)}(Z)}, \quad (5.57)$$

where c_O = replacement cost over time T, and c_N, c_Z and c_M are given in (5.6).

When $F(t) = 1 - \mathrm{e}^{-\lambda t}$ and $p(x) = 1 - \mathrm{e}^{-\theta x}$, $N \to \infty$ and $Z \to \infty$, and $c_O = c_N = c_Z$, the expected cost rate in (5.57) becomes

$$\frac{C_O(T)}{\lambda} \equiv \lim_{\substack{N\to\infty \\ Z\to\infty}} \frac{C_{OF}(T, N, Z)}{\lambda} = \frac{c_O + c_M \sum_{j=0}^{\infty} F^{(j)}(T)\{1 - [G^*(\theta)]^j\}}{\sum_{j=0}^{\infty} F^{(j)}(T)}. \quad (5.58)$$

We find optimum T_O^* to minimize $C_O(T)$. Differentiating $C_O(T)$ with respect to T and setting it equal to zero,

$$1 - (1 + \lambda T) G^*(\theta) \mathrm{e}^{-\lambda T[1 - G^*(\theta)]} + \frac{G^*(\theta)}{1 - G^*(\theta)} \left\{1 - \mathrm{e}^{-\lambda T[1-G^*(\theta)]}\right\} = \frac{c_O}{c_M}, \quad (5.59)$$

whose left-hand side increases strictly with T from $1 - G^*(\theta)$ to $1/[1 - G^*(\theta)]$. Thus, if $1/[1 - G^*(\theta)] > c_O/c_M$, then there exists a finite and unique T_O^* $(0 \le T_O^* < \infty)$ which satisfies (5.59), and the resulting cost rate is

$$\frac{C_O(T_O^*)}{\lambda} = c_M \left\{1 - G^*(\theta) \mathrm{e}^{-\lambda T_O^*[1-G^*(\theta)]}\right\}. \quad (5.60)$$

If $1 - G^*(\theta) \ge c_O/c_M$, then $T_O^* = 0$, i.e., the unit is replaced at the first shock, and $C_O(0)/\lambda = c_O$.

We compare replacement polices done at time T in (5.9) and over time T in (5.58) when $c_T = c_O$. Comparing the left-hand side of (5.11) and (5.59),

$$\begin{aligned} &1 - G^*(\theta)\{1 + (1 + \lambda T)[1 - G^*(\theta)]\} \mathrm{e}^{-\lambda T[1-G^*(\theta)]} \\ &> 1 - \{1 + \lambda T[1 - G^*(\theta)]\} \mathrm{e}^{-\lambda T[1-G^*(\theta)]}. \end{aligned} \quad (5.61)$$

which indicates that $T_O^* < T^*$. From the optimum cost rates in (5.12) and (5.60), $C_O(T_O^*) < C(T^*)$, i.e., we adopt replacement overtime to save the expected cost rate.

Table 5.6 Optimum λT_O^* and its cost rate $C_O(T_O^*)/(\lambda c_M)$

c_O/c_M	$\omega/\theta = 5.0$		$\omega/\theta = 10.0$		$\omega/\theta = 20.0$	
	λT_O^*	$C_O(T_O^*)/(\lambda c_M)$	λT_O^*	$C_O(T_O^*)/(\lambda c_M)$	λT_O^*	$C_O(T_O^*)/(\lambda c_M)$
0.1	0.000	0.100	0.105	0.100	0.833	0.085
0.2	0.220	0.197	0.941	0.165	1.841	0.128
0.5	1.561	0.358	2.509	0.276	3.824	0.206
1.0	3.161	0.508	4.380	0.389	6.165	0.290
2.0	5.959	0.691	7.401	0.536	9.774	0.402
5.0	18.287	0.960	15.829	0.784	18.456	0.604

When $F(t) = 1 - \mathrm{e}^{-\lambda t}$, $G(x) = 1 - \mathrm{e}^{-\omega x}$ and $p(x) = 1 - \mathrm{e}^{-\theta x}$, Table 5.6 presents optimum λT_O^* and its cost rate $C_O(T_O^*)/(\lambda c_M)$ for c_O/c_M and ω/θ. Comparing with λT^* and $C(T^*)/(\lambda c_M)$ in Table 5.1, it concludes that $\lambda T_O^* + 1 < \lambda T^*$ and $C_O(T_O^*) < C(T^*)$, which means replacement overtime in (5.58) saves more cost rate than that in (5.9) does. However, when $c_O = c_T$, the difference between $C_O(T_O^*)$ and $C(T^*)$ become smaller as c_O/c_M is larger.

We next compare replacement policies done at shock N in (5.13) and over time T in (5.58) when $c_N = c_O$. For this purpose, we consider the following policy of replacement overtime first: The unit is replaced at the forthcoming shock over time T $(0 \le T \le \infty)$ or at shock N $(N = 1, 2, \cdots)$, whichever occurs first. Putting that $Z \to \infty$ in (5.57), the expected cost rate is

$$\frac{C_{OF}(T, N)}{\lambda} \equiv \lim_{Z \to \infty} \frac{C_{OF}(T, N, Z)}{\lambda} = \frac{c_O + c_M \sum_{j=0}^{N-1} F^{(j)}(T)\{1 - [G^*(\theta)]^j\}}{\sum_{j=0}^{N-1} F^{(j)}(T)}. \tag{5.62}$$

When $N = 1$, i.e., the unit is replaced at the first shock,

$$C_{OF}(T, 1) = C_O(0) = c_O \lambda.$$

Forming the inequality $C_{OF}(T, N-1) - C_{OF}(T, N) > 0$ for $N \ge 2$,

$$\{1 - [G^*(\theta)]^{N-1}\} \sum_{j=0}^{N-1} F^{(j)}(T) - \sum_{j=0}^{N-1} F^{(j)}(T)\{1 - [G^*(\theta)]^j\} < \frac{c_O}{c_M}, \tag{5.63}$$

Differentiating $C_{OF}(T, N)$ with respect to T and setting it equal to zero,

$$\frac{\sum_{j=0}^{N-2}[(\lambda T)^j/j!]\{1-[G^*(\theta)]^{j+1}\}}{\sum_{j=0}^{N-2}[(\lambda T)^j/j!]}\sum_{j=0}^{N-1}F^{(j)}(T)$$
$$-\sum_{j=0}^{N-1}F^{(j)}(T)\{1-[G^*(\theta)]^j\}=\frac{c_O}{c_M}. \quad (5.64)$$

Substituting (5.63) for (5.64),

$$\frac{\sum_{j=0}^{N-2}[(\lambda T)^j/j!]\{1-[G^*(\theta)]^{j+1}\}}{\sum_{j=0}^{N-2}[(\lambda T)^j/j!]}>1-[G^*(\theta)]^{N-1}. \quad (5.65)$$

However, the above inequality (5.65) does not hold for any N, and hence, $T_O^*=\infty$. This concludes that if $c_N \le c_O$, then the optimum policy is $(T_O^*=\infty, N_O^*=N^*)$, where N^* is given in (5.14), i.e., replacement with shock N in (5.13) is better than replacement overtime in (5.58).

Next, we obtain optimum T_{OF}^* in (5.64) for given N to minimize $C_{OF}(T,N)$ in (5.62). The left-hand side of (5.64) increases strictly with T to

$$\sum_{j=0}^{N-1}\left\{[G^*(\theta)]^j-[G^*(\theta)]^{N-1}\right\}<\sum_{j=0}^{N-1}\left\{[G^*(\theta)]^j-[G^*(\theta)]^{N}\right\},$$

which agrees with the left-hand side of (5.14). Thus, if $N>N^*$ given in (5.14), then

$$\sum_{j=0}^{N-1}\{G^*(\theta)]^j-[G^*(\theta)]^{N-1}\}-\sum_{j=0}^{N^*}\{G^*(\theta)]^j-[G^*(\theta)]^{N^*}\}>0,$$

and hence, there exists a finite and unique T_{OF}^* $(0\le T_{OF}^*<\infty)$ which satisfies (5.64). Conversely, if $N\le N^*$, then $T_{OF}^*=\infty$.

5.4 Replacement Overtime Last

We give a counter model for the policy of replacement overtime first: The unit undergoes minimal repairs at failures and is replaced at the forthcoming shock over time T $(0\le T\le\infty)$, at shock N $(N=0,1,2,\cdots)$, or at damage Z $(0\le Z\le\infty)$, whichever occurs last, which is named as *replacement overtime last with minimal repair*.

The probabilities that the unit is replaced over time T, at shock N and at damage Z are obtained, respectively, in (5.32), (5.33) and (5.34). The mean time to replacement is (Problem 5.9)

$$\sum_{j=N}^{\infty}[1 - G^{(j)}(Z)] \int_0^T \left[\int_{T-t}^{\infty} (t + u)\mathrm{d}F(u)\right] \mathrm{d}F^{(j)}(t)$$
$$+ [1 - G^{(N)}(Z)] \int_T^{\infty} t\mathrm{d}F^{(N)}(t)$$
$$+ \sum_{j=N}^{\infty}[G^{(j)}(Z) - G^{(j+1)}(Z)] \int_T^{\infty} t\mathrm{d}F^{(j+1)}(t)$$
$$= \mu\left\{N + \sum_{j=N}^{\infty}[F^{(j)}(T) + G^{(j)}(Z) - F^{(j)}(T)G^{(j)}(Z)]\right\}. \tag{5.66}$$

The expected number of failures without replacement policies is

$$\sum_{j=0}^{\infty} \int_0^{\infty} p(x)\mathrm{d}G^{(j)}(x),$$

and the expected number of failures when replacement policies of T, N and Z have been done is

$$\sum_{j=N}^{\infty}[1 - F^{(j)}(T)] \int_Z^{\infty} p(x)\mathrm{d}G^{(j)}(x),$$

Thus, the expected number of failures until replacement is

$$\sum_{j=0}^{\infty} \int_0^{\infty} p(x)\mathrm{d}G^{(j)}(x) - \sum_{j=N}^{\infty}[1 - F^{(j)}(T)] \int_Z^{\infty} p(x)\mathrm{d}G^{(j)}(x). \tag{5.67}$$

Therefore, the expected cost rate is

$$C_{OL}(T, N, Z) = \frac{\begin{array}{l} c_Z - (c_Z - c_O)\sum_{j=N}^{\infty}[F^{(j)}(T) - F^{(j+1)}(T)][1 - G^{(j)}(Z)] \\ -(c_Z - c_N)[1 - F^{(N)}(T)][1 - G^{(N)}(Z)] \\ +c_M\{\sum_{j=0}^{\infty}\int_0^{\infty} p(x)\mathrm{d}G^{(j)}(x) - \sum_{j=N}^{\infty}[1 - F^{(j)}(T)]\int_Z^{\infty} p(x)\mathrm{d}G^{(j)}(x)\} \end{array}}{\mu\{N + \sum_{j=N}^{\infty}[F^{(j)}(T) + G^{(j)}(Z) - F^{(j)}(T)G^{(j)}(Z)]\}}. \tag{5.68}$$

In order to compare the optimum policy of T for given N in $C_{OF}(T, N)$ in (5.62) when $F(t) = 1 - \mathrm{e}^{-\lambda t}$ and $p(x) = 1 - \mathrm{e}^{-\theta x}$, we put $Z \to 0$ in (5.68) and obtain

$$\frac{C_{OL}(T,N)}{\lambda} \equiv \lim_{Z\to 0} \frac{C_{OL}(T,N,Z)}{\lambda}$$
$$= \frac{c_O + c_M\left(\sum_{j=0}^{N-1}\{1-[G^*(\theta)]^j\} + \sum_{j=N}^{\infty} F^{(j)}(T)\{1-[G^*(\theta)]^j\}\right)}{N + \sum_{j=N}^{\infty} F^{(j)}(T)}, \tag{5.69}$$

where $\lim_{T\to 0} C_{OL}(T,N) = \lim_{T\to\infty} C_{OF}(T,N)$, and $\lim_{N\to 0} C_{OL}(T,N) = \lim_{N\to\infty} C_{OF}(T,N)$.

Differentiating $C_{OL}(T,N)$ with respect to T and setting it equal to zero,

$$\frac{\sum_{j=N-1}^{\infty}[(\lambda T)^j/j!]\{1-[G^*(\theta)]^{j+1}\}}{\sum_{j=N-1}^{\infty}[(\lambda T)^j/j!]}\left[N + \sum_{j=N}^{\infty} F^{(j)}(T)\right]$$
$$-\left(\sum_{j=0}^{N-1}\{1-[G^*(\theta)]^j\} + \sum_{j=N}^{\infty} F^{(j)}(T)\{1-[G^*(\theta)]^j\}\right) = \frac{c_O}{c_M}, \tag{5.70}$$

whose left-hand side increases with T from

$$\sum_{j=0}^{N-1}\{[G^*(\theta)]^j - [G^*(\theta)]^N\},$$

which agrees with the left-hand side of (5.14). Thus, if $N < N^*$ given in (5.14), then there exists a finite and unique T_{OL}^* which satisfies (5.70). Conversely, if $N \geq N^*$, then $T_{OL}^* = 0$.

Therefore, when a finite N^* in (5.14) exists, we obtain the following comparative results for replacement overtime first and last:

1. If given N in (5.64) and (5.70) is less than N^*, then $T_{OF}^* = \infty$ and a finite T_{OL}^* $(0 < T_{OL}^* < \infty)$ exists. That is, we need to adopt replacement overtime last.
2. If given N is greater than N^*, then $T_{OL}^* = 0$ and a finite T_{OF}^* $(0 < T_{OF}^* < \infty)$ exists. That is, we need to adopt replacement overtime first.
3. If given N is equal to N^*, $C_{OF}(\infty, N^*) = C_{OL}(0, N^*) = C(N^*)$ in (5.13).

It would be interesting to note from the above comparisons that, replacement first and last policies discussed in Chap. 3, Sects. 5.2, and 5.4 show the same comparative results for optimum T for given N, which is also shown accordingly in Table 5.7. This indicates that when $N < N^*$, $0 < T_{OL}^* < \infty$ and $T_{OF}^* = \infty$, when $N = N^*$, $T_{OL}^* = 0$ and $T_{OF}^* = \infty$, and when $N > N^*$, $T_{OL}^* = 0$ and $0 < T_{OF}^* < \infty$.

Table 5.7 Optimum λT^*_{OL} and λT^*_{OF}, and their cost rates $C_{OL}(T^*_{OL}, N)/(\lambda c_M)$ and $C_{OF}(T^*_{OF}, N)/(\lambda c_M)$ when $\omega/\theta = 10.0$

c_O/c_M	λT^*_{OL}	$C_{OL}(T^*_{OL}, N)/(\lambda c_M)$	λT^*_{OF}	$C_{OF}(T^*_{OF}, N)/(\lambda c_M)$	N^*
	$N = 2$				
0.1	0.0	0.095	∞	0.095	2
0.2	0.0	0.145	∞	0.145	2
0.5	2.101	0.271	∞	0.295	4
1.0	4.294	0.389	∞	0.545	5
2.0	7.393	0.536	∞	1.045	8
5.0	15.829	0.784	∞	2.545	16
	$N = 5$				
0.1	0.0	0.186	0.105	0.100	2
0.2	0.0	0.206	0.991	0.165	2
0.5	0.0	0.266	4.344	0.264	4
1.0	0.0	0.366	∞	0.366	5
2.0	6.787	0.533	∞	0.566	8
5.0	15.826	0.784	∞	1.166	16
	$N = 10$				
0.1	0.0	0.334	0.105	0.100	2
0.2	0.0	0.344	0.941	0.165	2
0.5	0.0	0.374	2.515	0.276	4
1.0	0.0	0.424	4.552	0.388	5
2.0	0.0	0.524	11.495	0.524	8
5.0	15.435	0.784	∞	0.824	16

5.5 Replacement Middle Polices

Using the approach of whichever triggering event occurs middle for the replacement with three PR scenarios, and the same notations of t_N and t_Z, which have been introduced in Sect. 4.2, we obtain the expected cost rates for the following policies of *replacement middle with minimal repair*: The unit undergoes minimal repairs at failures and is replaced at time T $(0 \le T \le \infty)$, at shock N $(N = 0, 1, 2, \cdots)$, or at damage Z $(0 \le Z \le \infty)$, whichever occurs middle.

The probability that the unit is replaced at time T for $\{t_N < T < t_Z\}$ is

$$\sum_{j=N}^{\infty}[F^{(j)}(T) - F^{(j+1)}(T)]G^{(j)}(Z),$$

and the probability that it is replaced at time T for $t_Z < T < t_N$ is

$$\sum_{j=0}^{N-1}[F^{(j)}(T) - F^{(j+1)}(T)][1 - G^{(j)}(Z)].$$

The probability that the unit is replaced at shock N for $T < t_N < t_Z$ is

$$[1 - F^{(N)}(T)]G^{(N)}(Z),$$

and the probability that it is replaced at shock N for $t_Z < t_N < T$ is

$$F^{(N)}(T)[1 - G^{(N)}(Z)].$$

The probability that the unit is replaced at damage Z for $t_N < t_Z < T$ is

$$\sum_{j=N}^{\infty} F^{(j+1)}(T)[G^{(j)}(Z) - G^{(j+1)}(Z)],$$

and the probability that it is replaced at damage Z for $T < t_Z < t_N$ is

$$\sum_{j=0}^{N-1}[1 - F^{(j+1)}(T)][G^{(j)}(Z) - G^{(j+1)}(Z)].$$

The mean time to replacement is

$$\begin{aligned}
&T\left\{\sum_{j=N}^{\infty}[F^{(j)}(T) - F^{(j+1)}(T)]G^{(j)}(Z)\right.\\
&\left.+\sum_{j=0}^{N-1}[F^{(j)}(T) - F^{(j+1)}(T)][1 - G^{(j)}(Z)]\right\}\\
&+G^{(N)}(Z)\int_T^{\infty} t\mathrm{d}F^{(N)}(t) + [1 - G^{(N)}(Z)]\int_0^T t\mathrm{d}F^{(N)}(t)\\
&+\sum_{j=N}^{\infty}[G^{(j)}(Z) - G^{(j+1)}(Z)]\int_0^T t\mathrm{d}F^{(j+1)}(t)\\
&+\sum_{j=0}^{N-1}[G^{(j)}(Z) - G^{(j+1)}(Z)]\int_T^{\infty} t\mathrm{d}F^{(j+1)}(t)\\
&=\sum_{j=0}^{\infty} G^{(j)}(Z)\int_0^T [F^{(j)}(t) - F^{(j+1)}(t)]\mathrm{d}t
\end{aligned}$$

$$+\sum_{j=0}^{N-1}[1-G^{(j)}(Z)]\int_0^T[F^{(j)}(t)-F^{(j+1)}(t)]\mathrm{d}t$$
$$+\sum_{j=0}^{N-1}\int_T^\infty[F^{(j)}(t)-F^{(j+1)}(t)]\mathrm{d}t. \tag{5.71}$$

The expected number of shocks until replacement is

$$\sum_{j=N}^{\infty} j[F^{(j)}(T)-F^{(j+1)}(T)]G^{(j)}(Z)$$
$$+\sum_{j=0}^{N-1} j[F^{(j)}(T)-F^{(j+1)}(T)][1-G^{(j)}(Z)]$$
$$+N\left\{[1-F^{(N)}(T)]G^{(N)}(Z)+F^{(N)}(T)[1-G^{(N)}(Z)]\right\}$$
$$+\sum_{j=N}^{\infty}(j+1)F^{(j+1)}(T)[G^{(j)}(Z)-G^{(j+1)}(Z)]$$
$$+\sum_{j=0}^{N-1}(j+1)[1-F^{(j+1)}(T)][G^{(j)}(Z)-G^{(j+1)}(Z)]$$
$$=\sum_{j=0}^{\infty}F^{(j+1)}(T)G^{(j)}(Z)+\sum_{j=0}^{N-1}F^{(j+1)}(T)[1-G^{(j)}(Z)]$$
$$+\sum_{j=0}^{N-1}[1-F^{(j+1)}(T)]G^{(j)}(Z).$$

Thus, the expected number of failures until replacement is

$$\sum_{j=0}^{\infty}F^{(j+1)}(T)\int_0^Z p(x)\mathrm{d}G^{(j)}(x)+\sum_{j=0}^{N-1}F^{(j+1)}(T)\int_Z^\infty p(x)\mathrm{d}G^{(j)}(x)$$
$$+\sum_{j=0}^{N-1}[1-F^{(j+1)}(T)]\int_0^Z p(x)\mathrm{d}G^{(j)}(x). \tag{5.72}$$

Therefore, the expected cost rate is (Problem 5.10)

$$C_{MI}(T,N,Z)=$$

$$\frac{\begin{array}{l} c_Z - (c_Z - c_T)\{\sum_{j=N}^{\infty}[F^{(j)}(T) - F^{(j+1)}(T)]G^{(j)}(Z) \\ +\sum_{j=0}^{N-1}[F^{(j)}(T) - F^{(j+1)}(T)][1 - G^{(j)}(Z)]\} \\ -(c_Z - c_N)\{F^{(N)}(T) + G^{(N)}(Z) - 2F^{(N)}(T)G^{(N)}(Z) \\ +c_M\{\sum_{j=0}^{\infty} F^{(j+1)}(T) \int_0^Z p(x)\mathrm{d}G^{(j)}(x) \\ +\sum_{j=0}^{N-1} F^{(j+1)}(T) \int_Z^{\infty} p(x)\mathrm{d}G^{(j)}(x) \\ +\sum_{j=0}^{N-1}[1 - F^{(j+1)}(T)] \int_0^Z p(x)\mathrm{d}G^{(j)}(x)\} \end{array}}{\begin{array}{l} \sum_{j=0}^{\infty} G^{(j)}(Z) \int_0^T [F^{(j)}(t) - F^{(j+1)}(t)]\mathrm{d}t \\ +\sum_{j=0}^{N-1}[1 - G^{(j)}(Z)] \int_0^T [F^{(j)}(t) - F^{(j+1)}(t)]\mathrm{d}t \\ +\sum_{j=0}^{N-1} \int_T^{\infty}[F^{(j)}(t) - F^{(j+1)}(t)]\mathrm{d}t \end{array}}. \tag{5.73}$$

Next, the policy planned at time T in the above replacement middle is modified to be done at the forthcoming shock over time T. Then, the mean time to replacement is

$$\begin{aligned} &\sum_{j=N}^{\infty} G^{(j)}(Z) \int_0^T \left[\int_{T-t}^{\infty} (t+u)\mathrm{d}F(u)\right] \mathrm{d}F^{(j)}(t) \\ &+\sum_{j=0}^{N-1}[1 - G^{(j)}(Z)] \int_0^T \left[\int_{T-t}^{\infty} (t+u)\mathrm{d}F(u)\right] \mathrm{d}F^{(j)}(t) \\ &+ G^{(N)}(Z) \int_T^{\infty} t\mathrm{d}F^{(N)}(t) + [1 - G^{(N)}(Z)] \int_0^T t\mathrm{d}F^{(N)}(t) \\ &+\sum_{j=N}^{\infty}[G^{(j)}(Z) - G^{(j+1)}(Z)] \int_0^T t\mathrm{d}F^{(j+1)}(t) \\ &+\sum_{j=0}^{N-1}[G^{(j)}(Z) - G^{(j+1)}(Z)] \int_T^{\infty} t\mathrm{d}F^{(j+1)}(t) \\ &= \mu \left\{\sum_{j=0}^{\infty} F^{(j)}(T)G^{(j)}(Z) + \sum_{j=0}^{N-1} F^{(j)}(T)[1 - G^{(j)}(Z)] \right. \\ &\left. +\sum_{j=0}^{N-1}[1 - F^{(j)}(T)]G^{(j)}(Z)\right\}. \end{aligned} \tag{5.74}$$

The expected number of shocks until replacement is

$$\sum_{j=N}^{\infty}(j+1)[F^{(j)}(T) - F^{(j+1)}(T)]G^{(j)}(Z)$$

$$
\begin{aligned}
&+\sum_{j=0}^{N-1}(j+1)[F^{(j)}(T)-F^{(j+1)}(T)][1-G^{(j)}(Z)] \\
&+N\left\{[1-F^{(N)}(T)]G^{(N)}(Z)+F^{(N)}(T)[1-G^{(N)}(Z)]\right\} \\
&+\sum_{j=N}^{\infty}(j+1)F^{(j+1)}(T)[G^{(j)}(Z)-G^{(j+1)}(Z)] \\
&+\sum_{j=0}^{N-1}(j+1)[1-F^{(j+1)}(T)][G^{(j)}(Z)-G^{(j+1)}(Z)] \\
&=\sum_{j=0}^{\infty}F^{(j)}(T)G^{(j)}(Z)+\sum_{j=0}^{N-1}F^{(j)}(T)[1-G^{(j)}(Z)] \\
&\quad+\sum_{j=0}^{N-1}[1-F^{(j)}(T)]G^{(j)}(Z).
\end{aligned}
\tag{5.75}
$$

Thus, the expected number of failures until replacement is

$$
\begin{aligned}
&\sum_{j=0}^{\infty}F^{(j)}(T)\int_0^Z p(x)\mathrm{d}G^{(j)}(x)+\sum_{j=0}^{N-1}F^{(j)}(T)\int_Z^{\infty}p(x)\mathrm{d}G^{(j)}(x) \\
&+\sum_{j=0}^{N-1}[1-F^{(j)}(T)]\int_0^Z p(x)\mathrm{d}G^{(j)}(x),
\end{aligned}
\tag{5.76}
$$

where note that when $\mu = p(x) = 1$, both (5.74) and (5.76) are equal to the expected number of shocks in (5.75).

Therefore, the expected cost rate is (Problem 5.11)

$$
C_{MII}(T,N,Z)=
\frac{\begin{aligned}
&c_Z-(c_Z-c_T)\{\textstyle\sum_{j=N}^{\infty}[F^{(j)}(T)-F^{(j+1)}(T)]G^{(j)}(Z) \\
&+\textstyle\sum_{j=0}^{N-1}[F^{(j)}(T)-F^{(j+1)}(T)][1-G^{(j)}(Z)] \\
&-(c_Z-c_N)\{F^{(N)}(T)+G^{(N)}(Z)-2F^{(N)}(T)G^{(N)}(Z)\} \\
&+c_M\{\textstyle\sum_{j=0}^{\infty}F^{(j)}(T)\int_0^Z p(x)\mathrm{d}G^{(j)}(x) \\
&+\textstyle\sum_{j=0}^{N-1}F^{(j)}(T)\int_Z^{\infty}p(x)\mathrm{d}G^{(j)}(x) \\
&+\textstyle\sum_{j=0}^{N-1}[1-F^{(j)}(T)]\int_0^Z p(x)\mathrm{d}G^{(j)}(x)\}
\end{aligned}}{\begin{aligned}
&\mu\{\textstyle\sum_{j=0}^{\infty}F^{(j)}(T)G^{(j)}(Z)+\sum_{j=0}^{N-1}F^{(j)}(T)[1-G^{(j)}(Z)] \\
&+\textstyle\sum_{j=0}^{N-1}[1-F^{(j)}(T)]G^{(j)}(Z)\}
\end{aligned}}.
\tag{5.77}
$$

We could make similar discussions of obtaining optimum policies to minimize $C_{MI}(T, N, Z)$ and $C_{MII}(T, N, Z)$ by using the same methods in the previous sections.

5.6 Problem 5

5.1 Derive (5.5).
5.2 Derive (5.8) and discuss optimum policies to minimize $\widehat{C}_F(T, N, Z)$ when $p_j = 1 - q^j$ $(j = 1, 2, \cdots, 0 < q < 1)$.
5.3 Show that $T_F^* = \infty$ when the inequality (5.23) does not hold for any T.
5.4 Prove that when

$$\frac{\int_0^Z p(x)\mathrm{d}G^{(j)}(x)}{G^{(j)}(Z)}$$

increases strictly with j to $p(Z)$, Z_F^* decreases strictly with T from Z^* given in (5.17).
5.5 Derive (5.36).
5.6 Prove that the left-hand side of (5.40) decreases strictly with T from

$$\sum_{j=0}^{N-1}\{[G^*(\theta)]^j - [G^*(\theta)]^N\}.$$

5.7 Show that $T_L^* = 0$ when the inequality (5.42) does not hold for any T.
5.8 Prove that the left-hand side of (5.44) decreases strictly with T from that of (5.17).
5.9 Derive (5.66).
5.10 Derive $C_{MI}(T, N, Z)$ in (5.73) and discuss optimum policies to minimize it when $F(t) = 1 - \mathrm{e}^{-\lambda t}$ and $G(x) = 1 - \mathrm{e}^{-\omega x}$.
5.11 Derive $C_{MII}(T, N, Z)$ in (5.77), and show that when $\mu = p(x) = 1$, both (5.74) and (5.76) are equal to the expected number of shocks in (5.75).

Chapter 6
Replacement Policies with Maintenances

Maintenance actions, such as preventive and corrective maintenances [1], minor and major maintenances [39], routine and non-routine maintenances [40, 41], etc., have been proposed in pairs in literatures. Other maintenance plans, e.g., predictive maintenance [42, 43], time-based maintenance [44, 45], condition-based maintenance [45, 46], risk-based maintenance [47], etc., also have been discussed extensively. In this chapter, the notion of *reactive maintenance* [48] is modeled into replacement policies to preserve an operating unit. The reactive maintenance, also known as *breakdown maintenance*, which is distinguished from renewals for corrective maintenance, allows the unit to operate until failure and involves temporary repair or partial replacement for capital parts in order to restore the unit to its normal operating condition, so that preventive replacement actions can be put off until a later time.

In this chapter, a new unit with damage level 0 begins to operate at time 0 and degrades with damage produced by shocks, and the probability distributions $F(t)$ and $G(x)$ for shocks and damage have been supposed in Sect. 2.1. The unit fails when the total damage has exceeded a failure threshold K and can be quickly resumed to operation with undisturbed damage K by reactive maintenances at the following shocks including the shock that makes the unit fail. In order to save maintenance cost at failure, preventive replacement times are scheduled at time T, at shock N, and at failure M, where the approaches of *whichever triggering event occurs first*, *whichever triggering event occurs last* and *replacing over a planned measure* are modeled, respectively.

In Sect. 6.1, the models of replacement first and replacement overtime first for replacement scenarios planned at time T and at shock N are obtained. When the above replacement scenarios are modeled using the approach of whichever triggering event occurs last, replacement last and replacement overtime last are considered in Sect. 6.2. The number M of failures is considered into models in Sects. 6.3 and 6.4. In Sect. 6.5, replacement models obtained in Sects. 6.3 and 6.4 are given directly when shocks occur at a non-homogeneous Poisson process.

X. Zhao and T. Nakagawa, *Advanced Maintenance Policies for Shock and Damage Models*, Springer Series in Reliability Engineering,
https://doi.org/10.1007/978-3-319-70456-2_6

6.1 Replacement First with Shock Number

We consider two replacement models in which the unit undergoes reactive maintenances at shocks when the total damage has exceeded a failure threshold K $(0 < K < \infty)$, and is replaced preventively at time T or at shock N, and over time T or at shock N, whichever occurs first. We obtain their expected cost rates and derive optimum policies to minimize them analytically.

6.1.1 Replacement First

Suppose that the unit is replaced at time T $(0 < T \leq \infty)$ or at shock N $(N = 1, 2, \cdots)$, whichever occurs first, which is named as *replacement first with reactive maintenance*. Then, the probability that the unit is replaced at time T is

$$1 - F^{(N)}(T), \tag{6.1}$$

and the probability that it is replaced at shock N is

$$F^{(N)}(T). \tag{6.2}$$

The mean time to replacement is

$$T[1 - F^{(N)}(T)] + \int_0^T t\mathrm{d}F^{(N)}(t) = \int_0^T [1 - F^{(N)}(t)]\mathrm{d}t. \tag{6.3}$$

The expected number of shocks until replacement is

$$\sum_{j=1}^{N-1} j[F^{(j)}(T) - F^{(j+1)}(T)] + NF^{(N)}(T) = \sum_{j=1}^{N} F^{(j)}(T),$$

and the expected number of shocks without reactive maintenances is

$$\sum_{j=1}^{N-1}[F^{(j)}(T) - F^{(j+1)}(T)]\sum_{i=1}^{j} G^{(i)}(K) + F^{(N)}(T)\sum_{j=1}^{N} G^{(j)}(K)$$
$$= \sum_{j=1}^{N} F^{(j)}(T)G^{(j)}(K).$$

Thus, the expected number of reactive maintenances until replacement is (Problem 6.1)

$$\sum_{j=1}^{N} F^{(j)}(T) - \sum_{j=1}^{N} F^{(j)}(T)G^{(j)}(K) = \sum_{j=1}^{N} F^{(j)}(T)[1 - G^{(j)}(K)]. \tag{6.4}$$

Therefore, the expected replacement and maintenance cost rate is

$$C_F(T, N) = \frac{c_T + (c_N - c_T)F^{(N)}(T) + c_M \sum_{j=1}^{N} F^{(j)}(T)[1 - G^{(j)}(K)]}{\int_0^T [1 - F^{(N)}(t)]\mathrm{d}t}, \tag{6.5}$$

where c_T = replacement cost at time T, c_N = replacement cost at shock N, and c_M = cost for reactive maintenance.

In particular, when the unit is replaced only at time T,

$$C(T) \equiv \lim_{N\to\infty} C_F(T; N) = \frac{c_T + c_M \sum_{j=1}^{\infty} F^{(j)}(T)[1 - G^{(j)}(K)]}{T}, \tag{6.6}$$

and when the unit is replaced only at shock N,

$$C(N) \equiv \lim_{T\to\infty} C_F(T; N) = \frac{c_N + c_M \sum_{j=1}^{N}[1 - G^{(j)}(K)]}{N\mu} \quad (N = 1, 2, \cdots), \tag{6.7}$$

where $\mu \equiv \int_0^\infty \overline{F}(t)\mathrm{d}t < \infty$.

We find optimum T^* and N^* to minimize $C(T)$ and $C(N)$, respectively. Differentiating $C(T)$ with respect to T and setting it equal to zero,

$$\sum_{j=1}^{\infty}[1 - G^{(j)}(K)][Tf^{(j)}(T) - F^{(j)}(T)] = \frac{c_T}{c_M},$$

i.e.,

$$\sum_{j=1}^{\infty}[1 - G^{(j)}(K)]\int_0^T t\mathrm{d}f^{(j)}(t) = \frac{c_T}{c_M}. \tag{6.8}$$

Thus, if

$$\sum_{j=1}^{\infty}[1-G^{(j)}(K)]\int_0^{\infty} t\,\mathrm{d}f^{(j)}(t) > \frac{c_T}{c_M},$$

then there exists a finite and unique T^* $(0 < T^* < \infty)$ which satisfies (6.8).

When $F(t) = 1 - \mathrm{e}^{-\lambda t}$, we denote

$$p_j(t) \equiv F^{(j)}(t) - F^{(j+1)}(t) = \frac{(\lambda t)^j}{j!}\mathrm{e}^{-\lambda t} \quad (j = 0, 1, 2, \cdots),$$

then the expected cost rate $C(T)$ in (6.6) is

$$C(T) = \frac{c_T + c_M \sum_{j=1}^{\infty} p_j(T) \sum_{i=1}^{j}[1 - G^{(i)}(K)]}{T}, \tag{6.9}$$

and optimum T^* satisfies, from (6.8),

$$\sum_{j=1}^{\infty} p_j(T) \sum_{i=1}^{j}[G^{(i)}(K) - G^{(j)}(K)] = \frac{c_T}{c_M}, \tag{6.10}$$

whose left-hand side increases strictly with T from 0 to $M_G(K) \equiv \sum_{j=1}^{\infty} G^{(j)}(K)$ (Problem 6.2). Thus, if $M_G(K) > c_T/c_M$, then there exists a finite and unique T^* $(0 < T^* < \infty)$ which satisfies (6.10), and the resulting cost rate is

$$\begin{aligned}\frac{C(T^*)}{\lambda} &= c_M \sum_{j=0}^{\infty} p_j(T^*)[1 - G^{(j+1)}(K)] \\ &= \frac{c_T + c_M \sum_{j=0}^{\infty} F^{(j)}(T^*)[1 - G^{(j+1)}(K)]}{1 + \lambda T^*}.\end{aligned} \tag{6.11}$$

Forming the inequality $C(N+1) - C(N) \geq 0$,

$$\sum_{j=1}^{N}[G^{(j)}(K) - G^{(N+1)}(K)] \geq \frac{c_N}{c_M}, \tag{6.12}$$

whose left-hand side increases strictly with N from $G(K) - G^{(2)}(K)$ to $M_G(K)$. Thus, if $M_G(K) > c_N/c_M$, then there exists a finite and unique minimum N^* $(1 \leq N^* < \infty)$ which satisfies (6.12), and the resulting cost rate is

$$c_M \sum_{j=1}^{N^*-1} [1 - G^{(j)}(K)] < \frac{C(N^*)}{\mu} \le c_M \sum_{j=1}^{N^*} [1 - G^{(j)}(K)]. \tag{6.13}$$

Next, we find optimum (T_F^*, N_F^*) to minimize $C_F(T, N)$ when $c_T = c_N$ and $F(t) = 1 - \mathrm{e}^{-\lambda t}$. Forming the inequality $C_F(T, N-1) - C_F(T, N) > 0$,

$$\sum_{j=1}^{N} F^{(j)}(T)[G^{(j)}(K) - G^{(N)}(K)] < \frac{c_T}{c_M}, \tag{6.14}$$

and forming $C_F(T, N+1) - C_F(T, N) \ge 0$,

$$\sum_{j=1}^{N} F^{(j)}(T)[G^{(j)}(K) - G^{(N+1)}(K)] \ge \frac{c_T}{c_M}, \tag{6.15}$$

whose left-hand side increases strictly with T to that of (6.12). Thus, if N_F^* exists, it decreases strictly with T to N^* given in (6.12).

Differentiating $C_F(T, N)$ with respect to T and setting it equal to zero,

$$\sum_{j=1}^{N} F^{(j)}(T)\{Q_1(T, N) - [1 - G^{(j)}(K)]\} = \frac{c_T}{c_M}, \tag{6.16}$$

where

$$Q_1(T, N) \equiv \frac{\sum_{j=0}^{N-1} p_j(T)[1 - G^{(j+1)}(K)]}{\sum_{j=0}^{N-1} p_j(T)}.$$

Substituting (6.14) for (6.16),

$$Q_1(T, N) > 1 - G^{(N)}(K), \tag{6.17}$$

which does not hold for any N. Therefore, there does not exist any positive T_F^* which satisfies (6.16), i.e., the optimum policy is $(T_F^* = \infty, N_F^* = N^*)$.

Noting that $Q_1(T, N)$ increases strictly with T from $\overline{G}(K)$ to $1 - G^{(N)}(K)$, the left-hand side of (6.16) increases strictly with T from 0 to (Problem 6.3)

$$\sum_{j=1}^{N} [G^{(j)}(K) - G^{(N)}(K)] < \sum_{j=1}^{N} [G^{(j)}(K) - G^{(N+1)}(K)],$$

which agrees with the left-hand side of (6.12). Thus, if $N \le N^*$ given in (6.12), then $T_F^* = \infty$. Conversely, if $N > N^*$, then there exists a finite and unique T_F^* $(0 < T_F^* < \infty)$ which satisfies (6.16).

Table 6.1 Optimum λT^* and its cost rate $C(T^*)/(\lambda c_M)$

c_T/c_M	$\omega K = 5.0$		$\omega K = 10.0$		$\omega K = 20.0$	
	λT^*	$C(T^*)/(\lambda c_M)$	λT^*	$C(T^*)/(\lambda c_M)$	λT^*	$C(T^*)/(\lambda c_M)$
0.1	1.546	0.117	3.351	0.041	8.181	0.015
0.2	2.037	0.173	4.038	0.068	9.187	0.026
0.5	2.959	0.293	5.202	0.133	10.774	0.056
1.0	4.008	0.436	6.371	0.219	12.245	0.099
2.0	5.695	0.644	7.956	0.358	14.061	0.175
5.0	∞	1.000	11.512	0.668	17.386	0.365

Table 6.2 Optimum N^* and its cost rate $C(N^*)/(\lambda c_M)$

c_N/c_M	$\omega K = 5.0$		$\omega K = 10.0$		$\omega K = 20.0$	
	N^*	$C(N^*)/(\lambda c_M)$	N^*	$C(N^*)/(\lambda c_M)$	N^*	$C(N^*)/(\lambda c_M)$
0.1	2	0.074	4	0.028	11	0.011
0.2	2	0.124	5	0.049	11	0.020
0.5	3	0.224	6	0.102	13	0.045
1.0	4	0.359	7	0.177	14	0.082
2.0	5	0.575	8	0.308	15	0.150
5.0	∞	1.000	11	0.621	18	0.329

When $F(t) = 1 - \mathrm{e}^{-\lambda t}$ and $G(x) = 1 - \mathrm{e}^{-\omega x}$, Tables 6.1 and 6.2 present optimum λT^* in (6.10), N^* in (6.12) and their cost rates for c_i/c_M $(i = T, N)$ and ωK. When $c_T = c_N$, it is shown that $C(N^*) < C(T^*)$ and finite λT^* and N^* can be found and are almost equal to each other, which also agree with the above discussions of optimum (T_F^*, N_F^*).

6.1.2 Replacement Overtime First

Suppose that the unit is replaced at the forthcoming shock over time T $(0 \le T \le \infty)$ or at shock N $(N = 1, 2, \cdots)$, whichever occurs first, which is named as *replacement overtime first with reactive maintenance*. The mean time to replacement is

$$\sum_{j=1}^{N-1} \int_0^T \left[\int_{T-t}^{\infty} (t+u)\mathrm{d}F(u)\right] \mathrm{d}F^{(j)}(t) + \int_0^T t\mathrm{d}F^{(N)}(t) = \mu \sum_{j=0}^{N-1} \overline{F}^{(j)}(T). \quad (6.18)$$

The expected number of shocks until replacement is

$$\sum_{j=0}^{N-1}(j+1)[F^{(j)}(T)-F^{(j+1)}(T)]+NF^{(N)}(T)=\sum_{j=0}^{N-1}F^{(j)}(T),$$

and the expected number of shocks without reactive maintenances is

$$\sum_{j=0}^{N-1}[F^{(j)}(T)-F^{(j+1)}(T)]\sum_{i=1}^{j+1}G^{(i)}(K)+F^{(N)}(T)\sum_{i=1}^{N}G^{(i)}(K)$$
$$=\sum_{j=0}^{N-1}F^{(j)}(T)G^{(j+1)}(K).$$

Thus, the expected number of reactive maintenances until replacement is (Problem 6.4)

$$\sum_{j=0}^{N-1}F^{(j)}(T)-\sum_{j=0}^{N-1}F^{(j)}(T)G^{(j+1)}(K)=\sum_{j=0}^{N-1}F^{(j)}(T)[1-G^{(j+1)}(K)]. \quad (6.19)$$

Therefore, the expected cost rate is

$$C_{OF}(T,N)=$$
$$\frac{c_O+(c_N-c_O)F^{(N)}(T)+c_M\sum_{j=0}^{N-1}F^{(j)}(T)[1-G^{(j+1)}(K)]}{\mu\sum_{j=0}^{N-1}F^{(j)}(T)}, \quad (6.20)$$

where c_O = replacement cost over time T, and c_N and c_M are given in (6.5).

When $T\to\infty$, $\lim_{T\to\infty}C_{OF}(T,N)=C(N)$ in (6.7). When $N\to\infty$, replacement overtime first degrades into *replacement overtime*, and

$$C_O(T)\equiv\lim_{N\to\infty}C_{OF}(T;N)$$
$$=\frac{c_O+c_M\sum_{j=0}^{\infty}F^{(j)}(T)[1-G^{(j+1)}(K)]}{\mu\sum_{j=0}^{\infty}F^{(j)}(T)}. \quad (6.21)$$

We find optimum T_O^* to minimize $C_O(T)$ when $F(t)=1-e^{-\lambda t}$. Differentiating $C_O(T)$ with respect to T and setting it equal to zero,

$$\sum_{j=0}^{\infty}p_j(T)\left\{\sum_{i=1}^{j}[G^{(i)}(K)-G^{(j+1)}(K)]+[G^{(j+1)}(K)-G^{(j+2)}(K)]\right\}=\frac{c_O}{c_M}, \quad (6.22)$$

whose left-hand side increases strictly with T from $G(K) - G^{(2)}(K)$ to $M_G(K)$ (Problem 6.5). Therefore, if $M_G(K) > c_O/c_M$, then there exists a finite and unique T_O^* $(0 \le T_O^* < \infty)$ which satisfies (6.22), and the resulting cost rate is

$$\frac{C_O(T_O^*)}{\lambda} = c_M \sum_{j=0}^{\infty} p_j(T_O^*)[1 - G^{(j+2)}(K)]. \tag{6.23}$$

If $G(K) - G^{(2)}(K) \ge c_O/c_M$, then $T_O^* = 0$, i.e., the unit is replaced at the first shock.

By comparing T_O^* in (6.22) with T^* in (6.10) when $c_O = c_T$, we obtain $T_O^* < T^*$. Thus, from (6.11) and (6.21), $C_O(T_O^*) < C(T^*)$ given in (6.11), i.e., replacement overtime is better than that done at time T.

Next, we find optimum (T_{OF}^*, N_{OF}^*) to minimize $C_{OF}(T, N)$ in (6.20) when $c_N = c_O$ and $F(t) = 1 - \mathrm{e}^{-\lambda t}$. When $N = 1$,

$$\frac{C_{OF}(T, 1)}{\lambda} = c_O + c_M \overline{G}(K),$$

and the optimum policy is $T_{OF}^* = \infty$.

Forming the inequality $C_{OF}(T, N-1) - C_{OF}(T, N) > 0$ for $N \ge 2$,

$$\sum_{j=0}^{N-1} F^{(j)}(T)[G^{(j+1)}(K) - G^{(N)}(K)] < \frac{c_O}{c_M}. \tag{6.24}$$

Differentiating $C_{OF}(T, N)$ with respect to T and setting it equal to zero,

$$\sum_{j=0}^{N-1} F^{(j)}(T)G^{(j+1)}(K) - \frac{\sum_{j=0}^{N-2}[(\lambda T)^j/j!]G^{(j+2)}(K)}{\sum_{j=0}^{N-2}[(\lambda T)^j/j!]} \sum_{j=0}^{N-1} F^{(j)}(T) = \frac{c_O}{c_M}. \tag{6.25}$$

Substituting (6.24) for (6.25),

$$\frac{\sum_{j=0}^{N-2}[(\lambda T)^j/j!]G^{(j+2)}(K)}{\sum_{j=0}^{N-2}[(\lambda T)^j/j!]} < G^{(N)}(K), \tag{6.26}$$

which does not hold for any N. Therefore, there does not exist any positive T_{OF}^* which satisfies (6.25), i.e., the optimum policy is $(T_{OF}^* = \infty; N_{OF}^* = N^*)$, where N^* is given in (6.12).

Note that the left-hand side of (6.25) increases strictly with T to

$$\sum_{j=1}^{N}[G^{(j)}(K) - G^{(N)}(K)].$$

Table 6.3 Optimum λT_O^* and its cost rate $C_O(T_O^*)/(\lambda c_M)$

c_O/c_M	$\omega K = 5.0$		$\omega K = 10.0$		$\omega K = 20.0$	
	λT_O^*	$C_O(T_O^*)/(\lambda c_M)$	λT_O^*	$C_O(T_O^*)/(\lambda c_M)$	λT_O^*	$C_O(T_O^*)/(\lambda c_M)$
0.1	0.530	0.092	2.473	0.038	7.332	0.014
0.2	1.000	0.149	3.132	0.064	8.316	0.025
0.5	1.884	0.271	4.255	0.127	9.872	0.055
1.0	2.891	0.419	5.387	0.213	11.316	0.098
2.0	4.520	0.633	6.930	0.352	13.104	0.173
5.0	∞	1.000	10.408	0.665	16.382	0.363

Thus, by similar discussion above in Sect. 6.1.1, if $N \le N^*$ given in (6.12), then $T_{OF}^* = \infty$. Conversely, if $N > N^*$, then there exists a finite and unique T_{OF}^* $(0 < T_{OF}^* < \infty)$ which satisfies (6.25).

When $F(t) = 1 - e^{-\lambda t}$ and $G(x) = 1 - e^{-\omega x}$, Table 6.3 presents optimum λT_O^* in (6.22) and its cost rate $C_O(T_O^*)/(\lambda c_M)$ in (6.23) for c_O/c_M and ωK. When $c_O = c_T = c_N$, it is shown numerically that $C_O(T_O^*) < C(T^*)$ for $T_O^* < T^*$ and $C_O(T_O^*) = C(T^*)$ when $T_O^* = T^* = \infty$. Comparing Table 6.3 with Table 6.2, $C_O(T_O^*) > C(N^*)$, which also agrees with the above discussion of optimum (T_{OF}^*, N_{OF}^*).

6.2 Replacement Last with Shock Number

We consider two replacement models in which the unit is replaced at time T or at shock N, and over time T or at shock N, whichever occurs last. We obtain their expected cost rates and derive optimum policies to minimize them analytically.

6.2.1 Replacement Last

The unit is replaced at time T $(0 \le T < \infty)$ or at shock N $(N = 0, 1, 2, \cdots)$, whichever occurs last, which is named as *replacement last with reactive maintenance*. Then, the probabilities that the unit is replaced at time T and at shock N are $F^{(N)}(T)$ and $1 - F^{(N)}(T)$, respectively.

The mean time to replacement is

$$T F^{(N)}(T) + \int_T^\infty t \mathrm{d}F^{(N)}(t) = T + \int_T^\infty [1 - F^{(N)}(t)]\mathrm{d}t. \tag{6.27}$$

The expected number of shocks until replacement is

$$\sum_{j=N}^{\infty} j[F^{(j)}(T) - F^{(j+1)}(T)] + N[1 - F^{(N)}(T)] = N + \sum_{j=N+1}^{\infty} F^{(j)}(T),$$

and the expected number of shocks without reactive maintenances is

$$\sum_{j=N}^{\infty}[F^{(j)}(T) - F^{(j+1)}(T)]\sum_{i=1}^{j} G^{(i)}(K) + [1 - F^{(N+1)}(T)]\sum_{j=1}^{N} G^{(j)}(K)$$
$$= \sum_{j=N+1}^{\infty} F^{(j)}(T)G^{(j)}(K) + \sum_{j=1}^{N} G^{(j)}(K).$$

Thus, the expected number of reactive maintenances until replacement is (Problem 6.6)

$$N + \sum_{j=N+1}^{\infty} F^{(j)}(T) - \sum_{j=N+1}^{\infty} F^{(j)}(T)G^{(j)}(K) - \sum_{j=1}^{N} G^{(j)}(K)$$
$$= \sum_{j=1}^{N}[1 - G^{(j)}(K)] + \sum_{j=N+1}^{\infty} F^{(j)}(T)[1 - G^{(j)}(K)]. \tag{6.28}$$

Therefore, the expected cost rate is

$$C_L(T, N) = \frac{c_N - (c_N - c_T)F^{(N)}(T) + c_M\{\sum_{j=1}^{N}[1 - G^{(j)}(K)] + \sum_{j=N+1}^{\infty} F^{(j)}(T)[1 - G^{(j)}(K)]\}}{T + \int_T^{\infty}[1 - F^{(N)}(t)]\mathrm{d}t}. \tag{6.29}$$

Clearly,

$$\lim_{N\to 0} C_L(T, N) = \lim_{N\to\infty} C_F(T, N) = C(T),$$
$$\lim_{T\to 0} C_L(T, N) = \lim_{T\to\infty} C_F(T, N) = C(N),$$

where $C(T)$ and $C(N)$ are given in (6.6) and (6.7), respectively.

Next, we find optimum (T_L^*, N_L^*) to minimize $C_L(T, N)$ when $c_T = c_N$ and $p_j(t) = [(\lambda t)^j/j!]\mathrm{e}^{-\lambda t}$. Forming the inequality $C_L(T, N+1) - C_L(T, N) \geq 0$,

$$\lambda[1-G^{(N+1)}(K)]\left\{T+\int_T^{\infty}[1-F^{(N)}(t)]\mathrm{d}t\right\}$$
$$-\sum_{j=1}^{N}[1-G^{(j)}(K)]-\sum_{j=N+1}^{\infty}F^{(j)}(T)[1-G^{(j)}(K)]\geq\frac{c_T}{c_M}. \tag{6.30}$$

Differentiating $C_L(T,N)$ with respect to T and setting it equal to zero,

$$\lambda Q_2(T,N)\left\{T+\int_T^{\infty}[1-F^{(N)}(t)]\mathrm{d}t\right\}$$
$$-\sum_{j=1}^{N}[1-G^{(j)}(K)]-\sum_{j=N+1}^{\infty}F^{(j)}(T)[1-G^{(j)}(K)]=\frac{c_T}{c_M}. \tag{6.31}$$

where

$$Q_2(T,N)\equiv\frac{\sum_{j=N}^{\infty}p_j(T)[1-G^{(j+1)}(K)]}{\sum_{j=N}^{\infty}p_j(T)}.$$

Substituting (6.30) for (6.31),

$$Q_2(T,N)\leq 1-G^{(N+1)}(K), \tag{6.32}$$

which does not hold for any N. Therefore, there does not exist any positive T_L^* which satisfies (6.31), i.e., the optimum policy is $(T_L^*=0, N_L^*=N^*)$, where N^* is given in (6.12).

Noting that $Q_2(T,N)$ increases strictly with T from $1-G^{(N+1)}(K)$ to 1, the left-hand side of (6.31) increases strictly with T from

$$\sum_{j=1}^{N}[G^{(j)}(K)-G^{(N+1)}(K)],$$

which agrees with the left-hand side of (6.12). Thus, if $N\geq N^*$ given in (6.12), then $T_L^*=0$. Conversely, if $N<N^*$, then there exists a finite and unique T_L^* $(0<T_L^*<\infty)$ which satisfies (6.31).

From the above discussions, we should adopt replacement last in (6.29) for $N<N^*$, replacement first in (6.5) for $N>N^*$, and replacement with shock N in (6.7) for $N=N^*$.

When $F(t)=1-\mathrm{e}^{-\lambda t}$ and $G(x)=1-\mathrm{e}^{-\omega x}$, Table 6.4 presents optimum λT_L^* in (6.31), λT_F^* in (6.16), and their cost rates $C_L(T_L^*,N)/(\lambda c_M)$ and $C_F(T_F^*,N)/(\lambda c_M)$ for c_T/c_M and N when $\omega K=10.0$. Comparing given N with N^* in (6.12), optimum λT_L^* and λT_F^* can be determined in a finite value or not, i.e., the cases when replacement last or replacement first saves cost rate or not can also be determined.

Table 6.4 Optimum λT_L^* and λT_F^*, and their cost rates $C_L(T_L^*, N)/(\lambda c_M)$ and $C_F(T_F^*, N)/(\lambda c_M)$ when $\omega K = 10.0$

c_T/c_M	$N = 2$				
	λT_L^*	$C_L(T_L^*, N)/(\lambda c_M)$	λT_F^*	$C_F(T_F^*, N)/(\lambda c_M)$	N^*
0.1	3.021	0.039	∞	0.050	4
0.2	3.836	0.066	∞	0.100	5
0.5	5.111	0.132	∞	0.250	6
1.0	6.329	0.218	∞	0.500	7
2.0	7.942	0.358	∞	1.000	8
5.0	11.511	0.668	∞	2.500	11
c_T/c_M	$N = 5$				
	λT_L^*	$C_L(T_L^*, N)/(\lambda c_M)$	λT_F^*	$C_F(T_F^*, N)/(\lambda c_M)$	N^*
0.1	0.0	0.029	13.851	0.024	4
0.2	0.0	0.049	∞	0.049	5
0.5	2.467	0.108	∞	0.109	6
1.0	4.979	0.201	∞	0.209	7
2.0	7.342	0.350	∞	0.409	8
5.0	11.402	0.667	∞	1.009	11
c_T/c_M	$N = 10$				
	λT_L^*	$C_L(T_L^*, N)/(\lambda c_M)$	λT_F^*	$C_F(T_F^*, N)/(\lambda c_M)$	N^*
0.1	0.0	0.135	3.053	0.031	4
0.2	0.0	0.145	3.782	0.054	5
0.5	0.0	0.175	5.151	0.109	6
1.0	0.0	0.225	6.888	0.181	7
2.0	0.0	0.325	10.862	0.293	8
5.0	3.297	0.625	∞	0.625	11

6.2.2 Replacement Overtime Last

Suppose that the unit is replaced at the forthcoming shock over time T $(0 \le T \le \infty)$ or at shock N $(N = 1, 2, \cdots)$, whichever occurs last, which is named as *replacement overtime last with reactive maintenance*. The mean time to replacement is

$$\sum_{j=N}^{\infty} \int_0^T \left[\int_{T-t}^{\infty} (t+u)\mathrm{d}F(u)\right] \mathrm{d}F^{(j)}(t) + \int_T^{\infty} t\mathrm{d}F^{(N)}(t)$$

$$= T + \int_T^{\infty} [1 - F^{(N)}(t)]\mathrm{d}t + \sum_{j=N}^{\infty} \int_0^T \left[\int_{T-t}^{\infty} \overline{F}(u)\mathrm{d}u\right] \mathrm{d}F^{(j)}(t)$$

$$= \mu\left[N + \sum_{j=N}^{\infty} F^{(j)}(T)\right]. \tag{6.33}$$

The expected number of shocks until replacement is

$$\sum_{j=N}^{\infty}(j+1)[F^{(j)}(T)-F^{(j+1)}(T)]+N[1-F^{(N)}(T)]=N+\sum_{j=N}^{\infty}F^{(j)}(T),$$

and the expected number of shocks without reactive maintenances is

$$\sum_{j=N}^{\infty}[F^{(j)}(T)-F^{(j+1)}(T)]\sum_{i=1}^{j+1}G^{(i)}(K)+[1-F^{(N)}(T)]\sum_{j=1}^{N}G^{(j)}(K)$$
$$=\sum_{j=1}^{N}G^{(j)}(K)+\sum_{j=N}^{\infty}F^{(j)}(T)G^{(j+1)}(K).$$

Thus, the expected number of reactive maintenances until replacement is (Problem 6.6)

$$N+\sum_{j=N}^{\infty}F^{(j)}(T)-\sum_{j=1}^{N}G^{(j)}(K)-\sum_{j=N}^{\infty}F^{(j)}(T)G^{(j+1)}(K)$$
$$=\sum_{j=1}^{N}[1-G^{(j)}(K)]+\sum_{j=N}^{\infty}F^{(j)}(T)[1-G^{(j+1)}(K)]. \tag{6.34}$$

Therefore, the expected cost rate is

$$C_{OL}(T,N)=\frac{\begin{array}{l}c_N+(c_O-c_N)F^{(N)}(T)+c_M\{\sum_{j=1}^{N}[1-G^{(j)}(K)]\\ \quad+\sum_{j=N}^{\infty}F^{(j)}(T)[1-G^{(j+1)}(K)]\}\end{array}}{\mu[N+\sum_{j=N}^{\infty}F^{(j)}(T)]}. \tag{6.35}$$

Clearly,

$$\lim_{N\to 0}C_{OL}(T,N)=\lim_{N\to\infty}C_{OF}(T,N)=C_O(T),$$
$$\lim_{T\to 0}C_{OL}(T,N)=\lim_{T\to\infty}C_{OF}(T,N)=C(N),$$

where $C_O(T)$ and $C(N)$ are given in (6.21) and (6.7), respectively.

We find optimum (T_{OL}^*, N_{OL}^*) to minimize $C_{OL}(T,N)$ when $c_O=c_N$ and $F(t)=1-e^{-\lambda t}$. Forming the inequality $C_{OL}(T,N+1)-C_{OL}(T,N)\geq 0$,

$$[1-G^{(N+1)}(K)]\left[N+\sum_{j=N}^{\infty}F^{(j)}(T)\right]$$
$$-\sum_{j=1}^{N}[1-G^{(j)}(K)]-\sum_{j=N}^{\infty}F^{(j)}(T)[1-G^{(j+1)}(K)]\geq\frac{c_O}{c_M}. \quad (6.36)$$

Differentiating $C_{OL}(T, N)$ with respect to T and setting it equal to zero,

$$Q_3(T,N)\left[N+\sum_{j=N}^{\infty}F^{(j)}(T)\right]$$
$$-\sum_{j=1}^{N}[1-G^{(j)}(K)]-\sum_{j=N}^{\infty}F^{(j)}(T)[1-G^{(j+1)}(K)]=\frac{c_O}{c_M}, \quad (6.37)$$

where

$$Q_3(T,N)\equiv\frac{\sum_{j=N-1}^{\infty}p_j(T)[1-G^{(j+2)}(K)]}{\sum_{j=N-1}^{\infty}p_j(T)}.$$

Substituting (6.36) for (6.37),

$$Q_3(T,N)\leq 1-G^{(N+1)}(K), \quad (6.38)$$

which does not hold for any N. Therefore, there does not exist any positive T_L^* which satisfies (6.37), i.e., the optimum policy is ($T_{OL}^*=0$, $N_{OL}^*=N^*$), where N^* is given in (6.12).

Noting that $Q_3(T, N)$ increases strictly with T from $1-G^{(N+1)}(K)$ to 1, the left-hand side of (6.37) increases strictly with T from

$$\sum_{j=1}^{N}[G^{(j)}(K)-G^{(N+1)}(K)].$$

Thus, by similar discussion above in Sect. 6.2.1, if $N\geq N^*$ given in (6.12), then $T_{OL}^*=0$. Conversely, if $N<N^*$, then there exists a finite and unique T_{OL}^* ($0<T_{OL}^*<\infty$) which satisfies (6.37).

From the above discussions, it is also easily to obtain that we should adopt replacement overtime last in (6.35) for $N<N^*$, replacement overtime first in (6.20) for $N>N^*$, and replacement with shock N in (6.7) for $N=N^*$.

When $F(t)=1-e^{-\lambda t}$ and $G(x)=1-e^{-\omega x}$, Table 6.5 presents optimum λT_{OL}^* in (6.37), λT_{OF}^* in (6.25), and their cost rates $C_{OL}(T_{OL}^*, N)/(\lambda c_M)$ and $C_{OF}(T_{OF}^*, N)/(\lambda c_M)$ for c_O/c_M and N when $\omega K=10.0$, which agree with the above analytical discussions.

Table 6.5 Optimum λT_{OL}^* and λT_{OF}^*, and their cost rates $C_{OL}(T_{OL}^*, N)/(\lambda c_M)$ and $C_{OF}(T_{OF}^*, N)/(\lambda c_M)$ when $\omega K = 10.0$

c_O/c_M	$N = 2$				
	λT_{OL}^*	$C_{OL}(T_{OL}^*, N)/(\lambda c_M)$	λT_{OF}^*	$C_{OF}(T_{OF}^*, N)/(\lambda c_M)$	N^*
0.1	2.333	0.037	∞	0.050	4
0.2	3.051	0.063	∞	0.100	5
0.5	4.221	0.127	∞	0.250	6
1.0	5.373	0.213	∞	0.500	7
2.0	6.926	0.352	∞	1.000	8
5.0	10.408	0.665	∞	2.500	11
c_O/c_M	$N = 5$				
	λT_{OL}^*	$C_{OL}(T_{OL}^*, N)/(\lambda c_M)$	λT_{OF}^*	$C_{OF}(T_{OF}^*, N)/(\lambda c_M)$	N^*
0.1	0.0	0.029	∞	0.029	4
0.2	0.0	0.049	∞	0.049	5
0.5	2.102	0.108	∞	0.109	6
1.0	4.314	0.199	∞	0.209	7
2.0	6.471	0.346	∞	0.409	8
5.0	10.334	0.664	∞	1.009	11
c_O/c_M	$N = 10$				
	λT_{OL}^*	$C_{OL}(T_{OL}^*, N)/(\lambda c_M)$	λT_{OF}^*	$C_{OF}(T_{OF}^*, N)/(\lambda c_M)$	N^*
0.1	0.0	0.135	2.490	0.038	4
0.2	0.0	0.145	3.194	0.061	5
0.5	0.0	0.175	4.547	0.125	6
1.0	0.0	0.225	6.344	0.204	7
2.0	0.0	0.325	11.068	0.324	8
5.0	3.011	0.625	∞	0.625	11

6.3 Replacement Policies with Failure Number

We have supposed in the above sections that the unit can be quickly resumed to work with undisturbed damage K by reactive maintenances at failures, and the expected number of shocks is counted to balance the cost for maintenance and replacement. In this section, we take the expected number M ($M = 1, 2, \cdots$) of failures into consideration to save maintenance cost after failure, and the respective approaches of whichever triggering event occurs first and last are applied into replacement policies.

(1) Replacement First

Suppose that the unit is replaced at time T ($0 < T \leq \infty$) or at failure M ($M = 1, 2, \cdots$), whichever occurs first. Then, the probability that the unit is replaced at time T is

$$\sum_{j=0}^{\infty} G^{(j)}(K)[F^{(j)}(T) - F^{(j+1)}(T)] + \sum_{j=0}^{\infty}[G^{(j)}(K) - G^{(j+1)}(K)]$$
$$\times \sum_{i=0}^{M-2} \int_0^T \left[F^{(i)}(T-t) - F^{(i+1)}(T-t)\right] \mathrm{d}F^{(j+1)}(t)$$
$$= \sum_{j=0}^{\infty} G^{(j)}(K)[F^{(j)}(T) - F^{(j+1)}(T)]$$
$$+ \sum_{j=0}^{\infty}[G^{(j)}(K) - G^{(j+1)}(K)][F^{(j+1)}(T) - F^{(M+j)}(T)], \tag{6.39}$$

the probability that it is replaced at failure M is

$$\sum_{j=0}^{\infty}[G^{(j)}(K) - G^{(j+1)}(K)] \sum_{i=M-1}^{\infty} \int_0^T [F^{(i)}(T-t) - F^{(i+1)}(T-t)] \mathrm{d}F^{(j+1)}(t)$$
$$= \sum_{j=0}^{\infty}[G^{(j)}(K) - G^{(j+1)}(K)] F^{(M+j)}(T). \tag{6.40}$$

The mean time to replacement is

$$T\left\{1 - \sum_{j=0}^{\infty}[G^{(j)}(K) - G^{(j+1)}(K)] F^{(M+j)}(T)\right\}$$
$$+ \sum_{j=0}^{\infty}[G^{(j)}(K) - G^{(j+1)}(K)] \int_0^T t \mathrm{d}F^{(M+j)}(t)$$
$$= \sum_{j=0}^{\infty}[G^{(j)}(K) - G^{(j+1)}(K)] \int_0^T [1 - F^{(M+j)}(t)] \mathrm{d}t. \tag{6.41}$$

The expected number of failures until replacement is

$$\sum_{j=0}^{\infty}[G^{(j)}(K) - G^{(j+1)}(K)] \sum_{i=0}^{M-2} (i+1)[F^{(i+j+1)}(T) - F^{(i+j+2)}(T)]$$
$$+ M \sum_{j=0}^{\infty}[G^{(j)}(K) - G^{(j+1)}(K)] F^{(M+j)}(T)$$
$$= \sum_{j=0}^{\infty}[G^{(j)}(K) - G^{(j+1)}(K)] \sum_{i=1}^{M} F^{(i+j)}(T).$$

Therefore, the expected cost rate is

$$C_F(T,M)=\frac{\begin{array}{l}c_T+(c_F-c_T)\sum_{j=0}^{\infty}[G^{(j)}(K)-G^{(j+1)}(K)]F^{(M+j)}(T)\\ +c_M\sum_{j=0}^{\infty}[G^{(j)}(K)-G^{(j+1)}(K)]\sum_{i=1}^{M}F^{(i+j)}(T)\end{array}}{\sum_{j=0}^{\infty}[G^{(j)}(K)-G^{(j+1)}(K)]\int_0^T[1-F^{(M+j)}(t)]\mathrm{d}t}, \tag{6.42}$$

where c_F = replacement cost at failure M, and c_T and c_M are given in (6.5).

Clearly, $\lim_{M\to\infty} C_F(T,M)=C(T)$ in (6.6), $C_F(T,1)=C(T)$ in (2.9) when $c_F+c_M=c_K$, and (Problem 6.7)

$$C(M)\equiv\lim_{T\to\infty} C_F(T,M)=\frac{c_F+c_M M}{\mu[M+\sum_{j=1}^{\infty}G^{(j)}(K)]}. \tag{6.43}$$

We find optimum T_F^* to minimize $C_F(T,M)$ in (6.42) for given M ($M=1,2,\cdots$) when $c_F=c_T$ and $F(t)=1-\mathrm{e}^{-\lambda t}$. Differentiating $C_F(T,M)$ with respect to T and setting it equal to zero,

$$\begin{aligned}&Q_1(T,M)\sum_{j=0}^{\infty}[G^{(j)}(K)-G^{(j+1)}(K)]\sum_{i=1}^{M+j}F^{(i)}(T)\\&-\sum_{j=0}^{\infty}[G^{(j)}(K)-G^{(j+1)}(K)]\sum_{i=j+1}^{M+j}F^{(i)}(T)=\frac{c_T}{c_M},\end{aligned} \tag{6.44}$$

where

$$Q_1(T,M)=\frac{\sum_{j=0}^{\infty}[G^{(j)}(K)-G^{(j+1)}(K)][F^{(j)}(T)-F^{(M+j)}(T)]}{\sum_{j=0}^{\infty}[G^{(j)}(K)-G^{(j+1)}(K)][1-F^{(M+j)}(T)]},$$

which increases strictly with T from 0 to 1 (Problem 6.8). Thus, the left-hand side of (6.44) increases strictly with T from 0 to $M_G(K)$. Therefore, if $M_G(K)>c_T/c_M$, then there exists a finite and unique T_F^* ($0<T_F^*<\infty$) which satisfies (6.44), and the resulting cost rate is

$$\frac{C_F(T_F^*,M)}{\lambda}=c_M Q_1(T_F^*,M). \tag{6.45}$$

Furthermore, because $Q_1(T,M)$ increases with M, the left-hand side of (6.44) increases strictly to that of (6.10), and so that, T_F^* decreases with M to T^* given in (6.10).

(2) Replacement Last

Suppose that the unit is replaced at time T $(0 < T \leq \infty)$ or at failure M $(M = 1, 2, \cdots)$, whichever occurs last. Then, the probability that the unit is replaced at time T is

$$\sum_{j=0}^{\infty}[G^{(j)}(K) - G^{(j+1)}(K)]\sum_{i=M-1}^{\infty}\int_0^T [F^{(i)}(T-t) - F^{(i+1)}(T-t)]\mathrm{d}F^{(j+1)}(t)$$
$$= \sum_{j=0}^{\infty}[G^{(j)}(K) - G^{(j+1)}(K)]F^{(M+j)}(T), \tag{6.46}$$

and the probability that it is replaced at failure M is

$$\sum_{j=0}^{\infty}[G^{(j)}(K) - G^{(j+1)}(K)]\int_0^T [1 - F^{(M-1)}(T-t)]\mathrm{d}F^{(j+1)}(t)$$
$$+ \sum_{j=0}^{\infty}[G^{(j)}(K) - G^{(j+1)}(K)][1 - F^{(j+1)}(T)]$$
$$= \sum_{j=0}^{\infty}[G^{(j)}(K) - G^{(j+1)}(K)][1 - F^{(M+j)}(T)]. \tag{6.47}$$

The mean time to replacement is

$$T\sum_{j=0}^{\infty}[G^{(j)}(K) - G^{(j+1)}(K)]F^{(M+j)}(T)$$
$$+ \sum_{j=0}^{\infty}[G^{(j)}(K) - G^{(j+1)}(K)]\int_0^T \left[\int_{T-t}^{\infty}(t+u)\mathrm{d}F^{(M-1)}(u)\right]\mathrm{d}F^{(j+1)}(t)$$
$$+ \sum_{j=0}^{\infty}[G^{(j)}(K) - G^{(j+1)}(K)]\int_T^{\infty}\left[\int_0^{\infty}(t+u)\mathrm{d}F^{(M-1)}(u)\right]\mathrm{d}F^{(j+1)}(t)$$
$$= T + \sum_{j=0}^{\infty}[G^{(j)}(K) - G^{(j+1)}(K)]\int_T^{\infty}[1 - F^{(M+j)}(t)]\mathrm{d}t. \tag{6.48}$$

The expected number of failures until replacement is

$$\sum_{j=0}^{\infty}[G^{(j)}(K) - G^{(j+1)}(K)] \sum_{i=M-1}^{\infty} (i+1)[F^{(i+j+1)}(T) - F^{(i+j+2)}(T)]$$

$$+ M \sum_{j=0}^{\infty}[G^{(j)}(K) - G^{(j+1)}(K)][1 - F^{(M+j)}(T)]$$

$$= M + \sum_{j=0}^{\infty}[1 - G^{(j)}(K)]F^{(M+j)}(T). \tag{6.49}$$

Therefore, the expected cost rate is

$$C_L(T, M) = \frac{\begin{array}{c} c_F + (c_T - c_F)\sum_{j=0}^{\infty}[G^{(j)}(K) - G^{(j+1)}(K)]F^{(M+j)}(T) \\ +c_M\{M + \sum_{j=0}^{\infty}[1 - G^{(j)}(K)]F^{(M+j)}(T)\} \end{array}}{T + \sum_{j=0}^{\infty}[G^{(j)}(K) - G^{(j+1)}(K)]\int_T^{\infty}[1 - F^{(M+j)}(t)]\mathrm{d}t}. \tag{6.50}$$

Similarly, $\lim_{T\to 0} C_L(T, M) = \lim_{T\to\infty} C_F(T, M) = C(M)$ in (6.43).

When $F(t) = 1 - \mathrm{e}^{-\lambda t}$ and $c_T = c_F$, the expected cost rate is (Problem 6.9)

$$\frac{C_L(T, M)}{\lambda} = \frac{c_T - c_M M_G(K)}{M_G(K) + M + \sum_{j=0}^{\infty}[1 - G^{(j)}(K)]F^{(M+j)}(T)} + c_M. \tag{6.51}$$

It can be easily shown (Problem 6.7) that if $M_G(K) > c_T/c_M$, then $T_L^* = 0$, and the resulting cost rate is given in (6.43). Conversely, if $M_G(K) \le c_T/c_M$, then $T_L^* = \infty$, and the resulting cost rate is λc_M (Problem 6.10).

6.4 Replacement Overtime with Failure Number

We consider the following replacement models in which the unit is replaced over time T or at failure M, whichever occurs first and last, respectively.

(1) Replacement Overtime First

Suppose that the unit is replaced at the forthcoming shock over time T $(0 < T \le \infty)$ or at failure M $(M = 1, 2, \cdots)$, whichever occurs first. The mean time to replacement is

$$\sum_{j=0}^{\infty} G^{(j)}(K) \int_0^T \left[\int_{T-t}^{\infty} (t+u) \mathrm{d}F(u) \right] \mathrm{d}F^{(j)}(t)$$
$$+ \sum_{j=0}^{\infty} [G^{(j)}(K) - G^{(j+1)}(K)] \int_0^T t \mathrm{d}F^{(M+j)}(t) + \sum_{j=0}^{\infty} [G^{(j)}(K) - G^{(j+1)}(K)]$$
$$\times \sum_{i=0}^{M-2} \int_0^T \left\{ \int_0^{T-t} \left[\int_{T-t-u}^{\infty} (t+u+y) \mathrm{d}F(y) \right] \mathrm{d}F^{(i)}(u) \right\} \mathrm{d}F^{(j+1)}(t)$$
$$= \mu \left[\sum_{j=0}^{M-1} F^{(j)}(T) + \sum_{j=0}^{\infty} G^{(j+1)}(K) F^{(M+j)}(T) \right]$$
$$= \mu \sum_{j=0}^{\infty} [G^{(j)}(K) - G^{(j+1)}(K)] \sum_{i=0}^{M+j-1} F^{(i)}(T). \tag{6.52}$$

The expected number of failures until replacement is

$$\sum_{j=0}^{\infty} [G^{(j)}(K) - G^{(j+1)}(K)] \sum_{i=0}^{M-2} (i+1) [F^{(i+j+1)}(T) - F^{(i+j+2)}(T)]$$
$$+ M \sum_{j=0}^{\infty} [G^{(j)}(K) - G^{(j+1)}(K)] F^{(M+j)}(T)$$
$$= \sum_{j=0}^{\infty} [G^{(j)}(K) - G^{(j+1)}(K)] \sum_{i=0}^{M-1} F^{(i+j)}(T). \tag{6.53}$$

Therefore, the expected cost rate is

$$C_{OF}(T, M) = \frac{\begin{array}{c} c_O + (c_F - c_O) \sum_{j=0}^{\infty} [G^{(j)}(K) - G^{(j+1)}(K)] F^{(M+j)}(T) \\ + c_M \sum_{j=0}^{\infty} [G^{(j)}(K) - G^{(j+1)}(K)] \sum_{i=0}^{M-1} F^{(i+j)}(T) \end{array}}{\mu \sum_{j=0}^{\infty} [G^{(j)}(K) - G^{(j+1)}(K)] \sum_{i=0}^{M+j-1} F^{(i)}(T)}. \tag{6.54}$$

Clearly, $\lim_{T \to \infty} C_{OF}(T, M) = \lim_{T \to \infty} C_F(T, M) = C(M)$ in (6.43), and $\lim_{M \to \infty} C_{OF}(T, M) = C_O(T)$ in (6.21).

We find optimum T_{OF}^* to minimize $C_{OF}(T, M)$ for given M $(M = 1, 2, \cdots)$ when $F(t) = 1 - \mathrm{e}^{-\lambda t}$ and $c_F = c_O$. Differentiating $C_{OF}(T, M)$ with respect to T and setting it equal to zero,

$$Q_2(T,M)\sum_{j=0}^{\infty}[G^{(j)}(K)-G^{(j+1)}(K)]\sum_{i=0}^{M+j-1}F^{(i)}(T)$$
$$-\sum_{j=0}^{\infty}[G^{(j)}(K)-G^{(j+1)}(K)]\sum_{i=0}^{M-1}F^{(i+j)}(T)=\frac{c_O}{c_M}, \tag{6.55}$$

where

$$Q_2(T,M)\equiv\frac{\sum_{j=1}^{\infty}[G^{(j+1)}(K)-G^{(j+2)}(K)][F^{(j)}(T)-F^{(M+j)}(T)]}{\sum_{j=1}^{\infty}[G^{(j+1)}(K)-G^{(j+2)}(K)][1-F^{(M+j)}(T)]},$$

which increases strictly with T from 0 to 1 (Problem 6.8). Thus, the left-hand side of (6.55) increases strictly with T from 0 to $M_G(K)$. Therefore, if $M_G(K)>c_O/c_M$, then there exists a finite and unique T_{OF}^* $(0<T_{OF}^*<\infty)$ which satisfies (6.55), and the resulting cost rate is

$$\frac{C_{OF}(T_{OF}^*)}{\lambda}=c_M Q_2(T_{OF}^*,M). \tag{6.56}$$

Furthermore, because $Q_2(T,M)$ increases with M, the left-hand side of (6.55) increases strictly with M to that of (6.22), and so that, T_{OF}^* decreases with M to T_O^* given in (6.22).

(2) Replacement Overtime Last

Suppose that the unit is replaced at the forthcoming shock over time T $(0\le T\le\infty)$ or at failure M $(M=0,1,2,\cdots)$, whichever occurs last. The mean time to replacement is (Problem 6.11)

$$\mu\left\{M+M_G(K)+\sum_{j=0}^{\infty}[1-G^{(j+1)}(K)]F^{(M+j)}(T)\right\}. \tag{6.57}$$

The expected number of failures until replacement is

$$M+\sum_{j=0}^{\infty}[1-G^{(j+1)}(K)]F^{(M+j)}(T). \tag{6.58}$$

It can be easily shown that when $\mu=1$, the difference between (6.57) and (6.58) is $M_G(K)$, which represents the expected number of shocks before replacement.

Therefore, the expected cost rate is

$$C_{OL}(T, M) = \frac{\begin{array}{c} c_F + (c_O - c_F)\sum_{j=0}^{\infty}[G^{(j)}(K) - G^{(j+1)}(K)]F^{(M+j)}(T) \\ +c_M\{M + \sum_{j=0}^{\infty}[1 - G^{(j+1)}(K)]F^{(M+j)}(T)\} \end{array}}{\mu\{M + M_G(K) + \sum_{j=0}^{\infty}[1 - G^{(j+1)}(K)]F^{(M+j)}(T)\}}. \quad (6.59)$$

Clearly, $\lim_{T\to 0} C_{OL}(T, M) = \lim_{T\to\infty} C_{OF}(T, M) = C(M)$.

When $F(t) = 1 - \mathrm{e}^{-\lambda t}$ and $c_F = c_O$, the expected cost rate is

$$\frac{C_{OL}(T, M)}{\lambda} = \frac{c_O - c_M M_G(K)}{M + M_G(K) + \sum_{j=0}^{\infty}[1 - G^{(j+1)}(K)]F^{(M+j)}(T)} + c_M. \quad (6.60)$$

Thus, if $M_G(K) > c_O/c_M$, then $T_{OL}^* = 0$, and conversely, if $M_G(K) \le c_O/c_M$, then $T_L^* = \infty$. Comparing (6.60) with (6.51), if $M_G(K) > c_O/c_M$, then $C_{OL}(T, M) > C_L(T, M)$, and conversely, if $M_G(K) \le c_O/c_M$, then $C_{OL}(T, M) \le C_L(T, M)$.

6.5 Nonhomogeneous Poisson Shock Times

It is meaningless to assume that shocks occur at a Poisson process for replacement policies with failure number in Sects. 6.3 and 6.4, as shown in (6.43). We suppose that shocks occur at a nonhomogeneous Poisson process with mean value function $H(t) \equiv \int_0^t h(u)\mathrm{d}u$, i.e., the probability that shocks occur exactly j times in $[0, t]$ is

$$p_j(t) = \frac{H(t)^j}{j!}\mathrm{e}^{-H(t)} \quad (j = 0, 1, 2, \cdots).$$

Denote $P_j(t) \equiv \sum_{i=j}^{\infty} p_i(t)$ $(j = 0, 1, 2, \cdots)$ and $\overline{P}_j(t) \equiv 1 - P_j(t)$, and replace $F^{(j)}(t)$ with $P_j(t)$ formally in Sects. 6.3 and 6.4. For replacement first and last, (6.42) is

$$C_F(T, M) = \frac{\begin{array}{c} c_T + (c_F - c_T)\sum_{j=0}^{\infty}[G^{(j)}(K) - G^{(j+1)}(K)]P_{M+j}(T) \\ +c_M \sum_{j=0}^{\infty}[G^{(j)}(K) - G^{(j+1)}(K)]\sum_{i=1}^{M} P_{i+j}(T) \end{array}}{\sum_{j=0}^{\infty}[G^{(j)}(K) - G^{(j+1)}(K)]\int_0^T \overline{P}_{M+j}(t)\mathrm{d}t}, \quad (6.61)$$

and (6.50) is

$$C_L(T, M) = \frac{\begin{array}{c} c_F + (c_T - c_F)\sum_{j=0}^{\infty}[G^{(j)}(K) - G^{(j+1)}(K)]P_{M+j}(T) \\ +c_M\{M + \sum_{j=0}^{\infty}[1 - G^{(j)}(K)]P_{M+j}(T)\} \end{array}}{T + \sum_{j=0}^{\infty}[G^{(j)}(K) - G^{(j+1)}(K)]\int_T^{\infty} \overline{P}_{M+j}(t)\mathrm{d}t}. \quad (6.62)$$

For replacement overtime first and last, (6.54) is (Problem 6.12)

$$C_{OF}(T,M)=\frac{\begin{array}{c}c_O+(c_F-c_O)\sum_{j=0}^{\infty}[G^{(j)}(K)-G^{(j+1)}(K)]P_{M+j}(T)\\ +c_M\sum_{j=0}^{\infty}[G^{(j)}(K)-G^{(j+1)}(K)]\sum_{i=0}^{M-1}P_{i+j}(T)\end{array}}{\begin{array}{c}\sum_{j=0}^{\infty}[G^{(j)}(K)-G^{(j+1)}(K)]\{\int_0^T\overline{P}_{M+j}(t)\mathrm{d}t\\ +\sum_{i=0}^{M+j-1}p_i(T)\int_T^{\infty}\mathrm{e}^{-[H(t)-H(T)]}\mathrm{d}t\}\end{array}}, \tag{6.63}$$

and (6.59) is

$$C_{OL}(T,M)=\frac{\begin{array}{c}c_F+(c_O-c_F)\sum_{j=0}^{\infty}[G^{(j)}(K)-G^{(j+1)}(K)]P_{M+j}(T)\\ c_M\{M+\sum_{j=0}^{\infty}[G^{(j)}(K)-G^{(j+1)}(K)]\sum_{i=M}^{\infty}P_{i+j+1}(T)\}\end{array}}{\begin{array}{c}\sum_{j=0}^{\infty}[G^{(j)}(K)-G^{(j+1)}(K)]\{\int_T^{\infty}\overline{P}_{M+j}(t)\mathrm{d}t\\ +\sum_{i=M+j}^{\infty}p_i(T)\int_T^{\infty}\mathrm{e}^{-[H(t)-H(T)]}\mathrm{d}t\}\end{array}}. \tag{6.64}$$

In particular, when the unit is replaced only at failure M, the expected cost rate is

$$\begin{aligned}C(M)&\equiv\lim_{T\to\infty}C_F(T,M)=\lim_{T\to 0}C_L(T,M)\\&=\lim_{T\to\infty}C_{OF}(T,M)=\lim_{T\to 0}C_{OL}(T,M)\\&=\frac{c_F+c_M M}{\sum_{j=0}^{\infty}[G^{(j)}(K)-G^{(j+1)}(K)]\int_0^{\infty}\overline{P}_{M+j}(t)\mathrm{d}t}.\end{aligned} \tag{6.65}$$

We find optimum M^* to minimize $C(M)$ when $h(t)$ increases strictly to $h(\infty)$. Forming the inequality $C(M+1)-C(M)\geq 0$,

$$\frac{\sum_{j=0}^{\infty}[G^{(j)}(K)-G^{(j+1)}(K)]\int_0^{\infty}\overline{P}_{M+j}(t)\mathrm{d}t}{\sum_{j=0}^{\infty}[G^{(j)}(K)-G^{(j+1)}(K)]\int_0^{\infty}p_{M+j}(t)\mathrm{d}t}-M\geq\frac{c_F}{c_M}. \tag{6.66}$$

Noting that $\int_0^{\infty}p_{M+j}(t)\mathrm{d}t$ decreases strictly with M to $1/h(\infty)$ [1], and the left-hand side $L(M)$ of (6.66) increases strictly with M. Therefore, if $L(\infty)>c_F/c_M$, then there exists a finite and unique minimum M^* $(1\leq M^*<\infty)$ which satisfies (6.66).

6.6 Problem 6

6.1 Derive (6.4).

6.2 Derive (6.10) and prove that its left-hand side increases strictly with T from 0 to $M_G(K)$.

6.3 Prove that $Q_1(T, N)$ increases strictly with T from $\overline{G}(K)$ to $1 - G^{(N)}(K)$, and the left-hand side of (6.16) increases strictly with T from 0 to $\sum_{j=1}^{N}[G^{(j)}(K) - G^{(N)}(K)]$.
6.4 Derive (6.19).
6.5 Derive (6.22) and prove that its left-hand side increases strictly with T from $G(K) - G^{(2)}(K)$ to $M_G(K)$.
6.6 Derive (6.28) and (6.34).
6.7 Explain the physical meaning of $M^* = 1$ or $M^* = \infty$ to minimize $C(M)$ in (6.43).
6.8 Prove that when $F(t) = 1 - e^{-\lambda t}$,

$$Q_1(T, M) = \frac{\sum_{j=0}^{\infty}[G^{(j)}(K) - G^{(j+1)}(K)][F^{(j)}(T) - F^{(M+j)}(T)]}{\sum_{j=0}^{\infty}[G^{(j)}(K) - G^{(j+1)}(K)][1 - F^{(M+j)}(T)]} \quad (M = 1, 2, \cdots)$$

increases strictly with T from 0 to 1 and increases strictly with M to $\sum_{j=0}^{\infty}[G^{(j)}(K) - G^{(j+1)}(K)]F^{(j)}(T)$. Furthermore, show that when $M \to \infty$, (6.44) agrees with (6.10).
6.9 Derive (6.51).
6.10 Obtain the expected cost rate when the unit is replaced at shock N or failure M, whichever occurs first and last.
6.11 Derive (6.57) and (6.58).
6.12 Derive the mean times to replacement for replacement overtime first and last.

Chapter 7
Replacement Policies with Independent Damages

In general, an operating unit degrades gradually with *additive damage* cumulated by shocks in a stochastic way. In addition, *independent damage*, which occurs randomly and independently, can cause the unit to sudden crash and additional downtime. The independent damage was defined as non-additive damage that does no damage unless its amount exceeds a threshold δ of the mechanical strength [2]. A typical example of the failure mode caused by independent damage is semiconductor parts that fails due to over-current or fault voltage. However, both additive and independent damages should be considered in real cases to analyze the reliability and replacement models [18], e.g., corrosion, degrading the strength of pipelines with its age, is a predominant cause of pipeline leaks, and the third-part damage is a leading cause of pipeline ruptures, which occurs randomly in a statistical sense [49]. Three levels of damages such as small level of damage that is harmless, large level of damage resulting in failure and intermediate level of damage causing failure with probability are generalized in [50].

In this chapter, a new unit with damage level 0 begins to operate at time 0 and degrades with *additive damage* produced by shocks. The probability distributions $F(t)$ with mean μ and a density function $f(t)$, and $G(x)$ with mean $1/\omega$ for shocks and damage have been supposed in Sect. 2.1, and the unit fails when the total damage has exceeded a failure threshold K. We suppose that *independent damage* occurs at a nonhomogeneous Poisson process with and intensity function $h(t)$ and a mean-value function $H(t) \equiv \int_0^t h(u)\mathrm{d}u$, where $h(t)$ increases strictly with t to $h(\infty)$. Let $p_j(t)$ denote the probability that a number j of independent damages occurs exactly in $[0, t]$, then $p_j(t) = [H(t)^j/j!]\mathrm{e}^{-H(t)}$ $(j = 0, 1, 2, \ldots)$, and $P_j(t) \equiv \sum_{i=j}^{\infty} p_i(t)$ means at least number j of independent damages have occurred until t, where $\overline{P}_j(t) \equiv 1 - P_j(t)$. The most costly *corrective replacement* is done when the total additive damage exceeds K, *minimal repair* is made for the independent damage to let the unit return to operation, so that how to plan preventive replacement policies becomes the main focus to be solved in the following sections.

X. Zhao and T. Nakagawa, *Advanced Maintenance Policies for Shock and Damage Models*, Springer Series in Reliability Engineering,
https://doi.org/10.1007/978-3-319-70456-2_7

In Sect. 7.1, preventive replacement policies are planned at time T or at number N of independent damages, where the approaches of replacement first and last are modeled. In Sects. 7.2 and 7.3, replacement planned at time T is modified to be done over independent and additive damages, and replacement overtime first and replacement overtime last are discussed, respectively. In Sect. 7.4, both the numbers of additive and independent damages are modeled for the policies of replacement first, replacement last and replacement middle.

7.1 Replacement First and Last

We suppose that an operating unit is replaced preventively at time T or at number N of independent damages and model the policies of replacement first and last, using the approaches of whichever triggering event occurs first and whichever triggering event occurs last. Optimum replacement times to minimize their expected cost rates are derived analytically and computed numerically.

(1) Replacement First

Suppose that the unit undergoes minimal repairs at independent damages, and is replaced at time T ($0 < T \leq \infty$), at independent damage N ($N = 1, 2, \ldots$), or at failure K ($0 < K < \infty$) of additive damage, whichever occurs first, which is named as *replacement first with independent damage*. Then, the probability that the unit is replaced at time T is

$$\overline{P}_N(T) \sum_{j=0}^{\infty} G^{(j)}(K)[F^{(j)}(T) - F^{(j+1)}(T)], \tag{7.1}$$

the probability that it is replaced at independent damage N is

$$\sum_{j=0}^{\infty} G^{(j)}(K) \int_0^T [F^{(j)}(t) - F^{(j+1)}(t)] \mathrm{d}P_N(t), \tag{7.2}$$

and the probability that it is replaced at failure K is

$$\sum_{j=0}^{\infty} [G^{(j)}(K) - G^{(j+1)}(K)] \int_0^T \overline{P}_N(t) \mathrm{d}F^{(j+1)}(t), \tag{7.3}$$

where note that (7.1) + (7.2) + (7.3) = 1. Thus, the mean time to replacement is

$$
\begin{aligned}
&T\overline{P}_N(T)\sum_{j=0}^{\infty}G^{(j)}(K)[F^{(j)}(T)-F^{(j+1)}(T)] \\
&+\sum_{j=0}^{\infty}G^{(j)}(K)\int_0^T t[F^{(j)}(t)-F^{(j+1)}(t)]\mathrm{d}P_N(t) \\
&+\sum_{j=0}^{\infty}[G^{(j)}(K)-G^{(j+1)}(K)]\int_0^T t\overline{P}_N(t)\mathrm{d}F^{(j+1)}(t) \\
&=\sum_{j=0}^{\infty}G^{(j)}(K)\int_0^T [F^{(j)}(t)-F^{(j+1)}(t)]\overline{P}_N(t)\mathrm{d}t.
\end{aligned} \tag{7.4}
$$

The expected number of independent damages, i.e., minimal repairs, until replacement is (Problem 7.1)

$$
\begin{aligned}
&\sum_{i=0}^{N-1}ip_i(T)\sum_{j=0}^{\infty}G^{(j)}(K)[F^{(j)}(T)-F^{(j+1)}(T)] \\
&+N\sum_{j=0}^{\infty}G^{(j)}(K)\int_0^T [F^{(j)}(t)-F^{(j+1)}(t)]\mathrm{d}P_N(t) \\
&+\sum_{j=0}^{\infty}[G^{(j)}(K)-G^{(j+1)}(K)]\sum_{i=0}^{N-1}i\int_0^T p_i(t)\mathrm{d}F^{(j+1)}(t) \\
&=N-\sum_{i=0}^{N}(N-i)p_i(T)\sum_{j=0}^{\infty}G^{(j)}(K)[F^{(j)}(T)-F^{(j+1)}(T)] \\
&\quad-\sum_{j=0}^{\infty}[G^{(j)}(K)-G^{(j+1)}(K)]\sum_{i=0}^{N}(N-i)\int_0^T p_i(t)\mathrm{d}F^{(j+1)}(t) \\
&=\sum_{j=0}^{\infty}G^{(j)}(K)\int_0^T [F^{(j)}(t)-F^{(j+1)}(t)]\overline{P}_N(t)h(t)\mathrm{d}t,
\end{aligned} \tag{7.5}
$$

which agrees with (7.4) when $h(t)\equiv 1$.

Therefore, the expected repair and replacement cost rate is

$$
C_F(T,N)=
\frac{\begin{array}{l}
c_T+(c_N-c_T)\sum_{j=0}^{\infty}G^{(j)}(K)\int_0^T[F^{(j)}(t)-F^{(j+1)}(t)]\mathrm{d}P_N(t) \\
+(c_K-c_T)\sum_{j=0}^{\infty}[G^{(j)}(K)-G^{(j+1)}(K)]\int_0^T\overline{P}_N(t)\mathrm{d}F^{(j+1)}(t) \\
+c_M\sum_{j=0}^{\infty}G^{(j)}(K)\int_0^T[F^{(j)}(t)-F^{(j+1)}(t)]\overline{P}_N(t)h(t)\mathrm{d}t
\end{array}}{\sum_{j=0}^{\infty}G^{(j)}(K)\int_0^T[F^{(j)}(t)-F^{(j+1)}(t)]\overline{P}_N(t)\mathrm{d}t}, \tag{7.6}
$$

where c_T = replacement cost at time T, c_N = replacement cost at independent damage N, c_K = replacement cost at failure K, where $c_K > c_T$ and $c_K > c_N$, and c_M = cost for minimal repair at each independent damage. Clearly, when $h(t) = 0$, i.e., $\overline{P}_N(t) = 1$, $C_F(T, N)$ in (7.6) agrees with $C(T)$ in (2.9).

In particular, when the unit is replaced preventively only at time T,

$$C(T) \equiv \lim_{N\to\infty} C_F(T, N) = \frac{\begin{array}{c} c_T + (c_K - c_T)\sum_{j=0}^{\infty} F^{(j+1)}(T)[G^{(j)}(K) - G^{(j+1)}(K)] \\ +c_M \sum_{j=0}^{\infty} G^{(j)}(K) \int_0^T [F^{(j)}(t) - F^{(j+1)}(t)]h(t)\mathrm{d}t \end{array}}{\sum_{j=0}^{\infty} G^{(j)}(K) \int_0^T [F^{(j)}(t) - F^{(j+1)}(t)]\mathrm{d}t}. \tag{7.7}$$

We find optimum T^* to minimize $C(T)$. Differentiating $C(T)$ with respect to T and setting it equal to zero,

$$(c_K - c_T)\sum_{j=0}^{\infty} G^{(j)}(K) \int_0^T [F^{(j)}(t) - F^{(j+1)}(t)][Q_1(T) - Q_1(t)]\mathrm{d}t$$
$$+ c_M \sum_{j=0}^{\infty} G^{(j)}(K) \int_0^T [F^{(j)}(t) - F^{(j+1)}(t)][h(T) - h(t)]\mathrm{d}t = c_T, \tag{7.8}$$

where

$$Q_1(T) \equiv \frac{\sum_{j=0}^{\infty} f^{(j+1)}(T)[G^{(j)}(K) - G^{(j+1)}(K)]}{\sum_{j=0}^{\infty}[F^{(j)}(T) - F^{(j+1)}(T)]G^{(j)}(K)},$$

which is given in (2.10). Thus, if $Q_1(T)$ increases strictly with T from 0 to $Q_1(\infty)$, then the left-hand side $L_1(T)$ of (7.8) increases strictly from 0 to

$$L_1(\infty) =(c_K - c_T)\sum_{j=0}^{\infty} G^{(j)}(K) \int_0^{\infty} [F^{(j)}(t) - F^{(j+1)}(t)][Q_1(\infty) - Q_1(t)]\mathrm{d}t$$
$$+ c_M \sum_{j=0}^{\infty} G^{(j)}(K) \int_0^{\infty} [F^{(j)}(t) - F^{(j+1)}(t)][h(\infty) - h(t)]\mathrm{d}t.$$

Therefore, if $L_1(\infty) > c_T$, then there exists a finite and unique T^* $(0 < T^* < \infty)$ which satisfies (7.8), and the resulting cost rate is

$$C(T^*) = (c_K - c_T)Q_1(T^*) + c_M h(T^*). \tag{7.9}$$

Next, when the unit is replaced preventively only at independent damage N,

$$C(N) \equiv \lim_{T\to\infty} C_F(T,N)$$
$$= \frac{c_N + (c_K - c_N)\sum_{j=0}^{\infty}[G^{(j)}(K) - G^{(j+1)}(K)]\int_0^{\infty}\overline{P}_N(t)\mathrm{d}F^{(j+1)}(t) + c_M\sum_{j=0}^{\infty}G^{(j)}(K)\int_0^{\infty}[F^{(j)}(t) - F^{(j+1)}(t)]\overline{P}_N(t)h(t)\mathrm{d}t}{\sum_{j=0}^{\infty}G^{(j)}(K)\int_0^{\infty}[F^{(j)}(t) - F^{(j+1)}(t)]\overline{P}_N(t)\mathrm{d}t}. \tag{7.10}$$

We find optimum N^* to minimize $C(N)$. Forming the inequality $C(N+1) - C(N) \geq 0$,

$$(c_K - c_N)\left\{Q_2(N)\sum_{j=0}^{\infty}G^{(j)}(K)\int_0^{\infty}[F^{(j)}(t) - F^{(j+1)}(t)]\overline{P}_N(t)\mathrm{d}t - \sum_{j=0}^{\infty}[G^{(j)}(K) - G^{(j+1)}(K)]\int_0^{\infty}\overline{P}_N(t)\mathrm{d}F^{(j+1)}(t)\right\}$$
$$+ c_M\left\{Q_3(N)\sum_{j=0}^{\infty}G^{(j)}(K)\int_0^{\infty}[F^{(j)}(t) - F^{(j+1)}(t)]\overline{P}_N(t)\mathrm{d}t - \sum_{j=0}^{\infty}G^{(j)}(K)\int_0^{\infty}[F^{(j)}(t) - F^{(j+1)}(t)]\overline{P}_N(t)h(t)\mathrm{d}t\right\} \geq c_N, \tag{7.11}$$

where

$$Q_2(T,N) \equiv \frac{\sum_{j=0}^{\infty}[G^{(j)}(K) - G^{(j+1)}(K)]\int_0^{T}p_N(t)\mathrm{d}F^{(j+1)}(t)}{\sum_{j=0}^{\infty}G^{(j)}(K)\int_0^{T}[F^{(j)}(t) - F^{(j+1)}(t)]p_N(t)\mathrm{d}t},$$

$$Q_3(T,N) \equiv \frac{\sum_{j=0}^{\infty}G^{(j)}(K)\int_0^{T}[F^{(j)}(t) - F^{(j+1)}(t)]p_N(t)h(t)\mathrm{d}t}{\sum_{j=0}^{\infty}G^{(j)}(K)\int_0^{T}[F^{(j)}(t) - F^{(j+1)}(t)]p_N(t)\mathrm{d}t},$$

and $Q_2(N) \equiv \lim_{T\to\infty} Q_2(T,N)$ and $Q_3(N) \equiv \lim_{T\to\infty} Q_3(T,N)$. Thus, if both $Q_2(N)$ and $Q_3(N)$ increases strictly with N to $Q_2(\infty)$ and $Q_3(\infty)$, then the left-hand side $L_2(N)$ of (7.11) increases strictly with N to

$$L_2(\infty) = (c_K - c_N)\left[\mu Q_2(\infty)\sum_{j=0}^{\infty}G^{(j)}(K) - 1\right] + c_M\left\{\mu Q_3(\infty)\sum_{j=0}^{\infty}G^{(j)}(K) - \sum_{j=0}^{\infty}G^{(j)}(K)\int_0^{\infty}[F^{(j)}(t) - F^{(j+1)}(t)]h(t)\mathrm{d}t\right\}.$$

Therefore, if $L_2(\infty) > c_N$, then there exists a finite and unique minimum N^* $(1 \leq N^* < \infty)$ which satisfies (7.11), and the resulting cost rate is

$$(c_K - c_N)Q_2(N^* - 1) + c_M Q_3(N^* - 1) < C(N^*)$$
$$\leq (c_K - c_N)Q_2(N^*) + c_M Q_3(N^*). \tag{7.12}$$

When $c_T = c_N$ and $Q_1(T)$ increases strictly with T to $Q_1(\infty)$, we find optimum (T_F^*, N_F^*) to minimize $C_F(T, N)$ in (7.6). Differentiating $C_F(T, N)$ with respect to T and setting it equal to zero,

$$(c_K - c_T)\sum_{j=0}^{\infty} G^{(j)}(K)\int_0^T [F^{(j)}(t) - F^{(j+1)}(t)]\overline{P}_N(t)[Q_1(T) - Q_1(t)]\mathrm{d}t$$
$$+ c_M \sum_{j=0}^{\infty} G^{(j)}(K)\int_0^T [F^{(j)}(t) - F^{(j+1)}(t)]\overline{P}_N(t)[h(T) - h(t)]\mathrm{d}t = c_T. \tag{7.13}$$

Forming the inequality $C_F(T, N+1) - C_F(T, N) \geq 0$,

$$[(c_K - c_T)Q_2(T, N) + c_M Q_3(T, N)]\sum_{j=0}^{\infty} G^{(j)}(K)\int_0^T [F^{(j)}(t) - F^{(j+1)}(t)]\overline{P}_N(t)\mathrm{d}t$$
$$- (c_K - c_T)\sum_{j=0}^{\infty}[G^{(j)}(K) - G^{(j+1)}(K)]\int_0^T \overline{P}_N(t)\mathrm{d}F^{(j+1)}(t)$$
$$- c_M \sum_{j=0}^{\infty} G^{(j)}(K)\int_0^T [F^{(j)}(t) - F^{(j+1)}(t)]\overline{P}_N(t)h(t)\mathrm{d}t \geq c_T. \tag{7.14}$$

Substituting (7.13) for (7.14),

$$(c_K - c_T)Q_2(T, N) + c_M Q_3(T, N) \geq (c_K - c_T)Q_1(T) + c_M h(T). \tag{7.15}$$

The inequality (7.15) does not hold for any T because $Q_2(T, N) < Q_1(T)$ and $Q_3(T, N) < h(T)$ (Problem 7.2). Therefore, the optimum policy is $(T_F^* = T^*, N_F^* = \infty)$, where T^* is given in (7.8).

(2) Replacement Last

Suppose that the unit undergoes minimal repairs at independent damages, and is replaced preventively at time T $(0 \leq T \leq \infty)$ or at independent damage N $(N = 0, 1, 2, \ldots)$, whichever occurs last, which is named as *replacement last with independent damage*. Then, the probability that the unit is replaced at time T is

$$P_N(T)\sum_{j=0}^{\infty} G^{(j)}(K)[F^{(j)}(T) - F^{(j+1)}(T)], \tag{7.16}$$

the probability that it is replaced at independent damage N is

$$\sum_{j=0}^{\infty} G^{(j)}(K)\int_T^{\infty}[F^{(j)}(t)-F^{(j+1)}(t)]\mathrm{d}P_N(t), \tag{7.17}$$

and the probability that it is replaced at failure K is

$$\sum_{j=0}^{\infty} F^{(j+1)}(T)[G^{(j)}(K)-G^{(j+1)}(K)]$$
$$+\sum_{j=0}^{\infty}[G^{(j)}(K)-G^{(j+1)}(K)]\int_T^{\infty}\overline{P}_N(t)\mathrm{d}F^{(j+1)}(t), \tag{7.18}$$

where note that (7.16) + (7.17) + (7.18) = 1. Thus, the mean time to replacement is

$$TP_N(T)\sum_{j=0}^{\infty} G^{(j)}(K)[F^{(j)}(T)-F^{(j+1)}(T)]$$
$$+\sum_{j=0}^{\infty} G^{(j)}(K)\int_T^{\infty} t[F^{(j)}(t)-F^{(j+1)}(t)]\mathrm{d}P_N(t)$$
$$+\sum_{j=0}^{\infty}[G^{(j)}(K)-G^{(j+1)}(K)]\left[\int_0^T t\mathrm{d}F^{(j+1)}(t)+\int_T^{\infty} t\overline{P}_N(t)\mathrm{d}F^{(j+1)}(t)\right]$$
$$=\sum_{j=0}^{\infty} G^{(j)}(K)\left\{\int_0^T[F^{(j)}(t)-F^{(j+1)}(t)]\mathrm{d}t\right.$$
$$\left.+\int_T^{\infty}[F^{(j)}(t)-F^{(j+1)}(t)]\overline{P}_N(t)\mathrm{d}t\right\}. \tag{7.19}$$

The expected number of independent damages until replacement is

$$\sum_{i=N}^{\infty} ip_i(T)\sum_{j=0}^{\infty} G^{(j)}(K)[F^{(j)}(T)-F^{(j+1)}(T)]$$
$$+N\sum_{j=0}^{\infty} G^{(j)}(K)\int_T^{\infty}[F^{(j)}(t)-F^{(j+1)}(t)]\mathrm{d}P_N(t)$$
$$+\sum_{j=0}^{\infty}[G^{(j)}(K)-G^{(j+1)}(K)]\left[\int_0^T H(t)\mathrm{d}F^{(j+1)}(t)+\sum_{i=0}^{N-1} i\int_T^{\infty} p_i(t)\mathrm{d}F^{(j+1)}(t)\right]$$

$$
= \sum_{j=0}^{\infty} G^{(j)}(K) \left\{ \int_0^T [F^{(j)}(t) - F^{(j+1)}(t)] h(t) \mathrm{d}t \right.
$$

$$
\left. + \int_T^{\infty} [F^{(j)}(t) - F^{(j+1)}(t)] \overline{P}_N(t) h(t) \mathrm{d}t \right\}, \tag{7.20}
$$

which agrees with (7.19) when $h(t) \equiv 1$.

Therefore, the expected cost rate is

$$
C_L(T, N) = \frac{\begin{array}{l} c_T + (c_N - c_T) \sum_{j=0}^{\infty} G^{(j)}(K) \int_T^{\infty} [F^{(j)}(t) - F^{(j+1)}(t)] \mathrm{d}P_N(t) \\ +(c_K - c_T) \sum_{j=0}^{\infty} [G^{(j)}(K) - G^{(j+1)}(K)] \{F^{(j+1)}(T) \\ + \int_T^{\infty} \overline{P}_N(t) \mathrm{d}F^{(j+1)}(t)\} + c_M \sum_{j=0}^{\infty} G^{(j)}(K) \{\int_0^T [F^{(j)}(t) - F^{(j+1)}(t)] h(t) \mathrm{d}t \\ + \int_T^{\infty} [F^{(j)}(t) - F^{(j+1)}(t)] \overline{P}_N(t) h(t) \mathrm{d}t\} \end{array}}{\begin{array}{c} \sum_{j=0}^{\infty} G^{(j)}(K) \{\int_0^T [F^{(j)}(t) - F^{(j+1)}(t)] \mathrm{d}t \\ + \int_T^{\infty} [F^{(j)}(t) - F^{(j+1)}(t)] \overline{P}_N(t) \mathrm{d}t\} \end{array}}. \tag{7.21}
$$

Clearly, $\lim_{N \to 0} C_L(T, N) = \lim_{N \to \infty} C_F(T, N) = C(T)$ in (7.7), and $\lim_{T \to 0} C_L(T, N) = \lim_{T \to \infty} C_F(T, N) = C(N)$ in (7.10).

When $c_T = c_N$ and $Q_1(T)$ increases strictly with T to $Q_1(\infty)$, we find optimum (T_L^*, N_L^*) to minimize $C_L(T, N)$. Differentiating $C_L(T, N)$ with respect to T and setting it equal to zero,

$$
(c_K - c_T) \left(Q_1(T) \sum_{j=0}^{\infty} G^{(j)}(K) \left\{ \int_0^T [F^{(j)}(t) - F^{(j+1)}(t)] \mathrm{d}t \right. \right.
$$

$$
\left. + \int_T^{\infty} [F^{(j)}(t) - F^{(j+1)}(t)] \overline{P}_N(t) \mathrm{d}t \right\}
$$

$$
\left. - \sum_{j=0}^{\infty} [G^{(j)}(K) - G^{(j+1)}(K)] \left[F^{(j+1)}(T) + \int_T^{\infty} \overline{P}_N(t) \mathrm{d}F^{(j+1)}(t) \right] \right)
$$

$$
+ c_M \sum_{j=0}^{\infty} G^{(j)}(K) \left\{ \int_0^T [F^{(j)}(t) - F^{(j+1)}(t)][h(T) - h(t)] \mathrm{d}t \right.
$$

$$
\left. + \int_T^{\infty} [F^{(j)}(t) - F^{(j+1)}(t)] \overline{P}_N(t) [h(T) - h(t)] \mathrm{d}t \right\} = c_T, \tag{7.22}
$$

whose left-hand side $L_3(T)$ increases strictly with T to

$$L_3(\infty) \equiv (c_K - c_T)\left[Q_1(\infty)\mu\sum_{j=0}^{\infty}G^{(j)}(K) - 1\right]$$
$$+ c_M\sum_{j=0}^{\infty}G^{(j)}(K)\int_0^{\infty}[F^{(j)}(t) - F^{(j+1)}(t)][h(\infty) - h(t)]\mathrm{d}t.$$

Thus, if $L_3(\infty) > c_T$, then there exists a finite and unique T_L^* $(0 \leq T_L^* < \infty)$ which satisfies (7.22) for given N, and the resulting cost rate is

$$C_L(T_L^*, N) = (c_K - c_T)Q_1(T_L^*) + c_M h(T_L^*). \tag{7.23}$$

Forming the inequality $C_L(T, N-1) - C_L(T, N) > 0$ (Problem 2.3),

$$(c_K - c_T)\Bigg(Q_4(T, N-1)\sum_{j=0}^{\infty}G^{(j)}(K)\bigg\{\int_0^T[F^{(j)}(t) - F^{(j+1)}(t)]\mathrm{d}t$$
$$+ \int_T^{\infty}[F^{(j)}(t) - F^{(j+1)}(t)]\overline{P}_N(t)\mathrm{d}t\bigg\}$$
$$- \sum_{j=0}^{\infty}[G^{(j)}(K) - G^{(j+1)}(K)]\left[F^{(j+1)}(T) + \int_T^{\infty}\overline{P}_N(t)\mathrm{d}F^{(j+1)}(t)\right]\Bigg)$$
$$+ c_M\sum_{j=0}^{\infty}G^{(j)}(K)\bigg\{\int_0^T[F^{(j)}(t) - F^{(j+1)}(t)][Q_5(T, N-1) - h(t)]\mathrm{d}t$$
$$+ \int_T^{\infty}[F^{(j)}(t) - F^{(j+1)}(t)]\overline{P}_N(t)[Q_5(T, N-1) - h(t)]\mathrm{d}t\bigg\} < c_T, \tag{7.24}$$

where

$$Q_4(T, N) \equiv \frac{\sum_{j=0}^{\infty}[G^{(j)}(K) - G^{(j+1)}(K)]\int_T^{\infty}p_N(t)\mathrm{d}F^{(j+1)}(t)}{\sum_{j=0}^{\infty}G^{(j)}(K)\int_T^{\infty}[F^{(j)}(t) - F^{(j+1)}(t)]p_N(t)\mathrm{d}t},$$

$$Q_5(T, N) \equiv \frac{\sum_{j=0}^{\infty}G^{(j)}(K)\int_T^{\infty}[F^{(j)}(t) - F^{(j+1)}(t)]p_N(t)h(t)\mathrm{d}t}{\sum_{j=0}^{\infty}G^{(j)}(K)\int_T^{\infty}[F^{(j)}(t) - F^{(j+1)}(t)]p_N(t)\mathrm{d}t}.$$

Substituting (7.22) for (7.24),

$$(c_K - c_T)Q_4(T, N-1) + c_M Q_5(T, N-1) < (c_K - c_T)Q_1(T) + c_M h(T). \tag{7.25}$$

The inequality (7.25) does not hold for any T (Problem 7.2). Therefore, the optimum policy is $(T_L^* = T^*, N_L^* = 0)$, where T^* is given in (7.8).

(3) Poisson Shock Times

Suppose that shocks for additive damages occur at a Poisson process with rate λ, i.e., $F(t) = 1 - \mathrm{e}^{-\lambda t}$ and $F^{(j)}(t) = \sum_{i=j}^{\infty}[(\lambda t)^i/i!]\mathrm{e}^{-\lambda t}$ $(j = 0, 1, 2, \ldots)$, and independent damages occur at a nonhomogeneous Poisson process with an intensity function $h(t)$ that increases strictly with t from $h(0) = 0$ to $h(\infty)$. Then, the expected cost rate $C(T)$ in (7.7) is

$$\frac{C(T)}{\lambda} = \frac{c_T + (c_K - c_T)\sum_{j=0}^{\infty} F^{(j+1)}(T)[G^{(j)}(K) - G^{(j+1)}(K)] + c_M \sum_{j=0}^{\infty} G^{(j)}(K) \int_0^T [(\lambda t)^j/j!]\mathrm{e}^{-\lambda t}h(t)\mathrm{d}t}{\sum_{j=0}^{\infty} F^{(j+1)}(T)G^{(j)}(K)}. \tag{7.26}$$

From (7.8), optimum T^* satisfies

$$(c_K - c_T)\sum_{j=0}^{\infty} G^{(j)}(K)\int_0^T \frac{\lambda(\lambda t)^j}{j!}\mathrm{e}^{-\lambda t}[\widetilde{Q}_1(T) - \widetilde{Q}_1(t)]\mathrm{d}t + c_M\sum_{j=0}^{\infty} G^{(j)}(K)\int_0^T \frac{(\lambda t)^j}{j!}\mathrm{e}^{-\lambda t}[h(T) - h(t)]\mathrm{d}t = c_T, \tag{7.27}$$

where

$$\widetilde{Q}_1(T) = \frac{\sum_{j=0}^{\infty}[(\lambda T)^j/j!][G^{(j)}(K) - G^{(j+1)}(K)]}{\sum_{j=0}^{\infty}[(\lambda T)^j/j!]G^{(j)}(K)} \le 1.$$

Noting that when $[G^{(j)}(K) - G^{(j+1)}(K)]/G^{(j)}(K)$ increases with j to 1, $\widetilde{Q}_1(T)$ also increases with T to 1 (Problem 2.2), and the left-hand side $L_1(T)$ of (7.27) increases strictly with T from 0 to

$$L_1(\infty) = (c_K - c_T)M_G(K) + c_M\sum_{j=0}^{\infty} G^{(j)}(K)\int_0^{\infty} \frac{(\lambda t)^j}{j!}\mathrm{e}^{-\lambda t}[h(\infty) - h(t)]\mathrm{d}t,$$

where $M_G(K) \equiv \sum_{j=1}^{\infty} G^{(j)}(K)$. In this case, if $L_1(\infty) > c_T$, then a finite and unique T^* exists, and the expected cost rate is given in (7.9).

From (7.10) and (7.11),

$$C(N) = \frac{c_N + (c_K - c_N)\sum_{j=0}^{\infty}[G^{(j)}(K) - G^{(j+1)}(K)]\int_0^{\infty} \lambda[(\lambda t)^j/j!]\mathrm{e}^{-\lambda t}\overline{P}_N(t)\mathrm{d}t + c_M\sum_{j=0}^{\infty} G^{(j)}(K)\int_0^{\infty}[(\lambda t)^j/j!]\mathrm{e}^{-\lambda t}\overline{P}_N(t)h(t)\mathrm{d}t}{\sum_{j=0}^{\infty} G^{(j)}(K)\int_0^{\infty}[(\lambda t)^j/j!]\mathrm{e}^{-\lambda t}\overline{P}_N(t)\mathrm{d}t}, \tag{7.28}$$

and optimum N^* satisfies

$$(c_K - c_N)\left\{Q_2(N)\sum_{j=0}^{\infty} G^{(j)}(K)\int_0^{\infty}\frac{(\lambda t)^j}{j!}\mathrm{e}^{-\lambda t}\overline{P}_N(t)\mathrm{d}t \right.$$
$$\left. -\sum_{j=0}^{\infty}[G^{(j)}(K)-G^{(j+1)}(K)]\int_0^{\infty}\frac{\lambda(\lambda t)^j}{j!}\mathrm{e}^{-\lambda t}\overline{P}_N(t)\mathrm{d}t\right\}$$
$$+c_M\sum_{j=0}^{\infty} G^{(j)}(K)\int_0^{\infty}\frac{(\lambda t)^j}{j!}\mathrm{e}^{-\lambda t}\overline{P}_N(t)[Q_3(N)-h(t)]\mathrm{d}t \geq c_N, \quad (7.29)$$

where

$$Q_2(N) = \frac{\sum_{j=0}^{\infty}[G^{(j)}(K)-G^{(j+1)}(K)]\int_0^{\infty}[\lambda(\lambda t)^j/j!]\mathrm{e}^{-\lambda t}p_N(t)\mathrm{d}t}{\sum_{j=0}^{\infty}G^{(j)}(K)\int_0^{\infty}[(\lambda t)^j/j!]\mathrm{e}^{-\lambda t}p_N(t)\mathrm{d}t},$$

$$Q_3(N) = \frac{\sum_{j=0}^{\infty}G^{(j)}(K)\int_0^{\infty}[(\lambda t)^j/j!]\mathrm{e}^{-\lambda t}p_N(t)h(t)\mathrm{d}t}{\sum_{j=0}^{\infty}G^{(j)}(K)\int_0^{\infty}[(\lambda t)^j/j!]\mathrm{e}^{-\lambda t}p_N(t)\mathrm{d}t}.$$

Noting that when $r_{j+1}(K) = [G^{(j)}(K) - G^{(j+1)}(K)]/G^{(j)}(K)$ increases with j to 1, $Q_2(N)$ increases with N to λ, and when $h(t)$ increases with t to $h(\infty)$, $Q_3(N)$ increases with N to $h(\infty)$. In this case, the left-hand side $L_2(N)$ of (7.29) increases strictly with N to

$$L_2(\infty) = (c_K - c_N)M_G(K) + c_M\sum_{j=0}^{\infty}G^{(j)}(K)\int_0^{\infty}\frac{(\lambda t)^j}{j!}\mathrm{e}^{-\lambda t}[h(\infty)-h(t)]\mathrm{d}t,$$

which agrees with $L_1(\infty)$ when $c_T = c_N$. Thus, if $L_2(\infty) > c_N$, a finite and unique N^* exists.

It was shown that when $G(x) = 1 - \mathrm{e}^{-\omega x}$, $r_{j+1}(K)$ increases with j from $\mathrm{e}^{-\omega x}$ to 1 (Problem 2.4). So that, $\widetilde{Q}_1(T)$ increases strictly with T from $\mathrm{e}^{-\omega K}$ to 1, and $Q_2(N)$ increases strictly with N to λ.

When $F(t) = 1 - \mathrm{e}^{-\lambda t}$, $G(x) = 1 - \mathrm{e}^{-\omega x}$ and $H(t) = t^{\alpha}$ $(\alpha > 1)$, i.e., $h(t) = \alpha t^{\alpha-1}$, Tables 7.1 and 7.2 present optimum λT^* in (7.27) and N^* in (7.29), and their cost rates $C(T^*)/c_M$ and $C(N^*)/c_M$ for $\lambda c_T/c_M$, $\lambda c_N/c_M$ and ωK when $\alpha = 1.2$ and $\lambda c_K/c_M = 10.0$. This indicates that when $c_T = c_N$, $N^* \geq \lambda T^*$ and $C(T^*) < C(N^*)$, and the differences are becoming smaller as $\lambda c_T/c_M$ is larger. Thus, replacement with T is better than replacement with independent damage N, as shown in **(1)**.

Table 7.1 Optimum λT^* and its cost rate $C(T^*)/c_M$ when $\alpha = 1.2$ and $\lambda c_K/c_M = 10.0$

$\lambda c_T/c_M$	$\omega K = 5.0$		$\omega K = 10.0$		$\omega K = 20.0$	
	λT^*	$C(T^*)/c_M$	λT^*	$C(T^*)/c_M$	λT^*	$C(T^*)/c_M$
0.1	0.422	1.209	0.553	1.071	0.561	1.069
0.2	0.664	1.393	0.965	1.205	1.000	1.200
0.5	1.164	1.718	1.897	1.418	2.145	1.398
1.0	1.760	2.051	2.928	1.624	3.807	1.569
2.0	2.721	2.469	4.312	1.894	6.545	1.765
5.0	6.026	3.045	7.714	2.359	11.730	2.089

Table 7.2 Optimum N^* and its cost rate $C(N^*)/c_M$ when $\alpha = 1.2$ and $\lambda c_K/c_M = 10.0$

$\lambda c_N/c_M$	$\omega K = 5.0$		$\omega K = 10.0$		$\omega K = 20.0$	
	N^*	$C(N^*)/c_M$	N^*	$C(N^*)/c_M$	N^*	$C(N^*)/c_M$
0.1	1	1.506	1	1.184	1	1.165
0.2	1	1.611	1	1.291	1	1.272
0.5	2	1.921	2	1.483	3	1.431
1.0	2	2.200	4	1.681	5	1.591
2.0	4	2.571	6	1.945	10	1.780
5.0	10	3.068	13	2.386	19	2.102

7.2 Replacement Overtime First

We discuss two replacement overtime policies in which the unit is replaced preventively at the forthcoming independent damage or shock of additive damage over time T, or at number N of independent damage, using the approach of *whichever triggering event occurs first.*

7.2.1 Replacement Overtime for Independent Damage

Suppose that the unit is replaced at the forthcoming independent damage over time T $(0 < T \leq \infty)$, at independent damage N $(N = 1, 2, \ldots)$, or at failure K $(0 < K < \infty)$ of additive damage, whichever occurs first, which is named as *replacement overtime first with independent damage*. Then, the probability that the unit is replaced over time T is (Problem 7.3)

$$\overline{P}_N(T) \sum_{j=0}^{\infty} G^{(j)}(K) \int_T^{\infty} [F^{(j)}(t) - F^{(j+1)}(t)] h(t) \mathrm{e}^{-[H(t)-H(T)]} \mathrm{d}t, \tag{7.30}$$

the probability that it is replaced at independent damage N is

$$\sum_{j=0}^{\infty} G^{(j)}(K) \int_0^T [F^{(j)}(t) - F^{(j+1)}(t)] \mathrm{d}P_N(t), \tag{7.31}$$

and the probability that it is replaced at failure K is

$$\begin{aligned}
&\sum_{j=0}^{\infty}[G^{(j)}(K) - G^{(j+1)}(K)]\left[\int_0^T \overline{P}_N(t) \mathrm{d}F^{(j+1)}(t)\right. \\
&\left.+ \overline{P}_N(T) \int_T^{\infty} \mathrm{e}^{-[H(t)-H(T)]} \mathrm{d}F^{(j+1)}(t)\right],
\end{aligned} \tag{7.32}$$

where note that (7.30) + (7.31) + (7.32) = 1. Thus, the mean time to replacement is

$$\begin{aligned}
&\sum_{j=0}^{\infty} G^{(j)}(K)\left\{\overline{P}_N(T) \int_T^{\infty} t[F^{(j)}(t) - F^{(j+1)}(t)] h(t) \mathrm{e}^{-[H(t)-H(T)]} \mathrm{d}t\right. \\
&\left.+ \int_0^T t[F^{(j)}(t) - F^{(j+1)}(t)] \mathrm{d}P_N(t)\right\} \\
&+ \sum_{j=0}^{\infty}[G^{(j)}(K) - G^{(j+1)}(K)]\left[\int_0^T t\overline{P}_N(t) \mathrm{d}F^{(j+1)}(t)\right. \\
&\left.+ \overline{P}_N(T) \int_T^{\infty} t\mathrm{e}^{-[H(t)-H(T)]} \mathrm{d}F^{(j+1)}(t)\right] \\
&= \sum_{j=0}^{\infty} G^{(j)}(K)\left\{\int_0^T [F^{(j)}(t) - F^{(j+1)}(t)] \overline{P}_N(t) \mathrm{d}t\right. \\
&\left.+ \overline{P}_N(T) \int_T^{\infty} [F^{(j)}(t) - F^{(j+1)}(t)] \mathrm{e}^{-[H(t)-H(T)]} \mathrm{d}t\right\}.
\end{aligned} \tag{7.33}$$

The expected number of independent damages until replacement is

$$\begin{aligned}
&\sum_{j=0}^{\infty} G^{(j)}(K)\left\{\sum_{i=0}^{N-1} i p_i(T) \int_T^{\infty} [F^{(j)}(t) - F^{(j+1)}(t)] h(t) \mathrm{e}^{-[H(t)-H(T)]} \mathrm{d}t\right. \\
&\left.+ N \int_0^T [F^{(j)}(t) - F^{(j+1)}(t)] \mathrm{d}P_N(t)\right\} \\
&+ \sum_{j=0}^{\infty}[G^{(j)}(K) - G^{(j+1)}(K)] \sum_{i=0}^{N-1} i \left\{\int_0^T p_i(t) \mathrm{d}F^{(j+1)}(t)\right.
\end{aligned}$$

$$+p_i(T)\int_T^\infty e^{-[H(t)-H(T)]}dF^{(j+1)}(t)\Bigg\}$$

$$=\sum_{j=0}^\infty G^{(j)}(K)\Bigg\{\int_0^T [F^{(j)}(t)-F^{(j+1)}(t)]\overline{P}_N(t)h(t)dt$$

$$+\overline{P}_N(T)\int_T^\infty [F^{(j)}(t)-F^{(j+1)}(t)]e^{-[H(t)-H(T)]}h(t)dt\Bigg\}, \quad (7.34)$$

which agrees with (7.33) when $h(t)=1$.

Therefore, the expected cost rate is

$$C_{OF}(T,N)=$$

$$\frac{\begin{array}{l} c_O+(c_N-c_O)\sum_{j=0}^\infty G^{(j)}(K)\int_0^T[F^{(j)}(t)-F^{(j+1)}(t)]dP_N(t) \\ +(c_K-c_O)\sum_{j=0}^\infty[G^{(j)}(K)-G^{(j+1)}(K)]\{\int_0^T\overline{P}_N(t)dF^{(j+1)}(t) \\ +\overline{P}_N(T)\int_T^\infty e^{-[H(t)-H(T)]}dF^{(j+1)}(t)\} \\ +c_M\sum_{j=0}^\infty G^{(j)}(K)\{\int_0^T[F^{(j)}(t)-F^{(j+1)}(t)]\overline{P}_N(t)h(t)dt \\ +\overline{P}_N(T)\int_T^\infty[F^{(j)}(t)-F^{(j+1)}(t)]e^{-[H(t)-H(T)]}h(t)dt\} \end{array}}{\begin{array}{l} \sum_{j=0}^\infty G^{(j)}(K)\{\int_0^T[F^{(j)}(t)-F^{(j+1)}(t)]\overline{P}_N(t)dt \\ +\overline{P}_N(T)\int_T^\infty[F^{(j)}(t)-F^{(j+1)}(t)]e^{-[H(t)-H(T)]}dt\} \end{array}}, \quad (7.35)$$

where c_O = replacement cost over time T, and c_N, c_K and c_M are given in (7.6). Clearly, when $c_O=c_N$, $\lim_{T\to\infty}C_{OF}(T,N)=C(N)$ in (7.10).

In particular, when the unit is replaced preventively at the forthcoming independent damage over time T,

$$C_O(T)\equiv\lim_{N\to\infty}C_{OF}(T,N)$$

$$=\frac{\begin{array}{l} c_O+(c_K-c_O)\sum_{j=0}^\infty[G^{(j)}(K)-G^{(j+1)}(K)]\{F^{(j+1)}(T) \\ +\int_T^\infty e^{-[H(t)-H(T)]}dF^{(j+1)}(t)\} \\ +c_M\sum_{j=0}^\infty G^{(j)}(K)\{\int_0^T[F^{(j)}(t)-F^{(j+1)}(t)]h(t)dt \\ +\int_T^\infty[F^{(j)}(t)-F^{(j+1)}(t)]e^{-[H(t)-H(T)]}h(t)dt\} \end{array}}{\begin{array}{l} \sum_{j=0}^\infty G^{(j)}(K)\{\int_0^T[F^{(j)}(t)-F^{(j+1)}(t)]dt \\ +\int_T^\infty[F^{(j)}(t)-F^{(j+1)}(t)]e^{-[H(t)-H(T)]}dt\} \end{array}}. \quad (7.36)$$

We find optimum T_O^* to minimize $C_O(T)$ for $c_K>c_O$. Differentiating $C_O(T)$ with respect to T and setting it equal to zero,

$$[(c_K - c_O)Q_2(T) + c_M Q_3(T)] \sum_{j=0}^{\infty} G^{(j)}(K) \left\{ \int_0^T [F^{(j)}(t) - F^{(j+1)}(t)]\mathrm{d}t \right.$$
$$\left. + \int_T^{\infty} [F^{(j)}(t) - F^{(j+1)}(t)]\mathrm{e}^{-[H(t)-H(T)]}\mathrm{d}t \right\}$$
$$- (c_K - c_O) \sum_{j=0}^{\infty} [G^{(j)}(K) - G^{(j+1)}(K)] \left\{ \int_T^{\infty} \mathrm{e}^{-[H(t)-H(T)]}\mathrm{d}F^{(j+1)}(t) \right.$$
$$\left. + F^{(j+1)}(T) \right\} - c_M \sum_{j=0}^{\infty} G^{(j)}(K) \left\{ \int_0^T [F^{(j)}(t) - F^{(j+1)}(t)]h(t)\mathrm{d}t \right.$$
$$\left. + \int_T^{\infty} [F^{(j)}(t) - F^{(j+1)}(t)]\mathrm{e}^{-[H(t)-H(T)]}h(t)\mathrm{d}t \right\} = c_O, \tag{7.37}$$

where

$$Q_2(T) \equiv \frac{\sum_{j=0}^{\infty}[G^{(j)}(K) - G^{(j+1)}(K)] \int_T^{\infty} \mathrm{e}^{-H(t)}\mathrm{d}F^{(j+1)}(t)}{\sum_{j=0}^{\infty} G^{(j)}(K) \int_T^{\infty}[F^{(j)}(t) - F^{(j+1)}(t)]\mathrm{e}^{-H(t)}\mathrm{d}t},$$

$$Q_3(T) \equiv \frac{\sum_{j=0}^{\infty} G^{(j)}(K) \int_T^{\infty}[F^{(j)}(t) - F^{(j+1)}(t)]\mathrm{e}^{-H(t)}h(t)\mathrm{d}t}{\sum_{j=0}^{\infty} G^{(j)}(K) \int_T^{\infty}[F^{(j)}(t) - F^{(j+1)}(t)]\mathrm{e}^{-H(t)}\mathrm{d}t}.$$

When $F(t) = 1 - \mathrm{e}^{-\lambda t}$ and $r_{j+1}(K)$ increases with j to 1, $Q_2(T)$ increases strictly with T to λ, and when $F(t) = 1 - \mathrm{e}^{-\lambda t}$ and $h(t)$ increases strictly with t to $h(\infty)$, $Q_3(T)$ increases strictly with T to $h(\infty)$. In this case, (7.37) is

$$(c_K - c_O) \left\{ \frac{Q_2(T)}{\lambda} \sum_{j=0}^{\infty} F^{(j+1)}(T)G^{(j)}(K) \right.$$
$$\left. - \sum_{j=0}^{\infty} F^{(j+1)}(T)[G^{(j)}(K) - G^{(j+1)}(K)] \right\}$$
$$+ \frac{c_M}{\lambda} \sum_{j=0}^{\infty} G^{(j)}(K) \int_0^T [Q_3(T) - h(t)]\mathrm{d}F^{(j+1)}(t) = c_O, \tag{7.38}$$

whose left-hand side $L_4(T)$ increases strictly with T to

$$L_4(\infty) = (c_K - c_O)M_G(K) + c_M \sum_{j=0}^{\infty} G^{(j)}(K) \int_0^{\infty} \frac{(\lambda t)^j}{j!}\mathrm{e}^{-\lambda t}[h(\infty) - h(t)]\mathrm{d}t,$$

which agrees with $L_1(\infty)$ (Problem 7.4). Thus, if $L_4(\infty) > c_O$, then there exist a finite and unique T_O^* $(0 \le T_O^* < \infty)$ which satisfies (7.38), and the resulting cost rate is

$$C_O(T_O^*) = (c_K - c_O)Q_2(T_O^*) + c_M Q_3(T_O^*). \tag{7.39}$$

7.2.2 Replacement Overtime for Additive Damage

Suppose that the unit is replaced at the forthcoming shock over time T $(0 < T \le \infty)$, at independent damage N $(N = 1, 2, \ldots)$, or at failure K $(0 < K < \infty)$ of additive damage, whichever occurs first. Then, the probability that the unit is replaced over time T is

$$\sum_{j=0}^{\infty} G^{(j+1)}(K) \int_0^T \left[\int_{T-t}^{\infty} \overline{P}_N(t+u)\mathrm{d}F(u)\right] \mathrm{d}F^{(j)}(t), \tag{7.40}$$

the probability that it is replaced at independent damage N is

$$\sum_{j=0}^{\infty} G^{(j)}(K)\left\{\int_0^T [F^{(j)}(t) - F^{(j+1)}(t)]\mathrm{d}P_N(t) + \int_0^T \left[\int_T^{\infty} \overline{F}(u-t)\mathrm{d}P_N(u)\right] \mathrm{d}F^{(j)}(t)\right\}, \tag{7.41}$$

and the probability that it is replaced at failure K is

$$\sum_{j=0}^{\infty}[G^{(j)}(K) - G^{(j+1)}(K)]\left\{\int_0^T \overline{P}_N(t)\mathrm{d}F^{(j+1)}(t) + \int_0^T \left[\int_{T-t}^{\infty} \overline{P}_N(t+u)\mathrm{d}F(u)\right] \mathrm{d}F^{(j)}(t)\right\}, \tag{7.42}$$

where note that (7.40)+(7.41)+(7.42)=1. Thus, the mean time to replacement is (Problem 7.5)

$$\sum_{j=0}^{\infty} G^{(j)}(K)\left\{\int_0^T \left[\int_{T-t}^{\infty} (t+u)\overline{P}_N(t+u)\mathrm{d}F(u)\right] \mathrm{d}F^{(j)}(t) + \int_0^T t[F^{(j)}(t) - F^{(j+1)}(t)]\mathrm{d}P_N(t) + \int_0^T \left[\int_T^{\infty} u\overline{F}(u-t)\mathrm{d}P_N(u)\right] \mathrm{d}F^{(j)}(t)\right\}$$

$$+\sum_{j=0}^{\infty}[G^{(j)}(K)-G^{(j+1)}(K)]\int_0^T t\overline{P}_N(t)\mathrm{d}F^{(j+1)}(t)$$
$$=\sum_{j=0}^{\infty}G^{(j)}(K)\int_0^T\left[\int_0^{\infty}\overline{F}(u)\overline{P}_N(t+u)\mathrm{d}u\right]\mathrm{d}F^{(j)}(t). \tag{7.43}$$

The expected number of independent damages until replacement is

$$\sum_{j=0}^{\infty}G^{(j)}(K)\left\{\sum_{i=0}^{N-1}i\int_0^T\left[\int_{T-t}^{\infty}p_i(t+u)\mathrm{d}F(u)\right]\mathrm{d}F^{(j)}(t)\right.$$
$$\left.+N\int_0^T[F^{(j)}(t)-F^{(j+1)}(t)]\mathrm{d}P_N(t)+N\int_0^T\left[\int_T^{\infty}\overline{F}(u-t)\mathrm{d}P_N(u)\right]\mathrm{d}F^{(j)}(t)\right\}$$
$$+\sum_{j=0}^{\infty}[G^{(j)}(K)-G^{(j+1)}(K)]\sum_{i=0}^{N-1}i\int_0^T p_i(t)\mathrm{d}F^{(j+1)}(t)$$
$$=\sum_{j=0}^{\infty}G^{(j)}(K)\int_0^T\left[\int_0^{\infty}\overline{F}(u)\overline{P}_N(t+u)h(t+u)\mathrm{d}u\right]\mathrm{d}F^{(j)}(t), \tag{7.44}$$

which agrees with (7.43) when $h(t)\equiv 1$.

Therefore, the expected cost rate is

$$C_{OF}(T,N)=$$
$$\frac{\begin{array}{l}c_O+(c_N-c_O)\sum_{j=0}^{\infty}G^{(j)}(K)\{\int_0^T[F^{(j)}(t)-F^{(j+1)}(t)]\mathrm{d}P_N(t)\\ +\int_0^T[\int_T^{\infty}\overline{F}(u-t)\mathrm{d}P_N(u)]\mathrm{d}F^{(j)}(t)\}\\ +(c_K-c_O)\sum_{j=0}^{\infty}[G^{(j)}(K)-G^{(j+1)}(K)]\{\int_0^T\overline{P}_N(t)\mathrm{d}F^{(j+1)}(t)\\ +\int_0^T[\int_{T-t}^{\infty}\overline{P}_N(t+u)\mathrm{d}F(u)]\mathrm{d}F^{(j)}(t)\}\\ +c_M\sum_{j=0}^{\infty}G^{(j)}(K)\int_0^T[\int_0^{\infty}\overline{F}(u)\overline{P}_N(t+u)h(t+u)\mathrm{d}u]\mathrm{d}F^{(j)}(t)\end{array}}{\sum_{j=0}^{\infty}G^{(j)}(K)\int_0^T\left[\int_0^{\infty}\overline{F}(u)\overline{P}_N(t+u)\mathrm{d}u\right]\mathrm{d}F^{(j)}(t)}. \tag{7.45}$$

In particular, when the unit is replaced preventively at the forthcoming shock over time T,

$$C_O(T)\equiv\lim_{N\to\infty}C_{OF}(T;N)$$
$$=\frac{\begin{array}{l}c_O+(c_K-c_O)\sum_{j=0}^{\infty}F^{(j)}(T)[G^{(j)}(K)-G^{(j+1)}(K)]\\ +c_M\sum_{j=0}^{\infty}G^{(j)}(K)\int_0^T[\int_0^{\infty}\overline{F}(u)h(u+t)\mathrm{d}u]\mathrm{d}F^{(j)}(t)\end{array}}{\mu\sum_{j=0}^{\infty}G^{(j)}(K)F^{(j)}(T)}. \tag{7.46}$$

We find optimum T_O^* to minimize $C_O(T)$. Differentiating $C_O(T)$ with respect to T and setting it equal to zero,

$$(c_K - c_O)\left\{Q_4(T)\sum_{j=0}^{\infty} G^{(j)}(K)F^{(j)}(T) - \sum_{j=0}^{\infty}[G^{(j)}(K) - G^{(j+1)}(K)]F^{(j)}(T)\right\}$$
$$+ c_M \sum_{j=0}^{\infty} G^{(j)}(K)\int_0^T \left\{\int_0^{\infty} \overline{F}(u)[h(u+T) - h(u+t)]\mathrm{d}u\right\}\mathrm{d}F^{(j)}(t) = c_O, \tag{7.47}$$

where

$$Q_4(T) \equiv \frac{\sum_{j=0}^{\infty} f^{(j)}(T)[G^{(j)}(K) - G^{(j+1)}(K)]}{\sum_{j=0}^{\infty} f^{(j)}(T)G^{(j)}(K)}.$$

When $r_{j+1}(K)$ increases strictly with j to 1, and because

$$\int_0^{\infty}\left\{\int_0^{\infty} \overline{F}(u)[h(\infty) - h(u+t)]\mathrm{d}u\right\}\mathrm{d}F^{(j)}(t)$$
$$= \int_0^{\infty} \overline{F}(t)h(\infty)\mathrm{d}t - \int_0^{\infty}[F^{(j)}(t) - F^{(j+1)}(t)]h(t)\mathrm{d}t$$
$$= \int_0^{\infty}[F^{(j)}(t) - F^{(j+1)}(t)][h(\infty) - h(t)]\mathrm{d}t,$$

$Q_4(T)$ increases strictly to 1 and the left-hand side $L_5(T)$ of (7.47) increases strictly with T to

$$L_5(\infty) = (c_K - c_O)M_G(K)$$
$$+ c_M \sum_{j=0}^{\infty} G^{(j)}(K)\int_0^{\infty}[F^{(j)}(t) - F^{(j+1)}(t)][h(\infty) - h(t)]\mathrm{d}t,$$

which agrees with $L_1(\infty)$ in (7.27) when $c_O = c_T$ and $F(t) = 1 - \mathrm{e}^{-\lambda t}$.

Thus, if $L_5(\infty) > c_O$, then there exists a finite and unique T_O^* $(0 \le T_O^* < \infty)$ which satisfies (7.47), and the resulting cost rate is (Problem 7.6)

$$\mu C_O(T_O^*) = (c_K - c_O)Q_4(T_O^*) + c_M \int_0^{\infty} \overline{F}(t)h(t + T_O^*)\mathrm{d}t. \tag{7.48}$$

7.3 Replacement Overtime Last

In this section, the approach of *whichever triggering event occurs last* is used for the replacement polices given in Sect. 7.2.

7.3.1 Replacement Overtime for Independent Damage

Suppose that the unit is replaced preventively at the forthcoming independent damage over time T $(0 < T \leq \infty)$ or at independent damage N $(N = 1, 2, \ldots)$, whichever occurs last, which is named as *replacement overtime last with independent damage*. Then, the probability that the unit is replaced over time T is

$$\sum_{j=0}^{\infty} G^{(j)}(K) P_N(T) \int_T^{\infty} [F^{(j)}(t) - F^{(j+1)}(t)] h(t) \mathrm{e}^{-[H(t)-H(T)]} \mathrm{d}t, \tag{7.49}$$

the probability that it is replaced at independent damage N is

$$\sum_{j=0}^{\infty} G^{(j)}(K) \int_T^{\infty} [F^{(j)}(t) - F^{(j+1)}(t)] \mathrm{d}P_N(t), \tag{7.50}$$

and the probability that it is replaced at failure K is

$$\begin{aligned}
&\sum_{j=0}^{\infty} [G^{(j)}(K) - G^{(j+1)}(K)] \Bigg[F^{(j+1)}(T) \\
&+ P_N(T) \int_T^{\infty} \mathrm{e}^{-[H(t)-H(T)]} \mathrm{d}F^{(j+1)}(t) + \int_T^{\infty} \overline{P}_N(t) \mathrm{d}F^{(j+1)}(t) \Bigg],
\end{aligned} \tag{7.51}$$

where note that (7.49) + (7.50) + (7.51) = 1. Thus, the mean time to replacement is

$$\begin{aligned}
&P_N(T) \Bigg\{ \sum_{j=0}^{\infty} G^{(j)}(K) \int_T^{\infty} t [F^{(j)}(t) - F^{(j+1)}(t)] h(t) \mathrm{e}^{-[H(t)-H(T)]} \mathrm{d}t \\
&+ \sum_{j=0}^{\infty} [G^{(j)}(K) - G^{(j+1)}(K)] \int_T^{\infty} t \mathrm{e}^{-[H(t)-H(T)]} \mathrm{d}F^{(j+1)}(t) \Bigg\} \\
&+ \sum_{j=0}^{\infty} G^{(j)}(K) \int_T^{\infty} t [F^{(j)}(t) - F^{(j+1)}(t)] \mathrm{d}P_N(t)
\end{aligned}$$

$$+\sum_{j=0}^{\infty}[G^{(j)}(K)-G^{(j+1)}(K)]\left[\int_0^T t\mathrm{d}F^{(j+1)}(t)+\int_T^{\infty}t\overline{P}_N(t)\mathrm{d}F^{(j+1)}(t)\right]$$
$$=\sum_{j=0}^{\infty}G^{(j)}(K)\left\{P_N(T)\int_T^{\infty}[F^{(j)}(t)-F^{(j+1)}(t)]\mathrm{e}^{-[H(t)-H(T)]}\mathrm{d}t\right.$$
$$\left.+\int_0^T[F^{(j)}(t)-F^{(j+1)}(t)]\mathrm{d}t+\int_T^{\infty}[F^{(j)}(t)-F^{(j+1)}(t)]\overline{P}_N(t)\mathrm{d}t\right\}. \tag{7.52}$$

The expected number of independent damages until replacement is

$$\sum_{i=N}^{\infty}ip_i(T)\left\{\sum_{j=0}^{\infty}G^{(j)}(K)\int_T^{\infty}[F^{(j)}(t)-F^{(j+1)}(t)]h(t)\mathrm{e}^{-[H(t)-H(T)]}\mathrm{d}t\right.$$
$$\left.+\sum_{j=0}^{\infty}[G^{(j)}(K)-G^{(j+1)}(K)]\int_T^{\infty}H(t)\mathrm{e}^{-[H(t)-H(T)]}\mathrm{d}F^{(j+1)}(t)\right\}$$
$$+N\sum_{j=0}^{\infty}G^{(j)}(K)\int_T^{\infty}[F^{(j)}(t)-F^{(j+1)}(t)]\mathrm{d}P_N(t)$$
$$+\sum_{j=0}^{\infty}[G^{(j)}(K)-G^{(j+1)}(K)]\left[\int_0^T H(t)\mathrm{d}F^{(j+1)}(t)+\sum_{i=0}^{N-1}i\int_T^{\infty}p_i(t)\mathrm{d}F^{(j+1)}(t)\right]$$
$$=\sum_{j=0}^{\infty}G^{(j)}(K)\left\{P_N(T)\int_T^{\infty}[F^{(j)}(t)-F^{(j+1)}(t)]\mathrm{e}^{-[H(t)-H(T)]}h(t)\mathrm{d}t\right.$$
$$\left.+\int_0^T[F^{(j)}(t)-F^{(j+1)}(t)]h(t)\mathrm{d}t+\int_T^{\infty}[F^{(j)}(t)-F^{(j+1)}(t)]\overline{P}_N(t)h(t)\mathrm{d}t\right\}, \tag{7.53}$$

which agrees with (7.52) when $h(t)\equiv 1$.

Therefore, the expected cost rate is

$$C_{OL}(T,N)=$$
$$\frac{\begin{array}{l}c_O+(c_N-c_O)\sum_{j=0}^{\infty}G^{(j)}(K)\int_T^{\infty}[F^{(j)}(t)-F^{(j+1)}(t)]\mathrm{d}P_N(t)\\ +(c_K-c_O)\sum_{j=0}^{\infty}[G^{(j)}(K)-G^{(j+1)}(K)]\left\{F^{(j+1)}(T)\right.\\ \left.+\int_T^{\infty}\overline{P}_N(t)\mathrm{d}F^{(j+1)}(t)+\overline{P}_N(T)\int_T^{\infty}\mathrm{e}^{-[H(t)-H(T)]}\mathrm{d}F^{(j+1)}(t)\right\}\\ +c_M\sum_{j=0}^{\infty}G^{(j)}(K)\{P_N(T)\int_T^{\infty}[F^{(j)}(t)-F^{(j+1)}(t)]\mathrm{e}^{-[H(t)-H(T)]}h(t)\mathrm{d}t\\ +\int_0^T[F^{(j)}(t)-F^{(j+1)}(t)]h(t)\mathrm{d}t+\int_T^{\infty}[F^{(j)}(t)-F^{(j+1)}(t)]\overline{P}_N(t)h(t)\mathrm{d}t\}\end{array}}{\begin{array}{l}\sum_{j=0}^{\infty}G^{(j)}(K)\{P_N(T)\int_T^{\infty}[F^{(j)}(t)-F^{(j+1)}(t)]\mathrm{e}^{-[H(t)-H(T)]}\mathrm{d}t\\ +\int_0^T[F^{(j)}(t)-F^{(j+1)}(t)]\mathrm{d}t+\int_T^{\infty}[F^{(j)}(t)-F^{(j+1)}(t)]\overline{P}_N(t)\mathrm{d}t\}\end{array}}. \tag{7.54}$$

Clearly, $\lim_{T\to 0} C_{OL}(T,N) = \lim_{T\to\infty} C_{OF}(T,N) = C(N)$ in (7.10), and $\lim_{N\to 0} C_{OL}(T,N) = \lim_{N\to\infty} C_{OF}(T,N) = C_O(T)$ in (7.46).

7.3.2 Replacement Overtime for Additive Damage

Suppose that the unit is replaced preventively at the forthcoming shock over time T $(0 \le T \le \infty)$ or at independent damage N $(N = 1, 2, \ldots)$, whichever occurs last. Then, the probability that the unit is replaced over time T is

$$\sum_{j=0}^{\infty} G^{(j+1)}(K) \int_0^T \left[\int_{T-t}^{\infty} P_N(t+u)\mathrm{d}F(u)\right] \mathrm{d}F^{(j)}(t), \tag{7.55}$$

the probability that it is replaced at independent damage N is

$$\sum_{j=0}^{\infty} G^{(j)}(K) \left\{\int_T^{\infty} [F^{(j)}(t) - F^{(j+1)}(t)]\mathrm{d}P_N(t) - \int_0^T \left[\int_T^{\infty} \overline{F}(u-t)\mathrm{d}P_N(u)\right] \mathrm{d}F^{(j)}(t)\right\}, \tag{7.56}$$

and the probability that it is replaced at failure K is

$$\sum_{j=0}^{\infty} [G^{(j)}(K) - G^{(j+1)}(K)] \left\{F^{(j+1)}(T) + \int_0^T \left[\int_{T-t}^{\infty} P_N(t+u)\mathrm{d}F(u)\right] \mathrm{d}F^{(j)}(t) + \int_T^{\infty} \overline{P}_N(t)\mathrm{d}F^{(j+1)}(t)\right\}, \tag{7.57}$$

where note that (7.55) + (7.56) + (7.57) = 1. Thus, the mean time to replacement is

$$\sum_{j=0}^{\infty} G^{(j)}(K) \left\{\int_0^T \left[\int_{T-t}^{\infty} (t+u) P_N(t+u)\mathrm{d}F(u)\right] \mathrm{d}F^{(j)}(t) - \int_0^T \left[\int_{T-t}^{\infty} u\overline{F}(u-t)\mathrm{d}P_N(u)\right] \mathrm{d}F^{(j)}(t) + \int_T^{\infty} t[F^{(j)}(t) - F^{(j+1)}(t)]\mathrm{d}P_N(t)\right\}$$
$$+ \sum_{j=0}^{\infty} [G^{(j)}(K) - G^{(j+1)}(K)] \left[\int_0^T t\mathrm{d}F^{(j+1)}(t) + \int_T^{\infty} t\overline{P}_N(t)\mathrm{d}F^{(j+1)}(t)\right]$$

$$= \sum_{j=0}^{\infty} G^{(j)}(K) \left\{ \int_0^T \left[\int_0^{\infty} \overline{F}(u) P_N(t+u) \mathrm{d}u \right] \mathrm{d}F^{(j)}(t) \right.$$
$$\left. + \int_0^{\infty} [F^{(j)}(t) - F^{(j+1)}(t)] \overline{P}_N(t) \mathrm{d}t \right\}. \tag{7.58}$$

The expected number of independent damages until replacement is

$$\sum_{j=0}^{\infty} G^{(j)}(K) \sum_{i=N}^{\infty} i \int_0^T \left[\int_{T-t}^{\infty} p_i(t+u) \mathrm{d}F(u) \right] \mathrm{d}F^{(j)}(t)$$
$$+ N \sum_{j=0}^{\infty} G^{(j)}(K) \left\{ \int_T^{\infty} [F^{(j)}(t) - F^{(j+1)}(t)] P_N(t) \mathrm{d}t \right.$$
$$\left. - \int_0^T \left[\int_T^{\infty} \overline{F}(u-t) \mathrm{d}P_N(u) \right] \mathrm{d}F^{(j)}(t) \right\} + \sum_{j=0}^{\infty} [G^{(j)}(K) - G^{(j+1)}(K)]$$
$$\times \left\{ \int_0^T H(t) \mathrm{d}F^{(j+1)}(t) + \sum_{i=1}^{N-1} i \int_T^{\infty} p_i(t) \mathrm{d}F^{(j+1)}(t) \right\}$$
$$= \sum_{j=0}^{\infty} G^{(j)}(K) \left\{ \int_0^T \left[\int_0^{\infty} \overline{F}(u) P_N(t+u) h(t+u) \mathrm{d}u \right] \mathrm{d}F^{(j)}(t) \right.$$
$$\left. + \int_0^{\infty} [F^{(j)}(t) - F^{(j+1)}(t)] \overline{P}_N(t) h(t) \mathrm{d}t \right\}, \tag{7.59}$$

which agrees with (7.59) when $h(t) \equiv 1$.

Therefore, the expected cost rate is

$$C_{OL}(T, N) = \frac{\begin{array}{l} c_O + (c_N - c_O) \sum_{j=0}^{\infty} G^{(j)}(K) \{ \int_T^{\infty} [F^{(j)}(t) - F^{(j+1)}(t)] \mathrm{d}P_N(t) \\ - \int_0^T [\int_T^{\infty} \overline{F}(u-t) \mathrm{d}P_N(u)] \mathrm{d}F^{(j)}(t) \} \\ + (c_K - c_O) \sum_{j=0}^{\infty} [G^{(j)}(K) - G^{(j+1)}(K)] \{ F^{(j+1)}(T) \\ + \int_T^{\infty} \overline{P}_N(t) \mathrm{d}F^{(j+1)}(t) + \int_0^T [\int_{T-t}^{\infty} P_N(t+u) \mathrm{d}F(u)] \mathrm{d}F^{(j)}(t) \} \\ + c_M \sum_{j=0}^{\infty} G^{(j)}(K) \{ \int_0^T [\int_0^{\infty} \overline{F}(u) P_N(t+u) h(t+u) \mathrm{d}u] \mathrm{d}F^{(j)}(t) \\ + \int_0^{\infty} [F^{(j)}(t) - F^{(j+1)}(t)] \overline{P}_N(t) h(t) \mathrm{d}t \} \end{array}}{\begin{array}{l} \sum_{j=0}^{\infty} G^{(j)}(K) \{ \int_0^T \left[\int_0^{\infty} \overline{F}(u) P_N(t+u) \mathrm{d}u \right] \mathrm{d}F^{(j)}(t) \\ + \int_0^{\infty} [F^{(j)}(t) - F^{(j+1)}(t)] \overline{P}_N(t) \mathrm{d}t \} \end{array}}. \tag{7.60}$$

Clearly, $\lim_{T \to 0} C_{OL}(T, N) = \lim_{T \to \infty} C_{OF}(T, N) = C(N)$ in (7.10).

7.4 Additive and Independent Damages

When both of the number of independent damages and shocks of additive damage are considered, i.e., the unit is replaced preventively at planned time T, at number M of shocks of additive damage, or at number N of independent damages, we obtain the expected cost rates for replacement first, replacement last, and replacement middle.

(1) Replacement First

Suppose that the unit undergoes minimal repairs at independent damages, and is replaced at time T $(0 < T \leq \infty)$, at shock M $(M = 1, 2, \ldots)$, at independent damage N $(N = 1, 2, \ldots)$, or at failure K $(0 < K < \infty)$ of additive damage, whichever occurs first. Then, the probability that the unit is replaced at time T is

$$\overline{P}_N(T)\sum_{j=0}^{M-1} G^{(j)}(K)[F^{(j)}(T) - F^{(j+1)}(T)], \tag{7.61}$$

the probability that it is replaced at shock M is

$$G^{(M)}(K)\int_0^T \overline{P}_N(t)\mathrm{d}F^{(M)}(t), \tag{7.62}$$

the probability that it is replaced at independent damage N is

$$\sum_{j=0}^{M-1} G^{(j)}(K)\int_0^T [F^{(j)}(t) - F^{(j+1)}(t)]\mathrm{d}P_N(t), \tag{7.63}$$

and the probability that it is replaced at failure K is

$$\sum_{j=0}^{M-1}[G^{(j)}(K) - G^{(j+1)}(K)]\int_0^T \overline{P}_N(t)\mathrm{d}F^{(j+1)}(t), \tag{7.64}$$

where note that (7.61)+(7.62)+(7.63)+(7.64)=1. Thus, the mean time to replacement is

$$T\overline{P}_N(T)\sum_{j=0}^{M-1} G^{(j)}(K)[F^{(j)}(T) - F^{(j+1)}(T)]$$
$$+\sum_{j=0}^{M-1} G^{(j)}(K)\int_0^T t[F^{(j)}(t) - F^{(j+1)}(t)]\mathrm{d}P_N(t)$$
$$+G^{(M)}(K)\int_0^T t\overline{P}_N(t)\mathrm{d}F^{(M)}(t)$$

$$
+\sum_{j=0}^{M-1}[G^{(j)}(K)-G^{(j+1)}(K)]\int_0^T t\overline{P}_N(t)\mathrm{d}F^{(j+1)}(t)
$$
$$
=\sum_{j=0}^{M-1}G^{(j)}(K)\int_0^T [F^{(j)}(t)-F^{(j+1)}(t)]\overline{P}_N(t)\mathrm{d}t. \tag{7.65}
$$

The expected number of independent damages until replacement is

$$
\sum_{i=0}^{N-1} ip_i(T)\sum_{j=0}^{M-1}G^{(j)}(K)[F^{(j)}(T)-F^{(j+1)}(T)]
$$
$$
+N\sum_{j=0}^{M-1}G^{(j)}(K)\int_0^T [F^{(j)}(t)-F^{(j+1)}(t)]\mathrm{d}P_N(t)
$$
$$
+G^{(M)}(K)\sum_{i=0}^{N-1}i\int_0^T p_i(t)\mathrm{d}F^{(M)}(t)
$$
$$
+\sum_{j=0}^{M-1}[G^{(j)}(K)-G^{(j+1)}(K)]\sum_{i=0}^{N-1}i\int_0^T p_i(t)\mathrm{d}F^{(j+1)}(t)
$$
$$
=\sum_{j=0}^{M-1}G^{(j)}(K)\int_0^T [F^{(j)}(t)-F^{(j+1)}(t)]\overline{P}_N(t)h(t)\mathrm{d}t. \tag{7.66}
$$

Therefore, the expected cost rate is

$$
C_F(T,M,N)=
$$
$$
\frac{\begin{array}{l} c_T+(c_N-c_T)\sum_{j=0}^{M-1}G^{(j)}(K)\int_0^T[F^{(j)}(t)-F^{(j+1)}(t)]\mathrm{d}P_N(t) \\ +(c_R-c_T)G^{(M)}(K)\int_0^T\overline{P}_N(t)\mathrm{d}F^{(M)}(t) \\ +(c_K-c_T)\sum_{j=0}^{M-1}[G^{(j)}(K)-G^{(j+1)}(K)]\int_0^T\overline{P}_N(t)\mathrm{d}F^{(j+1)}(t) \\ +c_M\sum_{j=0}^{M-1}G^{(j)}(K)\int_0^T[F^{(j)}(t)-F^{(j+1)}(t)]\overline{P}_N(t)h(t)\mathrm{d}t \end{array}}{\sum_{j=0}^{M-1}G^{(j)}(K)\int_0^T[F^{(j)}(t)-F^{(j+1)}(t)]\overline{P}_N(t)\mathrm{d}t}, \tag{7.67}
$$

where $c_R =$ replacement cost at shock M, $c_N =$ replacement cost at independent damage N, and c_T, c_K and c_M are given in (7.6). Clearly, $\lim_{M\to\infty} C_F(T,M,N) = C_F(T,N)$ in (7.6).

(2) Replacement Last

Suppose that the unit undergoes minimal repairs at independent damages and is replaced preventively at time T $(0 \le T \le \infty)$, at shock M $(M = 0, 1, 2, \ldots)$, or at independent damage N $(N = 0, 1, 2, \ldots)$, whichever occurs last. Then, the probability that the unit is replaced at time T is

$$P_N(T)\sum_{j=M}^{\infty}G^{(j)}(K)[F^{(j)}(T)-F^{(j+1)}(T)], \tag{7.68}$$

the probability that it is replaced at shock M is

$$G^{(M)}(K)\int_T^{\infty}P_N(t)\mathrm{d}F^{(M)}(t), \tag{7.69}$$

the probability that it is replaced at independent damage N is

$$\sum_{j=M}^{\infty}G^{(j)}(K)\int_T^{\infty}[F^{(j)}(t)-F^{(j+1)}(t)]\mathrm{d}P_N(t), \tag{7.70}$$

and the probability that it is replaced at failure K is

$$\sum_{j=0}^{\infty}F^{(j+1)}(T)[G^{(j)}(K)-G^{(j+1)}(K)]$$
$$+\sum_{j=M}^{\infty}[G^{(j)}(K)-G^{(j+1)}(K)]\int_T^{\infty}\overline{P}_N(t)\mathrm{d}F^{(j+1)}(t)$$
$$+\sum_{j=0}^{M-1}[G^{(j)}(K)-G^{(j+1)}(K)]\left[\int_T^{\infty}P_N(t)\mathrm{d}F^{(j+1)}(t)+\int_T^{\infty}\overline{P}_N(t)\mathrm{d}F^{(j+1)}(t)\right], \tag{7.71}$$

where note that (7.68) + (7.69) + (7.70) + (7.71) = 1. Thus, the mean time to replacement is

$$TP_N(T)\sum_{j=M}^{\infty}G^{(j)}(K)[F^{(j)}(T)-F^{(j+1)}(T)]$$
$$+\sum_{j=M}^{\infty}G^{(j)}(K)\int_T^{\infty}t[F^{(j)}(t)-F^{(j+1)}(t)]\mathrm{d}P_N(t)$$
$$+G^{(M)}(K)\int_T^{\infty}tP_N(t)\mathrm{d}F^{(M)}(t)+\sum_{j=0}^{\infty}[G^{(j)}(K)-G^{(j+1)}(K)]\int_0^{T}t\mathrm{d}F^{(j+1)}(t)$$
$$+\sum_{j=M}^{\infty}[G^{(j)}(K)-G^{(j+1)}(K)]\int_T^{\infty}t\overline{P}_N(t)\mathrm{d}F^{(j+1)}(t)$$
$$+\sum_{j=0}^{M-1}[G^{(j)}(K)-G^{(j+1)}(K)]\int_T^{\infty}t\mathrm{d}F^{(j+1)}(t)$$

$$= \sum_{j=0}^{\infty} G^{(j)}(K) \int_0^T [F^{(j)}(t) - F^{(j+1)}(t)]\mathrm{d}t$$
$$+ \sum_{j=M}^{\infty} G^{(j)}(K) \int_T^{\infty} [F^{(j)}(t) - F^{(j+1)}(t)]\overline{P}_N(t)\mathrm{d}t$$
$$+ \sum_{j=0}^{M-1} G^{(j)}(K) \int_T^{\infty} [F^{(j)}(t) - F^{(j+1)}(t)]\mathrm{d}t. \tag{7.72}$$

The expected number of independent damages until replacement is

$$\sum_{i=N}^{\infty} i p_i(T) \sum_{j=M}^{\infty} G^{(j)}(K)[F^{(j)}(T) - F^{(j+1)}(T)]$$
$$+ N \sum_{j=M}^{\infty} G^{(j)}(K) \int_T^{\infty} [F^{(j)}(t) - F^{(j+1)}(t)]\mathrm{d}P_N(t)$$
$$+ G^{(M)}(K) \sum_{i=N}^{\infty} i \int_T^{\infty} p_i(t)\mathrm{d}F^{(M)}(t)$$
$$+ \sum_{j=0}^{\infty} [G^{(j)}(K) - G^{(j+1)}(K)] \int_0^T H(t)\mathrm{d}F^{(j+1)}(t)$$
$$+ \sum_{j=M}^{\infty} [G^{(j)}(K) - G^{(j+1)}(K)] \sum_{i=0}^{N-1} i \int_T^{\infty} p_i(t)\mathrm{d}F^{(j+1)}(t)$$
$$+ \sum_{j=0}^{M-1} [G^{(j)}(K) - G^{(j+1)}(K)] \int_T^{\infty} H(t)\mathrm{d}F^{(j+1)}(t)$$
$$= \sum_{j=0}^{\infty} G^{(j)}(K) \int_0^T [F^{(j)}(t) - F^{(j+1)}(t)]h(t)\mathrm{d}t$$
$$+ \sum_{j=M}^{\infty} G^{(j)}(K) \int_T^{\infty} [F^{(j)}(t) - F^{(j+1)}(t)]\overline{P}_N(t)h(t)\mathrm{d}t$$
$$+ \sum_{j=0}^{M-1} G^{(j)}(K) \int_T^{\infty} [F^{(j)}(t) - F^{(j+1)}(t)]h(t)\mathrm{d}t, \tag{7.73}$$

which agrees with (7.72) when $h(t) \equiv 1$.

Therefore, the expected cost rate is

$$C_L(T,M,N) = \frac{\begin{array}{l} c_T + (c_N - c_T)\sum_{j=M}^{\infty} G^{(j)}(K)\int_T^{\infty}[F^{(j)}(t) - F^{(j+1)}(t)]\mathrm{d}P_N(t) \\ +(c_R - c_T)G^{(M)}(K)\int_T^{\infty} P_N(t)\mathrm{d}F^{(M)}(t) \\ +(c_K - c_T)\{\sum_{j=0}^{\infty}[G^{(j)}(K) - G^{(j+1)}(K)]F^{(j+1)}(T) \\ +\sum_{j=M}^{\infty}[G^{(j)}(K) - G^{(j+1)}(K)]\int_T^{\infty}\overline{P}_N(t)\mathrm{d}F^{(j+1)}(t) \\ +\sum_{j=0}^{M-1}[G^{(j)}(K) - G^{(j+1)}(K)][1 - F^{(j+1)}(T)]\} \\ +c_M\{\sum_{j=0}^{\infty} G^{(j)}(K)\int_0^T[F^{(j)}(t) - F^{(j+1)}(t)]h(t)\mathrm{d}t \\ +\sum_{j=M}^{\infty} G^{(j)}(K)\int_T^{\infty}[F^{(j)}(t) - F^{(j+1)}(t)]\overline{P}_N(t)h(t)\mathrm{d}t \\ +\sum_{j=0}^{M-1} G^{(j)}(K)\int_T^{\infty}[F^{(j)}(t) - F^{(j+1)}(t)]h(t)\mathrm{d}t\} \end{array}}{\begin{array}{l} \sum_{j=0}^{\infty} G^{(j)}(K)\int_0^T[F^{(j)}(t) - F^{(j+1)}(t)]\mathrm{d}t \\ +\sum_{j=M}^{\infty} G^{(j)}(K)\int_T^{\infty}[F^{(j)}(t) - F^{(j+1)}(t)]\overline{P}_N(t)\mathrm{d}t \\ +\sum_{j=0}^{M-1} G^{(j)}(K)\int_T^{\infty}[F^{(j)}(t) - F^{(j+1)}(t)]\mathrm{d}t \end{array}}. \tag{7.74}$$

Clearly, $\lim_{M\to 0} C_L(T,M,N) = C_L(T,N)$ in (7.21).

(3) Replacement Middle

Suppose that the unit undergoes minimal repairs at independent damages, and is replaced preventively at time T $(0 < T \leq \infty)$, at shock M $(M = 1, 2, \ldots)$, or at independent damage N $(N = 1, 2, \ldots)$, whichever occurs middle. Then, the probability that the unit is replaced at time T is

$$P_N(T)\sum_{j=0}^{M-1} G^{(j)}(K)[F^{(j)}(T) - F^{(j+1)}(T)] + \overline{P}_N(T)\sum_{j=M}^{\infty} G^{(j)}(K)[F^{(j)}(T) - F^{(j+1)}(T)], \tag{7.75}$$

the probability that it is replaced at shock M is

$$G^{(M)}(K)\left[\int_T^{\infty}\overline{P}_N(t)\mathrm{d}F^{(M)}(t) + \int_0^T P_N(t)\mathrm{d}F^{(M)}(t)\right], \tag{7.76}$$

the probability that it is replaced at independent damage N is

$$\sum_{j=0}^{M-1} G^{(j)}(K) \int_T^{\infty} [F^{(j)}(t) - F^{(j+1)}(t)] \mathrm{d}P_N(t)$$
$$+ \sum_{j=M}^{\infty} G^{(j)}(K) \int_0^T [F^{(j)}(t) - F^{(j+1)}(t)] \mathrm{d}P_N(t), \tag{7.77}$$

and the probability that it is replaced at failure K is

$$\sum_{j=0}^{M-1} [G^{(j)}(K) - G^{(j+1)}(K)] \int_0^T \overline{P}_N(t) \mathrm{d}F^{(j+1)}(t)$$
$$+ \sum_{j=M}^{\infty} [G^{(j)}(K) - G^{(j+1)}(K)] \int_0^T \overline{P}_N(t) \mathrm{d}F^{(j+1)}(t)$$
$$+ \sum_{j=0}^{M-1} [G^{(j)}(K) - G^{(j+1)}(K)] \left[\int_0^T P_N(t) \mathrm{d}F^{(j+1)}(t) + \int_T^{\infty} \overline{P}_N(t) \mathrm{d}F^{(j+1)}(t) \right], \tag{7.78}$$

where note that (7.75) + (7.76) + (7.77) + (7.78) = 1. Thus, the mean time to replacement is

$$T \Bigg\{ P_N(T) \sum_{j=0}^{M-1} G^{(j)}(K)[F^{(j)}(T) - F^{(j+1)}(T)]$$
$$+ \overline{P}_N(T) \sum_{j=M}^{\infty} G^{(j)}(K)[F^{(j)}(T) - F^{(j+1)}(T)] \Bigg\}$$
$$+ \sum_{j=0}^{M-1} G^{(j)}(K) \int_T^{\infty} t[F^{(j)}(t) - F^{(j+1)}(t)] \mathrm{d}P_N(t)$$
$$+ \sum_{j=M}^{\infty} G^{(j)}(K) \int_0^T t[F^{(j)}(t) - F^{(j+1)}(t)] \mathrm{d}P_N(t)$$
$$+ G^{(M)}(K) \left[\int_T^{\infty} t\overline{P}_N(t) \mathrm{d}F^{(M)}(t) + \int_0^T t P_N(t) \mathrm{d}F^{(M)}(t) \right]$$
$$+ \sum_{j=0}^{M-1} [G^{(j)}(K) - G^{(j+1)}(K)] \left[\int_0^T t \mathrm{d}F^{(j+1)}(t) + \int_T^{\infty} t\overline{P}_N(t) \mathrm{d}F^{(j+1)}(t) \right]$$
$$+ \sum_{j=M}^{\infty} [G^{(j)}(K) - G^{(j+1)}(K)] \int_0^T t\overline{P}_N(t) \mathrm{d}F^{(j+1)}(t)$$

$$
= \sum_{j=0}^{M-1} G^{(j)}(K) \left\{ \int_0^T [F^{(j)}(t) - F^{(j+1)}(t)] \mathrm{d}t \right.
$$

$$
\left. + \int_T^\infty [F^{(j)}(t) - F^{(j+1)}(t)] \overline{P}_N(t) \mathrm{d}t \right\}
$$

$$
+ \sum_{j=M}^{\infty} G^{(j)}(K) \int_0^T [F^{(j)}(t) - F^{(j+1)}(t)] \overline{P}_N(t) \mathrm{d}t. \tag{7.79}
$$

The expected number of independent damages until replacement is

$$
\sum_{i=N}^{\infty} i p_i(T) \sum_{j=0}^{M-1} G^{(j)}(K)[F^{(j)}(T) - F^{(j+1)}(T)]
$$

$$
+ \sum_{i=0}^{N-1} i p_i(T) \sum_{j=M}^{\infty} G^{(j)}(K)[F^{(j)}(T) - F^{(j+1)}(T)]
$$

$$
+ N \left\{ \sum_{j=0}^{M-1} G^{(j)}(K) \int_T^\infty [F^{(j)}(t) - F^{(j+1)}(t)] \mathrm{d}P_N(t) \right.
$$

$$
\left. + \sum_{j=M}^{\infty} G^{(j)}(K) \int_0^T [F^{(j)}(t) - F^{(j+1)}(t)] \mathrm{d}P_N(t) \right\}
$$

$$
+ G^{(M)}(K) \left[\sum_{i=0}^{N-1} i \int_T^\infty p_i(t) \mathrm{d}F^{(M)}(t) + \sum_{i=N}^{\infty} i \int_0^T p_i(t) \mathrm{d}F^{(M)}(t) \right]
$$

$$
+ \sum_{j=0}^{M-1} [G^{(j)}(K) - G^{(j+1)}(K)] \left[\int_0^T H(t) \mathrm{d}F^{(j+1)}(t) + \sum_{i=0}^{N-1} i \int_T^\infty p_i(t) \mathrm{d}F^{(j+1)}(t) \right]
$$

$$
+ \sum_{j=M}^{\infty} [G^{(j)}(K) - G^{(j+1)}(K)] \sum_{i=0}^{N-1} i \int_0^T p_i(t) \mathrm{d}F^{(j+1)}(t)
$$

$$
= \sum_{j=0}^{M-1} G^{(j)}(K) \left\{ \int_0^T [F^{(j)}(t) - F^{(j+1)}(t)] h(t) \mathrm{d}t \right.
$$

$$
\left. + \int_T^\infty [F^{(j)}(t) - F^{(j+1)}(t)] \overline{P}_N(t) h(t) \mathrm{d}t \right\}
$$

$$
+ \sum_{j=M}^{\infty} G^{(j)}(K) \int_0^T [F^{(j)}(t) - F^{(j+1)}(t)] \overline{P}_N(t) h(t) \mathrm{d}t, \tag{7.80}
$$

which agrees with (7.79) when $h(t) \equiv 1$.

Therefore, the expected cost rate is (Problem 7.7)

$$C_M(T,M,N)=$$
$$\frac{\begin{array}{l} c_T+(c_N-c_T)\{\sum_{j=0}^{M-1}G^{(j)}(K)\int_T^\infty[F^{(j)}(t)-F^{(j+1)}(t)]\mathrm{d}P_N(t) \\ +\sum_{j=M}^\infty G^{(j)}(K)\int_0^T[F^{(j)}(t)-F^{(j+1)}(t)]\mathrm{d}P_N(t)\} \\ +(c_R-c_T)G^{(M)}(K)[\int_T^\infty \overline{P}_N(t)\mathrm{d}F^{(M)}(t)+\int_0^T P_N(t)\mathrm{d}F^{(M)}(t)] \\ +(c_K-c_T)\{\sum_{j=0}^{M-1}[G^{(j)}(K)-G^{(j+1)}(K)][F^{(j+1)}(T)+\int_T^\infty \overline{P}_N(t)\mathrm{d}F^{(j+1)}(t)] \\ +\sum_{j=M}^\infty[G^{(j)}(K)-G^{(j+1)}(K)]\int_0^T \overline{P}_N(t)\mathrm{d}F^{(j+1)}(t)\} \\ +c_M(\sum_{j=0}^{M-1}G^{(j)}(K)\{\int_0^T[F^{(j)}(t)-F^{(j+1)}(t)]h(t)\mathrm{d}t \\ +\int_T^\infty[F^{(j)}(t)-F^{(j+1)}(t)]\overline{P}_N(t)h(t)\mathrm{d}t\} \\ +\sum_{j=M}^\infty G^{(j)}(K)\int_0^T[F^{(j)}(t)-F^{(j+1)}(t)]\overline{P}_N(t)h(t)\mathrm{d}t) \end{array}}{\begin{array}{l} \sum_{j=0}^{M-1}G^{(j)}(K)\{\int_0^T[F^{(j)}(t)-F^{(j+1)}(t)]\mathrm{d}t \\ +\int_T^\infty[F^{(j)}(t)-F^{(j+1)}(t)]\overline{P}_N(t)\mathrm{d}t\} \\ +\sum_{j=M}^\infty G^{(j)}(K)\int_0^T[F^{(j)}(t)-F^{(j+1)}(t)]\overline{P}_N(t)\mathrm{d}t \end{array}}. \tag{7.81}$$

Clearly, comparing $C_F(T,M,N)$ in (7.67), $C_L(T,M,N)$ in (7.74) and $C_M(T,M,N)$ in (7.81),

$$\begin{aligned}
\lim_{T\to0}C_M(T,M,N)&=\lim_{T\to\infty}C_F(T,M,N),\\
\lim_{N\to0}C_M(T,M,N)&=\lim_{N\to\infty}C_F(T,M,N),\\
\lim_{M\to0}C_M(T,M,N)&=\lim_{M\to\infty}C_F(T,M,N),\\
\lim_{T\to\infty}C_M(T,M,N)&=\lim_{T\to0}C_L(T,M,N),\\
\lim_{N\to\infty}C_M(T,M,N)&=\lim_{N\to0}C_L(T,M,N),\\
\lim_{M\to\infty}C_M(T,M,N)&=\lim_{M\to0}C_L(T,M,N).
\end{aligned}$$

7.5 Problem 7

7.1 Derive (7.5).

7.2 Prove that $Q_2(T,N)<Q_1(T)$ and $Q_3(T,N)<h(T)$, and $Q_4(T,N)>Q_1(T)$ and $Q_5(T,N)>h(T)$ when both $Q_1(T)$ and $h(T)$ increase strictly with T.

7.3 Derive (7.30).

7.4 Prove that when $F(t)=1-\mathrm{e}^{-\lambda t}$ and $r_{j+1}(x)$ increases strictly with j to 1, $Q_2(T)$ increases strictly with T to λ and $Q_3(T)$ increases strictly with T to $h(\infty)$, and $Q_2(T)>Q_1(T)$ and $Q_3(T)>h(T)$ for $0\le T<\infty$. Furthermore, prove that the left-hand side of (7.38) increases strictly with T.

7.5 Derive (7.43) and (7.44).

7.6 Show that $T_O^* < T^*$.

7.7 Ascertain analytically that when the unit is replaced before failure K at $\max\{t_N, t_M\} < T$, at $\min\{t_N, t_M\} > T$, or at T in case of $\{t_N < T < t_M\}$ or $\{t_M < T < t_N\}$, the expected cost rate is given in (7.81). Furthermore, obtain the expected cost rates for replacement overtime policies with time T, shock M and independent damage N.

Chapter 8
Database Maintenance Models

Database, a set of data and the way they are organized [51], has now become the lifeblood for some organizations in this modern society. For instance, the transactional systems of commercial banks grind to halt within few minutes if something goes wrong with their databases. For the super critical databases in commercial airports and nuclear plants, it is expected to back up the data several times a day, or even to use database replication techniques [52] for real-time backups to achieve high data security. Normally, the database management system (DBMS) can be set to implement a hierarchy of daily, weekly and monthly defragmentations and backups [53, 54] to keep the database running efficiently and to achieve high data security, and their frequencies depend on the factors, such as rate of data update, database availability, criticality of data, and etc.

However, these periodic maintenance modes have not taken account of the busy states of the database systems, as we know that all defragmentation and backup processes will introduce some locks on the database and consume resources, e.g., the lockout time and backup window have certain amount of interferences with normal operations of database. In this chapter, we take up a database system that must be up and running 24 hours a day, 7 days a week, and try to consider the busy states into models, using the theory of random maintenances [15], that is, defragmentations and backups are scheduled in random ways, putting their lockout times and backup windows in non-busy states with user's convenience. The techniques of shock and damage models [2] are applied into modelings by replacing the random *shocks* with *database updates* in large volumes, and the amount of *damage* with the volumes of *fragmentation* and *updated data.*

We suppose in a database system that the data files updates in large volumes at a nonhomogeneous Poisson process with an intensity function $h(t)$ and a mean value function $H(t)$, where $H(t) \equiv \int_0^t h(u)\mathrm{d}u$ and $h(t)$ increases with t to $h(\infty)$ which might be infinity. Then, the probability that j updates, i.e., large volumes of updates, have arrived in $[0, t]$ is $p_j(t) \equiv [H(t)^j/j!]\mathrm{e}^{-H(t)}$ $(j = 0, 1, 2, \cdots)$. We denote

X. Zhao and T. Nakagawa, *Advanced Maintenance Policies for Shock and Damage Models*, Springer Series in Reliability Engineering,
https://doi.org/10.1007/978-3-319-70456-2_8

$P_j(t) \equiv \sum_{i=j}^{\infty} p_i(t)$ and $\overline{P}_j(t) \equiv 1 - P_j(t)$, then $P_{j+1}(t) = \int_0^t p_j(u)h(u)\mathrm{d}u$, and $P_0(t) = 1$, $\overline{P}_0(t) = 0$, $\lim_{j\to\infty} P_j(t) = 0$, $\lim_{j\to\infty} \overline{P}_j(t) = 1$, $\lim_{t\to\infty} P_j(t) = 1$, and $\lim_{t\to\infty} \overline{P}_j(t) = 0$. In Sect. 8.1, models of defragmentation first, last and overtime are discussed. In Sect. 8.2, optimum full backup times for incremental and differential backups are obtained. All discussions are analytically formulated and numerical examples are conducted with exponential distributions.

8.1 Defragmentation Models

Fragmentation takes place when the database management system cannot or will not allocate enough contiguous space for a complete file as a unit, but instead puts parts of it in small regions between existing files [54, 55]. The process of system fragmentation is sometimes called system aging and can be improved by periodically defragmenting and compacting the database.

In this section, we suppose that a volume W_j $(j = 1, 2, \cdots)$ of fragmentation due to the jth update is a random variable with an identical distribution $G(x) \equiv \Pr\{W_j \le x\}$ and finite mean $1/\omega \equiv \int_0^\infty \overline{G}(x)\mathrm{d}x$. The volumes of fragmentation are accumulated with updates and the total fragmentation $\sum_{i=1}^{j} W_i$ up to the jth update has distribution $G^{(j)}(x) = \Pr\{\sum_{i=1}^{j} W_i \le x\}$ $(j = 1, 2, \cdots)$, where $G^{(j)}(x)$ is the j-fold Stieltjes convolution of $G(x)$ with itself and $G^{(0)}(x) \equiv 1$ for $x \ge 0$. We suppose that when the total fragmentation has exceeded a level K $(0 < K < \infty)$ of upper limit, the database becomes useless for a long response time due to the lack of storage and should be defragmented immediately to get the database back to the normal running state.

8.1.1 Defragmentation First

In order to prevent the system from being useless states, the database is defragmented preventively at planned time T $(0 < T \le \infty)$, or at number N $(N = 1, 2, \cdots)$ of updates, whichever occurs first. Then, the probability that the database is defragmented at time T is

$$\sum_{j=0}^{N-1} p_j(T)G^{(j)}(K),$$

the probability that it is defragmented at update N is

$$P_N(T)G^{(N)}(K),$$

and the probability that it is defragmented at volume K is

$$\sum_{j=0}^{N-1} P_{j+1}(T)[G^{(j)}(K) - G^{(j+1)}(K)].$$

We introduce the following costs for defragmentations: c_T, c_N and c_K are constant costs defragmented at T, N and K, respectively, where $c_K > c_T$ and $c_K > c_N$. In addition, $c_0(x)$ is the variable cost of each defragmentation when a total volume of fragmentation is x $(0 \le x \le K)$. Then, the expected cost until defragmentation is [2]

$$\begin{aligned}
&\sum_{j=0}^{N-1} p_j(T) \int_0^K [c_T + c_0(x)]\mathrm{d}G^{(j)}(x) + P_N(T) \int_0^K [c_N + c_0(x)]\mathrm{d}G^{(N)}(x) \\
&+ [c_K + c_0(K)] \sum_{j=0}^{N-1} P_{j+1}(T)[G^{(j)}(K) - G^{(j+1)}(K)],
\end{aligned} \tag{8.1}$$

and the mean time to defragmentation is

$$\begin{aligned}
&T \sum_{j=0}^{N-1} p_j(T) G^{(j)}(K) + G^{(N)}(K) \int_0^T t\mathrm{d}P_N(t) \\
&+ \sum_{j=0}^{N-1} [G^{(j)}(K) - G^{(j+1)}(K)] \int_0^T t\mathrm{d}P_{j+1}(t) \\
&= \sum_{j=0}^{N-1} G^{(j)}(K) \int_0^T p_j(t)\mathrm{d}t.
\end{aligned} \tag{8.2}$$

Therefore, the expected cost rate is

$$C_F(T, N) = \frac{\begin{array}{l} \sum_{j=0}^{N-1} p_j(T) \int_0^K [c_T + c_0(x)]\mathrm{d}G^{(j)}(x) \\ +P_N(T) \int_0^K [c_N + c_0(x)]\mathrm{d}G^{(N)}(x) \\ +[c_K + c_0(K)] \sum_{j=0}^{N-1} P_{j+1}(T)[G^{(j)}(K) - G^{(j+1)}(K)] \end{array}}{\sum_{j=0}^{N-1} G^{(j)}(K) \int_0^T p_j(t)\mathrm{d}t}. \tag{8.3}$$

Clearly, when the database is defragmented only at volume K,

$$C(\infty) \equiv \lim_{\substack{T\to\infty \\ N\to\infty}} C_F(T, N) = \frac{c_K + c_0(K)}{\sum_{j=0}^{\infty} G^{(j)}(K) \int_0^{\infty} p_j(t)\mathrm{d}t}. \tag{8.4}$$

(1) Optimum T^*

Suppose that the database is defragmented preventively only at time T $(0 < T \leq \infty)$. Then, putting that $N \to \infty$ in (8.3),

$$C(T) \equiv \lim_{N\to\infty} C_F(T, N)$$
$$= \frac{\sum_{j=0}^{\infty} p_j(T) \int_0^K [c_T + c_0(x)]\mathrm{d}G^{(j)}(x) + [c_K + c_0(K)] \sum_{j=0}^{\infty} P_{j+1}(T)[G^{(j)}(K) - G^{(j+1)}(K)]}{\sum_{j=0}^{\infty} G^{(j)}(K) \int_0^T p_j(t)\mathrm{d}t}. \tag{8.5}$$

We find optimum T^* to minimize $C(T)$ when $c_0(x) \equiv c_0 x$. Differentiating $C(T)$ with respect to T and setting it equal to zero,

$$(c_K - c_T)\left\{h(T)Q_1(T)\sum_{j=0}^{\infty} G^{(j)}(K)\int_0^T p_j(t)\mathrm{d}t - \sum_{j=0}^{\infty} p_j(T)[1 - G^{(j)}(K)]\right\}$$
$$+ c_0\left\{h(T)Q_2(T)\sum_{j=0}^{\infty} G^{(j)}(K)\int_0^T p_j(t)\mathrm{d}t - \sum_{j=0}^{\infty} p_j(T)\int_0^K [1 - G^{(j)}(x)]\mathrm{d}x\right\}$$
$$= c_T, \tag{8.6}$$

where

$$Q_1(T) \equiv \frac{\sum_{j=0}^{\infty} p_j(T)[G^{(j)}(K) - G^{(j+1)}(K)]}{\sum_{j=0}^{\infty} p_j(T)G^{(j)}(K)},$$
$$Q_2(T) \equiv \frac{\sum_{j=0}^{\infty} p_j(T) \int_0^K [G^{(j)}(x) - G^{(j+1)}(x)]\mathrm{d}x}{\sum_{j=0}^{\infty} p_j(T)G^{(j)}(K)}.$$

In particular, when $c_0 = 0$, (8.6) becomes

$$h(T)Q_1(T)\sum_{j=0}^{\infty} G^{(j)}(K)\int_0^T p_j(t)\mathrm{d}t - \sum_{j=0}^{\infty} p_j(T)[1 - G^{(j)}(K)] = \frac{c_T}{c_K - c_T}. \tag{8.7}$$

If $Q_1(T)$ increases strictly with T to $Q_1(\infty)$ and

$$h(\infty)Q_1(\infty)\sum_{j=0}^{\infty} G^{(j)}(K)\int_0^{\infty} p_j(t)\mathrm{d}t > \frac{c_K}{c_K - c_T},$$

then there exists a finite and unique T^* $(0 < T^* < \infty)$ which satisfies (8.7).

(2) Optimum N^*

Suppose that the database is defragmented preventively only at updated N ($N = 1, 2, \cdots$). Then, putting that $T \to \infty$ in (8.3),

$$C(N) \equiv \lim_{T\to\infty} C_F(T, N) = \frac{[c_K + c_0(K)][1 - G^{(N)}(K)] + \int_0^K [c_N + c_0(x)]\mathrm{d}G^{(N)}(x)}{\sum_{j=0}^{N-1} G^{(j)}(K) \int_0^\infty p_j(t)\mathrm{d}t} \quad (N = 1, 2, \cdots). \tag{8.8}$$

We find optimum N^* to minimize $C(N)$ when $c_0(x) \equiv c_0 x$. Forming the inequality $C(N+1) - C(N) \geq 0$,

$$(c_K - c_N)\left\{Q_1(N)\sum_{j=0}^{N-1} G^{(j)}(K)\int_0^\infty p_j(t)\mathrm{d}t - [1 - G^{(N)}(K)]\right\} + c_0\left\{Q_2(N)\sum_{j=0}^{N-1} G^{(j)}(K)\int_0^\infty p_j(t)\mathrm{d}t - \int_0^K [1 - G^{(N)}(x)]\mathrm{d}x\right\} \geq c_N \quad (N = 1, 2, \cdots), \tag{8.9}$$

where

$$Q_1(N) \equiv \frac{r_{N+1}(K)}{\int_0^\infty p_N(t)\mathrm{d}t},$$

$$Q_2(N) \equiv \frac{\int_0^K [G^{(N)}(x) - G^{(N+1)}(x)]\mathrm{d}x}{G^{(N)}(K)\int_0^\infty p_N(t)\mathrm{d}t},$$

and $r_{N+1}(x)$ is given in (2.14).

In particular, when $c_0 = 0$, (8.9) becomes

$$Q_1(N)\sum_{j=0}^{N-1} G^{(j)}(K)\int_0^\infty p_j(t)\mathrm{d}t - [1 - G^{(N)}(K)] \geq \frac{c_N}{c_K - c_N}. \tag{8.10}$$

Note that when $h(t)$ increases with t, $\int_0^\infty p_N(t)\mathrm{d}t$ decreases with N to $1/h(\infty)$ from Theorem 4.2 of [1]. Thus, if $r_{N+1}(K)$ increases strictly with N to 1, and

$$h(\infty)\sum_{j=0}^{\infty} G^{(j)}(K)\int_0^\infty p_j(t)\mathrm{d}t \geq \frac{c_K}{c_K - c_N},$$

then there exists a finite and unique minimum N^* ($1 \leq N^* < \infty$) which satisfies (8.10).

We next compute optimum T^* and N^* when $c_K > c_T + c_0/\omega$, $c_0(x) = c_0 x$, $h(t) = \lambda$ and $G(x) = 1 - \mathrm{e}^{-\omega x}$, i.e., $P_j(t) = \sum_{i=j}^{\infty}[(\lambda t)^i/i!]\mathrm{e}^{-\lambda t}$ and $G^{(j)}(x) = \sum_{i=j}^{\infty}[(\omega x)^i/i!]\mathrm{e}^{-\omega x}$ $(j = 0, 1, 2, \cdots)$. Then, (8.6) is

$$Q_1(T)\sum_{j=0}^{\infty} G^{(j)}(K)P_{j+1}(T) - \sum_{j=0}^{\infty} p_j(T)[1 - G^{(j)}(K)] = \frac{c_T}{c_K - c_T - c_0/\omega}, \tag{8.11}$$

whose left-hand side increase strictly with T to ωK (Problem 2.2). Thus, if $\omega K > c_T/(c_K - c_T - c_0/\omega)$, then optimum T^* $(0 < T^* > \infty)$ to satisfy (8.11) exists.

Furthermore, (8.9) is

$$r_{N+1}(K)\sum_{j=0}^{N-1} G^{(j)}(K) - [1 - G^{(N)}(K)] \geq \frac{c_N}{c_K - c_N - c_0/\omega}, \tag{8.12}$$

whose left-hand side agrees with (2.14) and increases strictly with N to ωK. Thus, if $\omega K > c_T/(c_K - c_T - c_0/\omega)$, then optimum N^* $(1 \leq N^* < \infty)$ to satisfy (8.12) exists.

Tables 8.1 and 8.2 present optimum λT^* and N^*, and their cost rates $C(T^*)/(\lambda c_T)$ and $C(N^*)/(\lambda c_N)$ for ωK, c_K/c_T and c_K/c_N when $c_0 K/c_T = 1.0$ and $c_0 K/c_N = 1.0$. These indicate that λT^* are almost equal to N^*, however, $C(T^*) > C(N^*)$, which will be shown theoretically in **(3)**.

(3) Optimum T_F^* and N_F^*

When $c_T = c_N$, $c_0(x) = c_0 x$, $h(t) = \lambda$ and $G(x) = 1 - \mathrm{e}^{-\omega x}$, the expected cost rate in (8.3) is

$$\frac{C_F(T, N)}{\lambda} = \frac{c_T - (c_K - c_T - c_0/\omega)\sum_{j=1}^{N} P_j(T)G^{(j)}(K)}{\sum_{j=0}^{N-1} P_{j+1}(T)G^{(j)}(K)} + (c_K - c_T). \tag{8.13}$$

We find optimum T_F^* and N_F^* to minimize $C_F(T, N)$ when $c_K > c_T + c_0/\omega$ and $r_N(K)$ increases strictly with N to 1. Forming the inequality $C_F(T, N-1) - C_F(T, N) > 0$,

$$\begin{aligned} & r_N(K)\sum_{j=0}^{N-1} P_{j+1}(T)G^{(j)}(K) - \sum_{j=0}^{N-1} P_{j+1}(T)[G^{(j)}(K) - G^{(j+1)}(K)] \\ & < \frac{c_T}{c_K - c_T - c_0/\omega}, \end{aligned} \tag{8.14}$$

Table 8.1 Optimum λT^* and its cost rate $C(T^*)/(\lambda c_T)$ when $c_0 K/c_T = 1.0$

c_K/c_T	$\omega K = 5.0$		$\omega K = 10.0$		$\omega K = 20.0$	
	λT^*	$C(T^*)/(\lambda c_T)$	λT^*	$C(T^*)/(\lambda c_T)$	λT^*	$C(T^*)/(\lambda c_T)$
5	3.419	0.802	5.802	0.358	11.891	0.155
10	2.245	1.100	4.394	0.433	9.985	0.173
15	1.816	1.322	3.813	0.481	9.144	0.184
20	1.574	1.510	3.466	0.519	8.622	0.192
30	1.296	1.831	3.043	0.578	7.964	0.203
50	1.021	2.358	2.595	0.663	7.236	0.218

Table 8.2 Optimum N^* and its cost rate $C(N^*)/(\lambda c_N)$ when $c_0 K/c_N = 1.0$

c_K/c_N	$\omega K = 5.0$		$\omega K = 10.0$		$\omega K = 20.0$	
	N^*	$C(N^*)/(\lambda c_N)$	N^*	$C(N^*)/(\lambda c_N)$	λN^*	$C(N^*)/(\lambda c_N)$
5	2	0.779	5	0.323	12	0.141
10	2	0.880	4	0.373	11	0.150
15	1	1.293	4	0.386	10	0.157
20	1	1.327	3	0.451	10	0.160
30	1	1.394	3	0.460	9	0.168
50	1	1.529	3	0.479	8	0.180

whose left-hand side increases strictly with N to $\sum_{j=1}^{\infty} P_j(T)G^{(j)}(K)$. Thus, if

$$\sum_{j=1}^{\infty} P_j(T)G^{(j)}(K) > \frac{c_T}{c_K - c_T - c_0/\omega},$$

then there exists a finite and unique maximum N_F^* ($1 \le N_F^* < \infty$) for given T which satisfies (8.14). Furthermore, differentiating the left-hand side of (8.14) with respect to T,

$$\lambda \sum_{j=0}^{N-1} p_j(T)G^{(j)}(K)[r_N(K) - r_{j+1}(K)] > 0,$$

which follows that optimum N_F^* decreases strictly with T to N^*, where N^* is given in (8.12).

On the other hand, differentiating $C_F(T, N)$ with respect to T and setting it equal to zero,

$$Q_1(T,N)\sum_{j=0}^{N-1}P_{j+1}(T)G^{(j)}(K)-\sum_{j=0}^{N-1}P_{j+1}(T)[G^{(j)}(K)-G^{(j+1)}(K)]$$
$$=\frac{c_T}{c_K-c_T-c_O/\omega},\tag{8.15}$$

where

$$Q_1(T,N)\equiv\frac{\sum_{j=0}^{N-1}[(\lambda T)^j/j!][G^{(j)}(K)-G^{(j+1)}(K)]}{\sum_{j=0}^{N-1}[(\lambda T)^j/j!]G^{(j)}(K)}.$$

Substituting (8.14) for (8.15),

$$Q_1(T,N)>r_N(K),$$

which does not hold for any N. Thus, the optimum policy to minimize $C_F(T,N)$ is $(T_F^*=\infty, N_F^*=N^*)$, where N^* is given in (8.12).

For given N, the left-hand side of (8.15) increases strictly with T to the left-hand side of (8.14). Thus, if $N\le N^*$, then $T_F^*=\infty$, and conversely, if $N>N^*$, then optimum T_F^* $(0<T_F^*<\infty)$ to satisfy (8.15) exists.

8.1.2 Defragmentation Last

Suppose that the database is defragmented preventively at planned time T $(0\le T\le\infty)$, or at number N $(N=0,1,2,\cdots)$ of updates, whichever occurs last. Then, the probability that the database is defragmented at time T is

$$\sum_{j=N}^{\infty}p_j(T)G^{(j)}(K),$$

the probability that it is defragmented at update N is

$$\overline{P}_N(T)G^{(N)}(K),$$

and the probability that it is defragmented at volume K is

$$\sum_{j=0}^{\infty}P_{j+1}(T)[G^{(j)}(K)-G^{(j+1)}(K)]+\sum_{j=0}^{N-1}\overline{P}_{j+1}(T)[G^{(j)}(K)-G^{(j+1)}(K)]$$
$$=1-\sum_{j=N}^{\infty}\overline{P}_{j+1}(T)[G^{(j)}(K)-G^{(j+1)}(K)].$$

The mean time to defragmentation is

$$
\begin{aligned}
&T\sum_{j=N}^{\infty} p_j(T)G^{(j)}(K) + G^{(N)}(K)\int_T^{\infty} t\mathrm{d}P_N(t) \\
&+\sum_{j=0}^{\infty}[G^{(j)}(K) - G^{(j+1)}(K)]\int_0^T t\mathrm{d}P_{j+1}(t) \\
&+\sum_{j=0}^{N-1}[G^{(j)}(K) - G^{(j+1)}(K)]\int_T^{\infty} t\mathrm{d}P_{j+1}(t) \\
&=\sum_{j=0}^{\infty} G^{(j)}(K)\int_0^T p_j(t)\mathrm{d}t + \sum_{j=0}^{N-1} G^{(j)}(K)\int_T^{\infty} p_j(t)\mathrm{d}t. \qquad (8.16)
\end{aligned}
$$

Therefore, the expected cost rate is

$$
C_L(T,N) = \frac{\begin{array}{l}\sum_{j=N}^{\infty} p_j(T)\int_0^K [c_T + c_0(x)]\mathrm{d}G^{(j)}(x) \\ +\overline{P}_N(T)\int_0^K [c_N + c_0(x)]\mathrm{d}G^{(N)}(x) \\ +[c_K + c_0(K)]\{1 - \sum_{j=N}^{\infty}\overline{P}_{j+1}(T)[G^{(j)}(K) - G^{(j+1)}(K)]\}\end{array}}{\sum_{j=0}^{\infty} G^{(j)}(K)\int_0^T p_j(t)\mathrm{d}t + \sum_{j=0}^{N-1} G^{(j)}(K)\int_T^{\infty} p_j(t)\mathrm{d}t}. \qquad (8.17)
$$

Clearly, $\lim_{N\to 0} C_L(T,N) = C(T)$ in (8.5) and $\lim_{T\to 0} C_L(T,N) = C(N)$ in (8.8).

We find optimum T_L^* and N_L^* to minimize $C_L(T,N)$ when $c_K > c_T + c_0/\omega$, $c_T = c_N$, $c_0(x) = c_0 x$, $h(t) = \lambda$ and $G(x) = 1 - \mathrm{e}^{-\omega x}$. Then, the expected cost rate in (8.17) becomes

$$
\begin{aligned}
\frac{C_L(T,N)}{\lambda} &= (c_K - c_T) \\
&+ \frac{c_T - (c_K - c_T - c_0/\omega)[\sum_{j=1}^{\infty} P_j(T)G^{(j)}(K) + \sum_{j=1}^{N}\overline{P}_j(T)G^{(j)}(K)]}{\sum_{j=0}^{\infty} P_{j+1}(T)G^{(j)}(K) + \sum_{j=0}^{N-1}\overline{P}_{j+1}(T)G^{(j)}(K)}. \qquad (8.18)
\end{aligned}
$$

Forming the inequality $C_L(T,N+1) - C_L(T,N) \geq 0$,

$$
\begin{aligned}
&r_{N+1}(K)\left[\sum_{j=0}^{\infty} P_{j+1}(T)G^{(j)}(K) + \sum_{j=0}^{N-1}\overline{P}_{j+1}(T)G^{(j)}(K)\right] \\
&-\sum_{j=0}^{\infty} P_{j+1}(T)[G^{(j)}(K) - G^{(j+1)}(K)] - \sum_{j=0}^{N-1}\overline{P}_{j+1}(T)[G^{(j)}(K) - G^{(j+1)}(K)] \\
&\geq \frac{c_T}{c_K - c_T - c_0/\omega}, \qquad (8.19)
\end{aligned}
$$

whose left-hand side increases strictly with N to ωK. Thus, if $\omega K > c_T/(c_K - c_T - c_0/\omega)$, then there exists a finite and unique minimum N_L^* $(1 \le N_L^* < \infty)$ for given T which satisfies (8.19).

Furthermore, differentiating the left-hand side of (8.19) with respect to T,

$$\lambda \sum_{j=0}^{\infty} p_j(T) G^{(j)}(K)[r_{N+1}(K) - \widetilde{Q}_1(T, N)] < 0,$$

where

$$\widetilde{Q}_1(T, N)] \equiv \frac{\sum_{j=N}^{\infty}[(\lambda T)^j/j!][G^{(j)}(K) - G^{(j+1)}(K)]}{\sum_{j=N}^{\infty}[(\lambda T)^j/j!]G^{(j)}(K)} > r_{N+1}(K),$$

which increases strictly with T from $r_{N+1}(K)$ to 1 (Problem 3.2). Thus, optimum N_L^* increases strictly with T from N^*, where N^* is given in (8.12).

On the other hand, differentiating $C_L(T, N)$ with respect to T and setting it equal to zero,

$$\widetilde{Q}_1(T, N)] \left[\sum_{j=0}^{\infty} P_{j+1}(T) G^{(j)}(K) + \sum_{j=0}^{N-1} \overline{P}_{j+1}(T) G^{(j)}(K) \right]$$
$$- \sum_{j=0}^{\infty} P_{j+1}(T)[G^{(j)}(K) - G^{(j+1)}(K)] - \sum_{j=0}^{N-1} \overline{P}_{j+1}(T)[G^{(j)}(K) - G^{(j+1)}(K)]$$
$$= \frac{c_T}{c_K - c_T - c_0/\omega}. \tag{8.20}$$

Substituting (8.19) for (8.20),

$$\widetilde{Q}_1(T, N)] \le r_{N+1}(K),$$

which does not hold for any N. Thus, the optimum policy to minimize $C_L(T, N)$ is $(T_L^* = 0, N_L^* = N^*)$, where N^* is given in (8.12).

For given N, the left-hand side of (8.20) decreases strictly with T from the left-hand side of (8.12). Thus, if $N \le N^*$, then $T_L^* = 0$, and conversely, if $N > N^*$, then optimum T_L^* $(0 < T_L^* < \infty)$ to satisfy (8.20) exists.

8.1.3 Defragmentation Overtime

Suppose that the database is defragmented preventively at the forthcoming update over time T $(0 \le T < \infty)$. Then, the probability that the database is defragmented over time T is

$$\sum_{j=0}^{\infty} p_j(T) G^{(j+1)}(K),$$

and the probability that it is defragmented at volume K is

$$\sum_{j=0}^{\infty} P_j(T)[G^{(j)}(K) - G^{(j+1)}(K)] = \sum_{j=0}^{\infty} p_j(T)[1 - G^{(j+1)}(K)].$$

The mean time to defragmentation is (Problem 8.1)

$$\begin{aligned}
&\sum_{j=0}^{\infty}[G^{(j)}(K) - G^{(j+1)}(K)] \left\{ \int_0^T \left[\int_T^{\infty} u h(u) \mathrm{e}^{-H(u)+H(t)} \mathrm{d}u \right] \mathrm{d}P_j(t) \right. \\
&\left. + \int_0^T t \mathrm{d}P_{j+1}(t) \right\} + \sum_{j=0}^{\infty} G^{(j+1)}(K) \int_0^T \left[\int_T^{\infty} u h(u) \mathrm{e}^{-H(u)+H(t)} \mathrm{d}u \right] \mathrm{d}P_j(t) \\
&= \sum_{j=0}^{\infty} G^{(j)}(K) \left[\int_0^T p_j(t) \mathrm{d}t + p_j(T) \int_T^{\infty} \mathrm{e}^{-H(t)+H(T)} \mathrm{d}t \right]. \qquad (8.21)
\end{aligned}$$

Therefore, the expected cost rate is

$$C_O(T) = \frac{\begin{array}{c}\sum_{j=0}^{\infty} p_j(T) \int_0^K [c_{OV} + c_0(x)] \mathrm{d}G^{(j+1)}(x) \\ +[c_K + c_0(K)] \sum_{j=0}^{\infty} P_j(T)[G^{(j)}(K) - G^{(j+1)}(K)]\end{array}}{\sum_{j=0}^{\infty} G^{(j)}(K)[\int_0^T p_j(t)\mathrm{d}t + p_j(T) \int_T^{\infty} \mathrm{e}^{-H(t)+H(T)} \mathrm{d}t]}, \qquad (8.22)$$

where c_{OV} = defragmentation cost over time T.

We find optimum T_O^* to minimize $C_O(T)$ when $c_K > c_{OV} + c_0/\omega$ and $c_0(x) = c_0 x$. Differentiating $C_O(T)$ with respect to T and setting it equal to zero,

$$\begin{aligned}
&Q(T)[(c_K - c_{OV})Q_3(T) + c_0 Q_4(T)] \sum_{j=0}^{\infty} G^{(j)}(K) \left\{ \int_0^T p_j(t) \mathrm{d}t \right. \\
&\left. + p_j(T) \int_T^{\infty} \mathrm{e}^{-H(t)+H(T)} \mathrm{d}t \right\} - (c_K - c_{OV}) \sum_{j=0}^{\infty} P_j(T)[G^{(j)}(K) - G^{(j+1)}(K)] \\
&- c_0 \sum_{j=0}^{\infty} P_j(T) \int_0^K [G^{(j)}(x) - G^{(j+1)}(x)] \mathrm{d}x = c_{OV}, \qquad (8.23)
\end{aligned}$$

where

$$Q(T) \equiv \frac{\mathrm{e}^{-H(T)}}{\int_T^\infty \mathrm{e}^{-H(t)}\mathrm{d}t},$$

$$Q_3(T) \equiv \frac{\sum_{j=0}^\infty p_j(T)[G^{(j+1)}(K) - G^{(j+2)}(K)]}{\sum_{j=0}^\infty p_j(T)G^{(j+1)}(K)},$$

$$Q_4(T) \equiv \frac{\sum_{j=0}^\infty p_j(T)\int_0^K [G^{(j+1)}(x) - G^{(j+2)}(x)]\mathrm{d}x}{\sum_{j=0}^\infty p_j(T)G^{(j+1)}(K)}.$$

In particular, when $c_O = 0$, (8.23) becomes

$$Q(T)Q_3(T)\sum_{j=0}^\infty G^{(j)}(K)\left\{\int_0^T p_j(t)\mathrm{d}t + p_j(T)\int_T^\infty \mathrm{e}^{-H(t)+H(T)}\mathrm{d}t\right\}$$
$$-\sum_{j=0}^\infty P_j(T)[G^{(j)}(K) - G^{(j+1)}(K)] = \frac{c_{OV}}{c_K - c_{OV}}. \tag{8.24}$$

If $Q(T)Q_3(T)$ increases strictly with T to $Q(\infty)Q_3(\infty)$, and

$$Q(\infty)Q_3(\infty)\sum_{j=0}^\infty G^{(j)}(K)\int_0^\infty p_j(t)\mathrm{d}t > \frac{c_K}{c_K - c_{OV}},$$

then there exists a finite and unique T_O^* $(0 < T_O^* < \infty)$ which satisfies (8.24).

In addition, when $h(t) = \lambda$ and $G(x) = 1 - \mathrm{e}^{-\omega x}$, (8.23) is

$$Q_3(T)\sum_{j=0}^\infty P_j(T)G^{(j)}(K) - \sum_{j=0}^\infty P_j(T)[G^{(j)}(K) - G^{(j+1)}(K)]$$
$$= \frac{c_{OV}}{c_K - c_{OV} - c_O/\omega}, \tag{8.25}$$

where $Q(T) = \lambda$, $Q_4(T) = [1 - Q_3(T)]/\omega$ and

$$Q_3(T) = \frac{\sum_{j=0}^\infty[(\lambda T)^j/j!][G^{(j+1)}(K) - G^{(j+2)}(K)]}{\sum_{j=0}^\infty[(\lambda T)^j/j!]G^{(j+1)}(K)},$$

which increases strictly with T to 1. Thus, if $\omega K > c_{OV}/(c_K - c_{OV} - c_O/\omega)$, then there exist a finite and unique T_O^* $(0 < T_O^* < \infty)$ which satisfies (8.25).

Table 8.3 presents optimum λT_O^* and its cost rate $C_O(T_O^*)/(\lambda c_{OV})$ for ωK and c_K/c_{OV} when $c_O K/c_{OV} = 1.0$. It shows that $T_O^* < T^*$ and $C_O(T_O^*) < C(T^*)$ in Table 8.1 when $c_{OV} = c_T$.

Table 8.3 Optimum λT_O^* and its cost rate $C_O(T_O^*)/(\lambda c_{OV})$ when $c_O K/c_{OV} = 1.0$

c_K/c_{OV}	$\omega K = 5.0$		$\omega K = 10.0$		$\omega K = 20.0$	
	λT_O^*	$C_O(T_O^*)/(\lambda c_{OV})$	λT_O^*	$C_O(T_O^*)/(\lambda c_{OV})$	λT_O^*	$C_O(T_O^*)/(\lambda c_{OV})$
5	2.138	0.574	4.746	0.251	10.897	0.105
10	1.037	0.801	3.401	0.317	9.054	0.121
15	0.633	0.943	2.849	0.358	8.235	0.131
20	0.403	1.045	2.519	0.389	7.727	0.138
30	0.134	1.180	2.119	0.435	7.087	0.149
50	0.000	1.655	1.696	0.498	6.379	0.162

8.2 Database Backup Model

We use the following three widely accepted backup modes for our modelings and discussions, i.e., full backup, incremental backup and differential backup [53, 56–61]:

1. *Full backup*, a lazy but simple mode that exports all the data files updated since the last full backup. When a full recovery after database breakdown is needed, the data restoration needs only the last full backup. Obviously, This mode means that many periodic full backups will need to be implemented, requiring long periods of database downtime on a regular basis.
2. *Incremental backup*, the mode that exports only the data files updated since the last backup (a full or incremental backup). Incremental backups are much smaller and quicker than full backups, and the data restoration after breakdown needs the last full backup plus all the incremental backups until the point-in-time of breakdown.
3. *Differential backup*, the mode that exports all data files updated since the last full backup. The advantage to this mode is quicker recovery time, requiring only a full backup and the last differential backup to restore the entire updated data.

It is assumed that a volume W_j $(j = 1, 2, \cdots)$ of updated data due to the jth update is a random variable with an identical distribution $G(x) \equiv \Pr\{W_j \le x\}$ and finite mean $1/\omega \equiv \int_0^\infty \overline{G}(x)\mathrm{d}x$. In addition, the database breakdown occurs randomly with a general distribution $D(t)$ with a density function $d(t) \equiv \mathrm{d}D(t)/\mathrm{d}t$ and finite mean $1/\mu \equiv \int_0^\infty \overline{D}(t)\mathrm{d}t$. It is noted from the definition of failure rate [1] that the rate of breakdown $r(t) \equiv d(t)/\overline{D}(t)$ is supposed to be increasing with t strictly to $r(\infty)$ that might be infinity. A full backup should be made immediately after database breakdown as a renewal point for incremental and differential backups.

8.2.1 Backup First

In order to protect the security of data and prevent the enormous recovery cost due to breakdown, a full backup is scheduled preventively at planned time T $(0 < T \leq \infty)$ or at number N $(N = 1, 2, \cdots)$ of updates, i.e., at number N of incremental or differential backups, whichever occurs first. Then, the probability that a full backup is implemented at time T is

$$\overline{D}(T)[1 - F^{(N)}(T)],$$

the probability that it is implemented at update N is

$$\int_0^T \overline{D}(t)\mathrm{d}F^{(N)}(t),$$

and the probability that it is implemented at breakdown is

$$\int_0^T [1 - F^{(N)}(t)]\mathrm{d}D(t).$$

We introduce the following costs for backup and recovery schemes: c_F is the constant cost of full backup, $c_K + c_O x$ is the cost for incremental or differential backups when a total volume x of data has been updated, $c_R + c_O x + j c_N$ is the recovery cost after breakdown when a number j of incremental backups have been made, and $c_R + c_O x$ is the recovery cost after breakdown when the differential backup is scheduled. We denote for $j = 1, 2, \cdots, N$,

$$M_j \equiv \int_0^\infty (c_K + c_O x)\mathrm{d}G^{(j)}(x) = c_K + \frac{j c_O}{\omega},$$
$$N_j \equiv \int_0^\infty (c_R + c_O x)\mathrm{d}G^{(j)}(x) = c_R + \frac{j c_O}{\omega}.$$

Then, $j M_1$ is the expected cost of j incremental backups, $\sum_{i=1}^{j} M_i$ is the expected cost of j differential backups, $N_j + j c_N$ is the recovery cost when j incremental backups are implemented to import saved data files, and N_j is the recovery cost when the jth differential backup is implemented.

Thus, the expected cost until full backup for incremental backup is

$$\widetilde{C}_{IF}(T, N) = \sum_{j=0}^{N-1} (c_F + j M_1)\overline{D}(T)[F^{(j)}(T) - F^{(j+1)}(T)]$$
$$+ (c_F + N M_1) \int_0^T \overline{D}(t)\mathrm{d}F^{(N)}(t)$$

$$
\begin{aligned}
&+\sum_{j=0}^{N-1}(c_F + jM_1 + N_j + jc_N)\int_0^T [F^{(j)}(t) - F^{(j+1)}(t)]\mathrm{d}D(t) \\
&= c_F + \left(c_K + \frac{c_0}{\omega}\right)\sum_{j=1}^{N}\int_0^T \overline{D}(t)\mathrm{d}F^{(j)}(t) \\
&+\sum_{j=0}^{N-1}\left[c_R + j\left(c_N + \frac{c_0}{\omega}\right)\int_0^T [F^{(j)}(t) - F^{(j+1)}(t)]\mathrm{d}D(t)\right], \quad (8.26)
\end{aligned}
$$

and the expected cost until full backup for differential backup is

$$
\begin{aligned}
\widetilde{C}_{DF}(T, N) &= \sum_{j=0}^{N-1}\left(c_F + \sum_{i=1}^{j} M_i\right)\overline{D}(T)[F^{(j)}(T) - F^{(j+1)}(T)] \\
&+\left(c_F + \sum_{i=1}^{N} M_i\right)\int_0^T \overline{D}(t)\mathrm{d}F^{(N)}(t) \\
&+\sum_{j=0}^{N-1}\left(c_F + \sum_{i=1}^{j} M_i + N_j\right)\int_0^T [F^{(j)}(t) - F^{(j+1)}(t)]\mathrm{d}D(t) \\
&= c_F + \sum_{j=1}^{N}\left(c_K + \frac{jc_0}{\omega}\right)\int_0^T \overline{D}(t)\mathrm{d}F^{(j)}(t) \\
&+\sum_{j=0}^{N-1}\left(c_R + \frac{jc_0}{\omega}\right)\int_0^T [F^{(j)}(t) - F^{(j+1)}(t)]\mathrm{d}D(t). \quad (8.27)
\end{aligned}
$$

The difference of $\widetilde{C}_{DF}(T, N)$ and $\widetilde{C}_{IF}(T, N)$ is

$$
\begin{aligned}
\widetilde{C}_{DF}(T, N) - \widetilde{C}_{IF}(T, N) &= \frac{c_0}{\omega}\sum_{j=0}^{N-1} j\int_0^T \overline{D}(t)\mathrm{d}F^{(j+1)}(t) \\
&- c_N\sum_{j=0}^{N-1} j\int_0^T [F^{(j)}(t) - F^{(j+1)}(t)]\mathrm{d}D(t),
\end{aligned}
$$

that is, if

$$
\frac{c_0}{\omega}\sum_{j=0}^{N-1} j\int_0^T \overline{D}(t)\mathrm{d}F^{(j+1)}(t) > c_N\sum_{j=0}^{N-1} j\int_0^T [F^{(j)}(t) - F^{(j+1)}(t)]\mathrm{d}D(t)
$$

holds, then incremental backup would save more cost than differential backup.

The mean time to full backup is

$$T\overline{D}(T)[1-F^{(N)}(T)]+\int_0^T t\overline{D}(t)\mathrm{d}F^{(N)}(t)+\int_0^T t[1-F^{(N)}(t)]\mathrm{d}D(t)$$
$$=\int_0^T \overline{D}(t)[1-F^{(N)}(t)]\mathrm{d}t. \tag{8.28}$$

Therefore, the expected cost rate for incremental backup is

$$C_{IF}(T,N)=\frac{c_F+(c_K+c_O/\omega)\sum_{j=1}^{N}\int_0^T\overline{D}(t)\mathrm{d}F^{(j)}(t)+\sum_{j=0}^{N-1}[c_R+j(c_N+c_O/\omega)]\int_0^T[F^{(j)}(t)-F^{(j+1)}(t)]\mathrm{d}D(t)}{\int_0^T\overline{D}(t)[1-F^{(N)}(t)]\mathrm{d}t}, \tag{8.29}$$

and the expected cost rate for differential backup is

$$C_{DF}(T,N)=\frac{c_F+\sum_{j=1}^{N}(c_K+jc_O/\omega)\int_0^T\overline{D}(t)\mathrm{d}F^{(j)}(t)+\sum_{j=0}^{N-1}(c_R+jc_O/\omega)\int_0^T[F^{(j)}(t)-F^{(j+1)}(t)]\mathrm{d}D(t)}{\int_0^T\overline{D}(t)[1-F^{(N)}(t)]\mathrm{d}t}. \tag{8.30}$$

8.2.1.1 Optimum Incremental Backup Policies

We find optimum full backup times T_I^* and N_I^* to minimize $C_I(T)\equiv\lim_{N\to\infty}C_{IF}(T,N)$ and $C_I(N)\equiv\lim_{T\to\infty}C_{IF}(T,N)$, respectively.

(1) Optimum T_I^*

Suppose that a full backup is scheduled preventively only at time T $(0<T\leq\infty)$. Then, putting that $N\to\infty$ in (8.29),

$$C_I(T)\equiv\lim_{N\to\infty}C_{IF}(T,N)=\frac{c_F+c_RD(T)+(c_K+c_O/\omega)\int_0^T\overline{D}(t)\mathrm{d}M_F(t)+(c_N+c_O/\omega)\int_0^T M_F(t)\mathrm{d}D(t)}{\int_0^T\overline{D}(t)\mathrm{d}t}, \tag{8.31}$$

where $M_F(t)=\sum_{j=1}^{\infty}F^{(j)}(t)$.

When $F(t)=1-\mathrm{e}^{-\lambda t}$, i.e., $M_F(t)=\lambda t$, differentiating $C_I(T)$ with respect to T and setting it equal to zero,

$$c_R\int_0^T\overline{D}(t)[r(T)-r(t)]\mathrm{d}t+\left(c_N+\frac{c_O}{\omega}\right)\lambda\int_0^T\overline{D}(t)[r(T)T-r(t)t]\mathrm{d}t=c_F, \tag{8.32}$$

whose left-hand side increases strictly with T to ∞. Thus, there exists a finite and unique T_I^* $(0 < T_I^* < \infty)$ which satisfies (8.32), and the resulting cost rate is

$$\frac{C_I(T_I^*)}{\lambda} = \left[\frac{c_R}{\lambda} + \left(c_N + \frac{c_O}{\omega}\right) T_I^*\right] r(T_I^*) + \left(c_K + \frac{c_O}{\omega}\right). \tag{8.33}$$

Furthermore, when $D(t) = 1 - \mathrm{e}^{-\mu t}$, optimum T_I^* satisfies

$$\frac{\lambda}{\mu}[\mu T - (1 - \mathrm{e}^{-\mu T})] = \frac{c_F}{c_N + c_O/\omega}, \tag{8.34}$$

and (8.33) becomes

$$\frac{C_I(T_I^*)}{\mu} = \left(c_N + \frac{c_O}{\omega}\right)\lambda T_I^* + c_R + \frac{\lambda}{\mu}\left(c_K + \frac{c_O}{\omega}\right). \tag{8.35}$$

(2) Optimum N_I^*

Suppose that a full backup is scheduled preventively only at update N $(N = 1, 2, \cdots)$. Then, putting that $T \to \infty$ in (8.29),

$$C_I(N) \equiv \lim_{T\to\infty} C_{IF}(T, N)$$
$$= \frac{c_F + (c_K + c_O/\omega)\sum_{j=1}^{N}\int_0^\infty F^{(j)}(t)\mathrm{d}D(t) + \sum_{j=0}^{N-1}[c_R + j(c_N + c_O/\omega)]\int_0^\infty [F^{(j)}(t) - F^{(j+1)}(t)]\mathrm{d}D(t)}{\int_0^\infty \overline{D}(t)[1 - F^{(N)}(t)]\mathrm{d}t}. \tag{8.36}$$

When $D(t) = 1 - \mathrm{e}^{-\mu t}$, i.e., $\int_0^\infty F^{(j)}(t)\mathrm{d}D(t) = [F^*(\mu)]^j$, where $F^*(s) \equiv \int_0^\infty \mathrm{e}^{-st}\mathrm{d}F(t)$, forming the inequality $C_I(N+1) - C_I(N) \geq 0$,

$$\sum_{j=1}^{N}\{1 - [F^*(\mu)]^j\} \geq \frac{c_F}{c_N + c_O/\omega}, \tag{8.37}$$

whose left-hand side increases strictly with N to ∞. Thus, there exists a finite and unique minimum N_I^* $(1 \leq N_I^* < \infty)$ which satisfies (8.37).

Furthermore, when $F(t) = 1 - \mathrm{e}^{-\lambda t}$, optimum N_I^* satisfies

$$\sum_{j=1}^{N}\left[1 - \left(\frac{\lambda}{\lambda + \mu}\right)^j\right] \geq \frac{c_F}{c_N + c_O/\omega}. \tag{8.38}$$

8.2.1.2 Optimum Differential Backup Policies

We find optimum full backup times T_D^* and N_D^* to minimize $C_D(T) \equiv \lim_{N\to\infty} C_{DF}(T,N)$ and $C_D(N) \equiv \lim_{T\to\infty} C_{DF}(T,N)$, respectively.

(1) Optimum T_D^*

Suppose that a full backup is scheduled preventively only at time T $(0 < T \le \infty)$. Then, putting that $N \to \infty$ in (8.30),

$$C_D(T) \equiv \lim_{N\to\infty} C_{DF}(T,N)$$
$$= \frac{\begin{array}{c} c_F + c_R D(T) + (c_0/\omega)\int_0^T M_F(t)\mathrm{d}D(t) \\ + \sum_{j=1}^{\infty}(c_K + jc_0/\omega)\int_0^T \overline{D}(t)\mathrm{d}F^{(j)}(t) \end{array}}{\int_0^T \overline{D}(t)\mathrm{d}t}. \tag{8.39}$$

When $F(t) = 1 - \mathrm{e}^{-\lambda t}$, differentiating $C_D(T)$ with respect to T and setting it equal to zero,

$$c_R\int_0^T \overline{D}(t)[r(T) - r(t)]\mathrm{d}t + \frac{c_0\lambda}{\omega}\int_0^T \overline{D}(t)[\lambda(T-t) + r(T)T - r(t)t]\mathrm{d}t = c_F, \tag{8.40}$$

whose left-hand side increases strictly with T to ∞. Thus, there exists a finite and unique T_D^* $(0 < T_D^* < \infty)$ which satisfies (8.40), and the resulting cost rate is

$$\frac{C_D(T_D^*)}{\lambda} = \left(\frac{c_R}{\lambda} + \frac{c_0 T_D^*}{\omega}\right) r(T_D^*) + c_K + \frac{c_0}{\omega}(1 + \lambda T_D^*). \tag{8.41}$$

Furthermore, when $D(t) = 1 - \mathrm{e}^{-\mu t}$, optimum T_D^* satisfies

$$\frac{\lambda}{\mu}\left(1 + \frac{\lambda}{\mu}\right)[\mu T - (1 - \mathrm{e}^{-\mu T})] = \frac{c_F}{c_0/\omega}, \tag{8.42}$$

and (8.41) becomes

$$\frac{C_D(T_D^*)}{\mu} = \frac{c_0}{\omega}\left(1 + \frac{\lambda}{\mu}\right)\lambda T_D^* + c_R + \frac{\lambda}{\mu}\left(c_K + \frac{c_0}{\omega}\right). \tag{8.43}$$

(2) Optimum N_D^*

Suppose that a full backup is scheduled preventively only at update N $(N = 1, 2, \cdots)$. Then, putting that $T \to \infty$ in (8.30),

$$C_D(N) \equiv \lim_{T\to\infty} C_{DF}(T,N)$$
$$= \frac{c_F + \sum_{j=1}^{N}(c_K + jc_O/\omega)\int_0^\infty F^{(j)}(t)\mathrm{d}D(t) + \sum_{j=0}^{N-1}(c_R + jc_O/\omega)\int_0^\infty [F^{(j)}(t) - F^{(j+1)}(t)]\mathrm{d}D(t)}{\int_0^\infty \overline{D}(t)[1 - F^{(N)}(t)]\mathrm{d}t}. \quad (8.44)$$

When $D(t) = 1 - \mathrm{e}^{-\mu t}$, forming the inequality $C_D(N+1) - C_D(N) \geq 0$,

$$\frac{1}{1 - F^*(\mu)}\sum_{j=1}^{N}\{1 - [F^*(\mu)]^j\} \geq \frac{c_F}{c_O/\omega}, \quad (8.45)$$

whose left-hand side increases strictly with N to ∞. Thus, there exists a finite and unique minimum N_D^* $(1 \leq N_D^* < \infty)$ which satisfies (8.45).

Furthermore, when $F(t) = 1 - \mathrm{e}^{-\lambda t}$, optimum N_D^* satisfies

$$\sum_{j=0}^{N}(N - j)\left(\frac{\lambda}{\lambda + \mu}\right)^j \geq \frac{c_F}{c_O/\omega}. \quad (8.46)$$

8.2.1.3 Numerical Examples

Table 8.4 presents optimum λT_I^* in (8.34), λT_D^* in (8.42) and their cost rates $C_I(T_I^*)/\mu$ and $C_D(T_D^*)/\mu$ for λ/μ and c_O/ω when $c_K = 0.1$, $c_R = 0.2$, $c_N = 0.5$ and $c_F = 10.0$. Obviously, both λT_I^* and λT_D^* decrease with c_O/ω, λT_I^* increases with λ/μ and λT_D^* decreases λ/μ, respectively. Table 8.4 can be explained physically that when the database is busy with updates, more incremental backups can be implemented and less differential backups are implemented to save backup costs.

Table 8.5 presents N_I^* in (8.38), N_D^* in (8.46) and their cost rates for λ/μ and c_O/ω when $c_K = 0.1$, $c_R = 0.2$, $c_N = 0.5$ and $c_F = 10.0$. Optimum N_i^* $(i = I, D)$ are

Table 8.4 Optimum λT_I^*, λT_D^* and their cost rates when $c_K = 0.1$, $c_R = 0.2$, $c_N = 0.5$ and $c_F = 10.0$

λ/μ	$c_O/\omega = 0.5$				$c_O/\omega = 1.0$			
	λT_I^*	$C_I(T_I^*)/\mu$	λT_D^*	$C_D(T_D^*)/\mu$	λT_I^*	$C_I(T_I^*)/\mu$	λT_D^*	$C_D(T_D^*)/\mu$
1.0	10.999	11.800	10.999	11.800	7.666	12.799	5.997	13.294
2.0	11.996	13.396	8.641	14.361	8.641	15.361	5.184	17.952
5.0	14.738	17.938	7.133	24.600	11.126	22.389	4.723	34.035
10.0	18.414	24.614	6.702	43.062	14.265	32.597	4.590	61.691
20.0	23.967	36.167	6.507	80.527	18.889	50.534	4.529	117.312
50.0	35.339	65.537	6.396	193.301	28.247	97.571	4.494	284.400

Table 8.5 Optimum N_I^*, N_D^* and their cost rates when $c_K = 0.1, c_R = 0.2, c_N = 0.5$ and $c_F = 10.0$

λ/μ	$c_0/\omega = 0.5$				$c_0/\omega = 1.0$			
	N_I^*	$C_I(N_I^*)/\mu$	N_D^*	$C_D(N_D^*)/\mu$	N_I^*	$C_I(N_I^*)/\mu$	N_D^*	$C_D(N_D^*)/\mu$
1.0	11	11.800	11	11.800	8	12.792	6	13.268
2.0	12	13.384	9	14.307	9	15.307	6	17.630
5.0	15	17.853	7	23.942	11	22.189	5	32.262
10.0	19	24.441	7	41.159	15	32.266	5	57.491
20.0	24	35.908	7	76.219	19	50.084	5	108.348
50.0	35	65.197	6	181.740	28	97.017	5	261.260

almost the same in Table 8.4. However, it is of great interest that $C_I(N_I^*) < C_I(T_I^*)$ and $C_D(N_D^*) < C_D(T_D^*)$.

8.2.2 Backup Last

Suppose that a full backup is scheduled preventively at planned time T $(0 \le T \le \infty)$ or at number N $(N = 0, 1, 2, \cdots)$ of updates, whichever occurs last. Then, the probability that the full backup is implemented at time T is

$$\overline{D}(T)F^{(N)}(T),$$

the probability that it is implemented at update N is

$$\int_T^\infty \overline{D}(t)\mathrm{d}F^{(N)}(t),$$

and the probability that it is implemented at breakdown is

$$D(T) + \int_T^\infty [1 - F^{(N)}(t)]\mathrm{d}D(t).$$

Therefore, the expected cost until full backup for incremental backup is

$$\widetilde{C}_{IL}(T, N) =$$

$$\sum_{j=N}^{\infty}(c_F + jM_1)\overline{D}(T)[F^{(j)}(T) - F^{(j+1)}(T)]$$

$$+ (c_F + NM_1)\int_T^\infty \overline{D}(t)\mathrm{d}F^{(N)}(t)$$

$$
\begin{aligned}
&+\sum_{j=0}^{\infty}(c_F + jM_1 + N_j + jc_N)\int_0^T [F^{(j)}(t) - F^{(j+1)}(t)]\mathrm{d}D(t) \\
&+\sum_{j=0}^{N-1}(c_F + jM_1 + N_j + jc_N)\int_T^{\infty} [F^{(j)}(t) - F^{(j+1)}(t)]\mathrm{d}D(t) \\
&= c_F + \left(c_K + \frac{c_0}{\omega}\right)\left[\sum_{j=1}^{\infty}\int_0^T \overline{D}(t)\mathrm{d}F^{(j)}(t) + \sum_{j=1}^{N}\int_T^{\infty}\overline{D}(t)\mathrm{d}F^{(j)}(t)\right] \\
&+ c_R\left\{D(T) + \int_T^{\infty}[1 - F^{(N)}(t)]\mathrm{d}D(t)\right\} \\
&+ \left(c_N + \frac{c_0}{\omega}\right)\left\{\sum_{j=1}^{\infty}\int_0^T F^{(j)}(t)\mathrm{d}D(t)\right. \\
&\left. + \sum_{j=0}^{N-1} j\int_T^{\infty}[F^{(j)}(t) - F^{(j+1)}(t)]\mathrm{d}D(t)\right\}, \qquad (8.47)
\end{aligned}
$$

and the expected cost until full backup for differential backup is

$$
\begin{aligned}
&\widetilde{C}_{DL}(T,N) = \\
&\sum_{j=N}^{\infty}\left(c_F + \sum_{i=1}^{j} M_i\right)\overline{D}(T)[F^{(j)}(T) - F^{(j+1)}(T)] \\
&+\left(c_F + \sum_{i=1}^{N} M_i\right)\int_T^{\infty}\overline{D}(t)\mathrm{d}F^{(N)}(t) \\
&+\sum_{j=0}^{\infty}\left(c_F + \sum_{i=1}^{j} M_i + N_j\right)\int_0^T [F^{(j)}(t) - F^{(j+1)}(t)]\mathrm{d}D(t) \\
&+\sum_{j=0}^{N-1}\left(c_F + \sum_{i=1}^{j} M_i + N_j\right)\int_T^{\infty} [F^{(j)}(t) - F^{(j+1)}(t)]\mathrm{d}D(t) \\
&= c_F + \sum_{j=1}^{\infty}\left(c_K + \frac{jc_0}{\omega}\right)\int_0^T \overline{D}(t)\mathrm{d}F^{(j)}(t) \\
&+\sum_{j=1}^{N}\left(c_K + \frac{jc_0}{\omega}\right)\int_T^{\infty}\overline{D}(t)\mathrm{d}F^{(j)}(t)
\end{aligned}
$$

$$+ c_R \left\{ D(T) + \int_T^{\infty} [1 - F^{(N)}(t)] \mathrm{d}D(t) \right\}$$
$$+ \frac{c_O}{\omega} \left\{ \sum_{j=1}^{\infty} \int_0^T F^{(j)}(t) \mathrm{d}D(t) + \sum_{j=0}^{N-1} j \int_T^{\infty} [F^{(j)}(t) - F^{(j+1)}(t)] \mathrm{d}D(t) \right\}. \tag{8.48}$$

The mean time to full backup is

$$T\overline{D}(T)F^{(N)}(T) + \int_T^{\infty} t\overline{D}(t) \mathrm{d}F^{(N)}(t) + \int_0^T t \mathrm{d}D(t) + \int_T^{\infty} t[1 - F^{(N)}(t)] \mathrm{d}D(t)$$
$$= \int_0^T \overline{D}(t) \mathrm{d}t + \int_T^{\infty} \overline{D}(t)[1 - F^{(N)}(t)] \mathrm{d}t. \tag{8.49}$$

Therefore, the expected cost rate for incremental backup is

$$C_{IL}(T, N) = \frac{\begin{array}{l} c_F + (c_K + c_O/\omega)[\sum_{j=1}^{\infty} \int_0^T \overline{D}(t) \mathrm{d}F^{(j)}(t) \\ + \sum_{j=1}^{N} \int_T^{\infty} \overline{D}(t) \mathrm{d}F^{(j)}(t)] \\ + c_R\{D(T) + \int_T^{\infty} [1 - F^{(N)}(t)] \mathrm{d}D(t)\} \\ + (c_N + c_O/\omega)\{\sum_{j=1}^{\infty} \int_0^T F^{(j)}(t) \mathrm{d}D(t) \\ + \sum_{j=0}^{N-1} j \int_T^{\infty} [F^{(j)}(t) - F^{(j+1)}(t)] \mathrm{d}D(t)\} \end{array}}{\int_0^T \overline{D}(t) \mathrm{d}t + \int_T^{\infty} \overline{D}(t)[1 - F^{(N)}(t)] \mathrm{d}t}, \tag{8.50}$$

and the expected cost rate for differential backup is

$$C_{DL}(T, N) = \frac{\begin{array}{l} c_F + \sum_{j=1}^{\infty} (c_K + jc_O/\omega) \int_0^T \overline{D}(t) \mathrm{d}F^{(j)}(t) \\ + \sum_{j=1}^{N} (c_K + jc_O/\omega) \int_T^{\infty} \overline{D}(t) \mathrm{d}F^{(j)}(t) \\ + c_R\{D(T) + \int_T^{\infty} [1 - F^{(N)}(t)] \mathrm{d}D(t)\} \\ + (c_O/\omega)\{\sum_{j=1}^{\infty} \int_0^T F^{(j)}(t) \mathrm{d}D(t) \\ + \sum_{j=0}^{N-1} j \int_T^{\infty} [F^{(j)}(t) - F^{(j+1)}(t)] \mathrm{d}D(t)\} \end{array}}{\int_0^T \overline{D}(t) \mathrm{d}t + \int_T^{\infty} \overline{D}(t)[1 - F^{(N)}(t)] \mathrm{d}t}. \tag{8.51}$$

Clearly,

$$C_{IL}(T, 0) = C_{IF}(T, \infty) = C_I(T),$$
$$C_{IL}(0, N) = C_{IF}(\infty, N) = C_I(N)$$

in (8.31) and (8.36), respectively, and

$$C_{DL}(T,0)=C_{DF}(T,\infty)=C_D(T),$$
$$C_{DL}(0,N)=C_{DF}(\infty,N)=C_D(N)$$

in (8.39) and (8.44), respectively.

8.2.3 *Backup Overtime*

Suppose that a full backup is scheduled at the completion of the forthcoming update over time T $(0 \le T < \infty)$ or at database breakdown, whichever occurs first. Then, the probability that the full backup is implemented over time T is

$$\sum_{j=0}^{\infty}\int_0^T\left[\int_{T-t}^{\infty}\overline{D}(t+u)\mathrm{d}F(u)\right]\mathrm{d}F^{(j)}(t),$$

and the probability that it is implemented at breakdown is

$$D(T)+\sum_{j=0}^{\infty}\int_0^T\left[\int_{T-t}^{\infty}\overline{F}(u)\mathrm{d}D(t+u)\right]\mathrm{d}F^{(j)}(t)$$
$$=\sum_{j=0}^{\infty}\int_0^T\left[\int_{T-t}^{\infty}D(t+u)\mathrm{d}F(u)\right]\mathrm{d}F^{(j)}(t).$$

Therefore, the expected cost until full backup for incremental backup is

$$\begin{aligned}\widetilde{C}_{OI}(T)=&\\ &\sum_{j=0}^{\infty}[c_F+(j+1)M_1]\int_0^T\left[\int_{T-t}^{\infty}\overline{D}(t+u)\mathrm{d}F(u)\right]\mathrm{d}F^{(j)}(t)\\ &+\sum_{j=0}^{\infty}(c_F+jM_1+N_j+jc_N)\left\{\int_0^T[F^{(j)}(t)-F^{(j+1)}(t)]\mathrm{d}D(t)\right.\\ &\left.+\int_0^T\left[\int_{T-t}^{\infty}\overline{F}(u)\mathrm{d}D(t+u)\right]\mathrm{d}F^{(j)}(t)\right\}\end{aligned}$$

$$
= c_F + \left(c_K + \frac{c_0}{\omega}\right) \sum_{j=0}^{\infty} \int_0^T \left[\int_0^{\infty} \overline{D}(t+u)\mathrm{d}F(u) \right] \mathrm{d}F^{(j)}(t)
$$

$$
+ \sum_{j=0}^{\infty} \left[c_R + j\left(c_N + \frac{c_0}{\omega}\right) \right] \int_0^T \left\{ \int_0^{\infty} [D(t+u) - D(t)]\mathrm{d}F(u) \right\} \mathrm{d}F^{(j)}(t),
\tag{8.52}
$$

and the expected cost until full backup for differential backup is

$$
\widetilde{C}_{OD}(T) =
$$

$$
\sum_{j=0}^{\infty} \left(c_F + \sum_{i=1}^{j+1} M_i \right) \int_0^T \left[\int_{T-t}^{\infty} \overline{D}(t+u)\mathrm{d}F(u) \right] \mathrm{d}F^{(j)}(t)
$$

$$
+ \sum_{j=0}^{\infty} \left(c_F + \sum_{i=1}^{j} M_i + N_j \right) \left\{ \int_0^T [F^{(j)}(t) - F^{(j+1)}(t)]\mathrm{d}D(t) \right.
$$

$$
\left. + \int_0^T \left[\int_{T-t}^{\infty} \overline{F}(u)\mathrm{d}D(t+u) \right] \mathrm{d}F^{(j)}(t) \right\}
$$

$$
= c_F + \sum_{j=0}^{\infty} \left[c_K + \frac{(j+1)c_0}{\omega} \right] \int_0^T \left[\int_0^{\infty} \overline{D}(t+u)\mathrm{d}F(u) \right] \mathrm{d}F^{(j)}(t)
$$

$$
+ \sum_{j=0}^{\infty} \left(c_R + \frac{jc_0}{\omega} \right) \int_0^T \left\{ \int_0^{\infty} [D(t+u) - D(t)]\mathrm{d}F(u) \right\} \mathrm{d}F^{(j)}(t). \tag{8.53}
$$

The difference of $\widetilde{C}_{OD}(T)$ and $\widetilde{C}_{OI}(T)$ is

$$
\widetilde{C}_{OD}(T) - \widetilde{C}_{OI}(T) = \frac{c_0}{\omega} \sum_{j=0}^{\infty} j \int_0^T \left[\int_0^{\infty} \overline{D}(t+u)\mathrm{d}F(u) \right] \mathrm{d}F^{(j)}(t)
$$

$$
- c_N \sum_{j=0}^{\infty} j \int_0^T \left\{ \int_0^{\infty} [D(t+u) - D(t)]\mathrm{d}F(u) \right\} \mathrm{d}F^{(j)}(t),
$$

that is, if

$$
\left(c_N + \frac{c_0}{\omega}\right) \sum_{j=0}^{\infty} j \int_0^T \left[\int_0^{\infty} \overline{D}(t+u)\mathrm{d}F(u) \right] \mathrm{d}F^{(j)}(t)
$$

$$
> c_N \sum_{j=0}^{\infty} j \int_0^T \overline{D}(t)\mathrm{d}F^{(j)}(t)
$$

holds, then incremental backup would save more cost than differential backup.

The mean time to full backup is (Problem 8.2)

$$\sum_{j=0}^{\infty}\int_0^T\left[\int_{T-t}^{\infty}(t+u)\overline{D}(t+u)\mathrm{d}F(u)\right]\mathrm{d}F^{(j)}(t)+\int_0^T t\mathrm{d}D(t)$$
$$+\sum_{j=0}^{\infty}\int_0^T\left[\int_{T-t}^{\infty}(t+u)\overline{F}(u)\mathrm{d}D(t+u)\right]\mathrm{d}F^{(j)}(t)$$
$$=\sum_{j=0}^{\infty}\int_0^T\left[\int_0^{\infty}\overline{D}(t+u)\overline{F}(u)\mathrm{d}u\right]\mathrm{d}F^{(j)}(t). \tag{8.54}$$

Therefore, the expected cost rate for incremental backup is

$$C_{OI}(T)=$$
$$\frac{c_F+(c_K+c_O/\omega)\sum_{j=0}^{\infty}[\int_0^{\infty}\overline{D}(t+u)\mathrm{d}F(u)]\mathrm{d}F^{(j)}(t)+\sum_{j=0}^{\infty}[c_R+j(c_N+c_O/\omega)]\int_0^T\{\int_0^{\infty}[D(t+u)-D(t)]\mathrm{d}F(u)\}\mathrm{d}F^{(j)}(t)}{\sum_{j=0}^{\infty}\int_0^T[\int_0^{\infty}\overline{D}(t+u)\overline{F}(u)\mathrm{d}u]\mathrm{d}F^{(j)}(t)}, \tag{8.55}$$

and the expected cost rate for differential backup is

$$C_{OD}(T)=\frac{c_F+\sum_{j=0}^{\infty}[c_K+(j+1)c_O/\omega]\int_0^T[\int_0^{\infty}\overline{D}(t+u)\mathrm{d}F(u)]\mathrm{d}F^{(j)}(t)+\sum_{j=0}^{\infty}(c_R+jc_O/\omega)\int_0^T\{\int_0^{\infty}[D(t+u)-D(t)]\mathrm{d}F(u)\}\mathrm{d}F^{(j)}(t)}{\sum_{j=0}^{\infty}\int_0^T[\int_0^{\infty}\overline{D}(t+u)\overline{F}(u)\mathrm{d}u]\mathrm{d}F^{(j)}(t)}. \tag{8.56}$$

When $F(t)=1-\mathrm{e}^{-\lambda t}$, we next find optimum T_{OI}^* and T_{OD}^* to minimize $C_{OI}(T)$ and $C_{OD}(T)$, respectively. The expected cost rate in (8.55) is

$$C_{OI}(T)=$$
$$\frac{c_F+c_R[D(T)+\int_T^{\infty}\mathrm{e}^{-\lambda(t-T)}\mathrm{d}D(t)]+(c_N+c_O/\omega)[\int_0^T\lambda t\mathrm{d}D(t)+\lambda T\int_T^{\infty}\mathrm{e}^{-\lambda(t-T)}\mathrm{d}D(t)]}{\int_0^T\overline{D}(t)\mathrm{d}t+\int_T^{\infty}\mathrm{e}^{-\lambda(t-T)}\overline{D}(t)\mathrm{d}t}+\lambda\left(c_K+\frac{c_O}{\omega}\right). \tag{8.57}$$

Differentiating $C_{OI}(T)$ with respect to T and setting it equal to zero,

$$\left[c_R + \left(c_N + \frac{c_O}{\omega}\right)(1+\lambda T)\right] \frac{\int_T^\infty \mathrm{e}^{-\lambda t}\mathrm{d}D(t)}{\int_T^\infty \mathrm{e}^{-\lambda t}\overline{D}(t)\mathrm{d}t}$$
$$\times \left[\int_0^T \overline{D}(t)\mathrm{d}t + \int_0^\infty \mathrm{e}^{-\lambda t}\overline{D}(t+T)\mathrm{d}t\right] - c_R\left[D(T) + \int_0^\infty \mathrm{e}^{-\lambda t}\mathrm{d}D(t+T)\right]$$
$$-\left(c_N + \frac{c_O}{\omega}\right)\left[\int_0^T \lambda t\mathrm{d}D(t) + \lambda T\int_0^\infty \mathrm{e}^{-\lambda t}\mathrm{d}D(t+T)\right] = c_F, \tag{8.58}$$

whose left-hand side increases strictly with T to ∞. Therefore, there exists a finite and unique T_{OI}^* $(0 \le T_{OI}^* < \infty)$ which satisfies (8.58) (Problem 8.3), and the resulting cost rate is

$$C_{OI}(T_{OI}^*) = \left[c_R + \left(c_N + \frac{c_O}{\omega}\right)(1+\lambda T_{OI}^*)\right] \frac{\int_{T_{OI}^*}^\infty \mathrm{e}^{-\lambda t}\mathrm{d}D(t)}{\int_{T_{OI}^*}^\infty \mathrm{e}^{-\lambda t}\overline{D}(t)\mathrm{d}t} + \lambda\left(c_K + \frac{c_O}{\omega}\right). \tag{8.59}$$

In particular, when $D(t) = 1 - \mathrm{e}^{-\mu t}$, (8.58) becomes

$$\frac{\mu}{\lambda+\mu} + \lambda T - \frac{\lambda^2}{(\lambda+\mu)\mu}(1 - \mathrm{e}^{-\mu T}) = \frac{c_F}{c_N + c_O/\omega}, \tag{8.60}$$

whose left-hand side increases strictly with T from $\mu/(\lambda+\mu)$ to ∞. Thus, if $[\mu/(\lambda+\mu)] > c_F/(c_N + c_O/\omega)$, then $T_{OI}^* = 0$. Furthermore, comparing the left-hand sides of (8.60) and (8.34), it can be easily shown that $T_{OI}^* < T_I^*$.

The expected cost rate in (8.56) is

$$C_{OD}(T) = \frac{\begin{array}{c} c_F + c_R[D(T) + \int_T^\infty \mathrm{e}^{-\lambda(t-T)}\mathrm{d}D(t)] \\ +(c_O/\omega)\lambda\int_0^T (1+\lambda t)\overline{D}(t)\mathrm{d}t \end{array}}{\int_0^T \overline{D}(t)\mathrm{d}t + \int_T^\infty \overline{D}(t)\mathrm{e}^{-\lambda(t-T)}\mathrm{d}t} + \lambda\left(c_K + \frac{c_O}{\omega}\right). \tag{8.61}$$

Differentiating $C_{OD}(T)$ with respect to T and setting it equal to zero,

$$c_R\left[\frac{\int_T^\infty \mathrm{e}^{-\lambda t}\mathrm{d}D(t)}{\int_T^\infty \overline{D}(t)\mathrm{e}^{-\lambda t}\mathrm{d}t}\int_0^T \overline{D}(t)\mathrm{d}t - D(T)\right]$$
$$+\frac{c_O}{\omega}\left\{\frac{(1+\lambda T)\mathrm{e}^{-\lambda T}\overline{D}(T)}{\int_T^\infty \overline{D}(t)\mathrm{e}^{-\lambda t}\mathrm{d}t}\left[\int_0^T \overline{D}(t)\mathrm{d}t + \int_T^\infty \overline{D}(t)\mathrm{e}^{-\lambda(t-T)}\mathrm{d}t\right]\right.$$
$$\left.-\lambda\int_0^T (1+\lambda t)\overline{D}(t)\mathrm{d}t\right\} = c_F, \tag{8.62}$$

whose left-hand side increase strictly with T to ∞. Thus, there exists a finite and unique T_{OD}^* $(0 \le T_{OD}^* < \infty)$ which satisfies (8.62) (Problem 8.4), and the resulting cost rate is

$$C_{OD}(T_{OD}^*) = \lambda\left(c_K + \frac{c_O}{\omega}\right) + \frac{c_R \int_{T_{OD}^*}^{\infty} \mathrm{e}^{-\lambda t}\mathrm{d}D(t) + (c_O/\omega)(1 + \lambda T_{OD}^*)\mathrm{e}^{-\lambda T_{OD}^*}\overline{D}(T_{OD}^*)}{\int_{T_{OD}^*}^{\infty} \overline{D}(t)\mathrm{e}^{-\lambda t}\mathrm{d}t}. \tag{8.63}$$

In particular, when $D(t) = 1 - \mathrm{e}^{-\mu t}$, (8.62) becomes

$$1 + \lambda T + \frac{\lambda^2 T}{\mu} - \left(\frac{\lambda}{\mu}\right)^2 (1 - \mathrm{e}^{-\mu T}) = \frac{c_F}{c_O/\omega}, \tag{8.64}$$

whose left-hand side increases strictly with T from 1 to ∞. Thus, if $c_O/\omega \geq c_F$, then $T_{OD}^* = 0$. Furthermore, comparing the left-hand sides of (8.64) and (8.42), it can be shown that $T_{OD}^* < T_D^*$.

Table 8.6 presents λT_{OI}^* in (8.60), λT_{OD}^* in (8.64) and their cost rates for λ/μ and c_O/ω when $c_K = 0.1$, $c_R = 0.2$, $c_N = 0.5$ and $c_F = 10.0$. It shows that $T_{OI}^* \geq T_{OD}^*$ and $C_{OI}(T_{OI}^*) \leq C_{OD}(T_{OD}^*)$. Comparing with Table 8.4, $T_{OI}^* < T_I^*$ and $T_{OD}^* < T_D^*$, and $C_{OI}(T_{OI}^*) \leq C_I(T_I^*)$ and $C_{OD}(T_{OD}^*) \leq C_D(T_D^*)$.

8.2.4 Backup Overtime First and Last

We finally obtain the expected cost rates for incremental and differential backup overtime first and last policies without discussing their optimum full backup times.

(1) Backup Overtime First

Suppose that a full backup is scheduled preventively at the completion of the forthcoming update over time T ($0 \leq T \leq \infty$) or at update N ($N = 1, 2, \cdots$), whichever occurs first. Then, the probability that the full backup is implemented over time T is

$$\sum_{j=0}^{N-1} \int_0^T \left[\int_{T-t}^{\infty} \overline{D}(t+u)\mathrm{d}F(u)\right] \mathrm{d}F^{(j)}(t),$$

the probability that it is implemented at update N is

$$\int_0^T \overline{D}(t)\mathrm{d}F^{(N)}(t),$$

and the probability that it is implemented at breakdown is

$$\int_0^T [1 - F^{(N)}(t)]\mathrm{d}D(t) + \sum_{j=0}^{N-1} \int_0^T \left[\int_{T-t}^{\infty} \overline{F}(u)\mathrm{d}D(t+u)\right] \mathrm{d}F^{(j)}(t).$$

Table 8.6 Optimum λT_{OI}^*, λT_{OD}^* and their cost rates when $c_K = 0.1$, $c_R = 0.2$, $c_N = 0.5$ and $c_F = 10.0$

λ/μ	$c_0/\omega = 0.5$				$c_0/\omega = 1.0$			
	λT_{OI}^*	$C_{OI}(T_{OI}^*)/\mu$	λT_{OD}^*	$C_{OD}(T_{OD}^*)/\mu$	λT_{OI}^*	$C_{OI}(T_{OI}^*)/\mu$	λT_{OD}^*	$C_{OD}(T_{OD}^*)/\mu$
1.0	10.001	11.800	10.001	11.800	6.666	12.799	4.997	13.293
2.0	10.995	13.395	7.637	14.356	7.637	15.356	4.167	17.900
5.0	13.733	17.933	6.104	24.511	10.116	22.375	3.664	33.685
10.0	17.405	24.605	5.653	42.791	13.250	32.576	3.507	60.781
20.0	22.955	36.155	5.443	79.850	17.874	50.511	3.433	115.287
50.0	34.328	65.528	5.323	191.425	27.234	97.552	3.389	279.016

Therefore, the expected cost until full backup for incremental backup is

$$
\begin{aligned}
&\widetilde{C}_{OIF}(T,N) = \\
&\sum_{j=0}^{N-1}[c_F + (j+1)M_1]\int_0^T\left[\int_{T-t}^{\infty}\overline{D}(t+u)\mathrm{d}F(u)\right]\mathrm{d}F^{(j)}(t) \\
&+ (c_F + NM_1)\int_0^T\overline{D}(t)\mathrm{d}F^{(N)}(t) \\
&+\sum_{j=0}^{N-1}(c_F + jM_1 + N_j + jc_N)\left\{\int_0^T[F^{(j)}(t) - F^{(j+1)}(t)]\mathrm{d}D(t)\right. \\
&\left.+\int_0^T\left[\int_{T-t}^{\infty}\overline{F}(u)\mathrm{d}D(t+u)\right]\mathrm{d}F^{(j)}(t)\right\} \\
&= c_F + \left(c_K + \frac{c_0}{\omega}\right)\sum_{j=0}^{N-1}\int_0^T\left[\int_0^{\infty}\overline{D}(t+u)\overline{F}(u)\right]\mathrm{d}F^{(j)}(t) \\
&+\sum_{j=0}^{N-1}\left[c_R + j\left(c_N + \frac{c_0}{\omega}\right)\right]\int_0^T\left\{\int_0^{\infty}[D(t+u) - D(t)]\mathrm{d}F(u)\right\}\mathrm{d}F^{(j)}(t),
\end{aligned}
\tag{8.65}
$$

and the expected cost until full backup for differential backup is

$$
\begin{aligned}
&\widetilde{C}_{ODF}(T,N) = \\
&\sum_{j=0}^{N-1}\left(c_F + \sum_{i=1}^{j+1}M_i\right)\int_0^T\left[\int_{T-t}^{\infty}\overline{D}(t+u)\mathrm{d}F(u)\right]\mathrm{d}F^{(j)}(t) \\
&+\left(c_F + \sum_{i=1}^{N}M_i\right)\int_0^T\overline{D}(t)\mathrm{d}F^{(N)}(t) \\
&+\sum_{j=0}^{N-1}\left(c_F + \sum_{i=1}^{j}M_i + N_j\right)\left\{\int_0^T[F^{(j)}(t) - F^{(j+1)}(t)]\mathrm{d}D(t)\right. \\
&\left.+\int_0^T\left[\int_{T-t}^{\infty}\overline{F}(u)\mathrm{d}D(t+u)\right]\mathrm{d}F^{(j)}(t)\right\} \\
&= c_F + \sum_{j=0}^{N-1}\left[c_K + \frac{(j+1)c_0}{\omega}\right]\int_0^T\left[\int_0^{\infty}\overline{D}(t+u)\mathrm{d}F(u)\right]\mathrm{d}F^{(j)}(t) \\
&+\sum_{j=0}^{N-1}\left(c_R + \frac{jc_0}{\omega}\right)\int_0^T\left\{\int_0^{\infty}[D(t+u) - D(t)]\mathrm{d}F(u)\right\}\mathrm{d}F^{(j)}(t).
\end{aligned}
\tag{8.66}
$$

The mean time to full backup is (Problem 8.5)

$$\sum_{j=0}^{N-1}\int_0^T\left[\int_0^\infty \overline{D}(t+u)\overline{F}(u)\mathrm{d}u\right]\mathrm{d}F^{(j)}(t). \tag{8.67}$$

Therefore, the expected cost rate for incremental backup is

$$C_{OIF}(T,N) = \frac{\begin{array}{l} c_F + (c_K + c_O/\omega)\sum_{j=0}^{N-1}\int_0^T[\int_0^\infty \overline{D}(t+u)\mathrm{d}F(u)]\mathrm{d}F^{(j)}(t) \\ +\sum_{j=0}^{N-1}[c_R + j(c_N + c_O/\omega)]\int_0^T\{\int_0^\infty[D(t+u)-D(t)]\mathrm{d}F(u)\}\mathrm{d}F^{(j)}(t)\end{array}}{\sum_{j=0}^{N-1}\int_0^T[\int_0^\infty \overline{D}(t+u)\overline{F}(u)\mathrm{d}u]\mathrm{d}F^{(j)}(t)}, \tag{8.68}$$

and the expected cost rate for differential backup is

$$C_{ODF}(T,N) = \frac{\begin{array}{l} c_F + \sum_{j=0}^{N-1}[c_K + (j+1)c_O/\omega]\int_0^T[\int_0^\infty \overline{D}(t+u)\mathrm{d}F(u)]\mathrm{d}F^{(j)}(t) \\ +\sum_{j=0}^{N-1}(c_R + jc_O/\omega)\int_0^T\{\int_0^\infty[D(t+u)-D(t)]\mathrm{d}F(u)\}\mathrm{d}F^{(j)}(t)\end{array}}{\sum_{j=0}^{N-1}\int_0^T[\int_0^\infty \overline{D}(t+u)\overline{F}(u)\mathrm{d}u]\mathrm{d}F^{(j)}(t)}. \tag{8.69}$$

(2) Backup Overtime Last

Suppose that a full backup is scheduled preventively at the completion of the forthcoming update over time T $(0 \le T < \infty)$ or at update N $(N = 1, 2, \cdots)$, whichever occurs last. Then, the probability that the full backup is implemented over time T is

$$\sum_{j=N}^{\infty}\int_0^T\left[\int_{T-t}^\infty \overline{D}(t+u)\mathrm{d}F(u)\right]\mathrm{d}F^{(j)}(t),$$

the probability that it is implemented at update N is

$$\int_T^\infty \overline{D}(t)\mathrm{d}F^{(N)}(t),$$

and the probability that it is implemented at breakdown is

$$D(T) + \int_T^\infty[1-F^{(N)}(t)]\mathrm{d}D(t) + \sum_{j=N}^{\infty}\int_0^T\left[\int_{T-t}^\infty \overline{F}(u)\mathrm{d}D(t+u)\right]\mathrm{d}F^{(j)}(t).$$

Therefore, the expected cost until full backup for incremental backup is

$$
\begin{aligned}
\widetilde{C}_{OIL} &= \sum_{j=N}^{\infty}[c_F + (j+1)M_1]\int_0^T \left[\int_{T-t}^{\infty} \overline{D}(t+u)\mathrm{d}F(u)\right]\mathrm{d}F^{(j)}(t) \\
&+ (c_F + NM_1)\int_T^{\infty} \overline{D}(t)\mathrm{d}F^{(N)}(t) \\
&+ \sum_{j=0}^{N-1}(c_F + jM_1 + N_j + jc_N)\int_0^{\infty}[F^{(j)}(t) - F^{(j+1)}(t)]\mathrm{d}D(t) \\
&+ \sum_{j=N}^{\infty}(c_F + jM_1 + N_j + jc_N)\int_0^T \left[\int_0^{\infty} \overline{F}(t)\mathrm{d}D(t+u)\right]\mathrm{d}F^{(j)}(t) \\
&= c_F + \left(c_K + \frac{c_0}{\omega}\right)\left\{\sum_{j=1}^{N}\int_0^{\infty}\overline{D}(t)\mathrm{d}F^{(j)}(t)\right. \\
&\left. + \sum_{j=N}^{\infty}\int_0^T \left[\int_0^{\infty}\overline{D}(t+u)\mathrm{d}F(u)\right]\mathrm{d}F^{(j)}(t)\right\} \\
&+ \sum_{j=N}^{\infty}\left[c_R + j\left(c_N + \frac{c_0}{\omega}\right)\right]\left(\int_0^{\infty}[1 - F^{(N)}(t)]\mathrm{d}D(t)\right. \\
&\left. + \int_0^T \left\{\int_0^{\infty}[D(t+u) - D(t)]\mathrm{d}F(u)\right\}\mathrm{d}F^{(j)}(t)\right),
\end{aligned} \tag{8.70}
$$

and the expected cost until full backup for differential backup is

$$
\begin{aligned}
\widetilde{C}_{ODL}(T, N) &= \sum_{j=N}^{\infty}\left(c_F + \sum_{i=1}^{j+1} M_i\right)\sum_{j=N}^{\infty}\int_0^T \left[\int_{T-t}^{\infty}\overline{D}(t+u)\mathrm{d}F(u)\right]\mathrm{d}F^{(j)}(t) \\
&+ \left(c_F + \sum_{i=1}^{N} M_i\right)\int_T^{\infty}\overline{D}(t)\mathrm{d}F^{(N)}(t) \\
&+ \sum_{j=N}^{\infty}\left(c_F + \sum_{i=1}^{j} M_i + N_j\right)\left\{\int_0^{\infty}[F^{(j)}(t) - F^{(j+1)}(t)]\mathrm{d}D(t)\right. \\
&\left. + \int_0^T \left[\int_0^{\infty}\overline{F}(u)\mathrm{d}D(t+u)\right]\mathrm{d}F^{(j)}(t)\right\} \\
&= c_F + \sum_{j=1}^{N-1}\left[c_K + \frac{(j+1)c_0}{\omega}\right]\int_0^{\infty}\overline{D}(t)\mathrm{d}F^{(j)}(t) \\
&+ \sum_{j=N}^{\infty}\left[c_K + \frac{(j+1)c_0}{\omega}\right]\int_0^T \left[\overline{D}(t+u)\mathrm{d}F(u)\right]\mathrm{d}F^{(j)}(t)
\end{aligned}
$$

$$+\sum_{j=N}^{\infty}\left(c_R+\frac{jc_O}{\omega}\right)\left(\int_0^{\infty}[1-F^{(N)}(t)]\mathrm{d}D(t)\right.$$
$$\left.+\int_0^T\left\{\int_0^{\infty}[D(t+u)-D(t)]\mathrm{d}F(u)\right\}\mathrm{d}F^{(j)}(t)\right). \tag{8.71}$$

The mean time to full backup is (Problem 8.5)

$$\int_0^{\infty}\overline{D}(t)[1-F^{(N)}(t)]\mathrm{d}t+\sum_{j=N}^{\infty}\int_0^T\left[\int_0^{\infty}\overline{D}(t+u)\overline{F}(u)\mathrm{d}u\right]\mathrm{d}F^{(j)}(t). \tag{8.72}$$

Therefore, the expected cost rate for incremental backup is

$$C_{OIL}(T,N)=$$
$$\frac{\begin{array}{l}c_F+(c_K+c_O/\omega)\{\sum_{j=1}^{N}\int_0^{\infty}\overline{D}(t)\mathrm{d}F^{(j)}(t)\\ +\sum_{j=N}^{\infty}\int_0^T[\int_0^{\infty}\overline{D}(t+u)\mathrm{d}F(u)]\mathrm{d}F^{(j)}(t)\}\\ +\sum_{j=N}^{\infty}[c_R+j(c_N+c_O/\omega)](\int_0^{\infty}[1-F^{(N)}(t)]\mathrm{d}D(t)\\ +\int_0^T\{\int_0^{\infty}[D(t+u)-D(t)]\mathrm{d}F(u)\}\mathrm{d}F^{(j)}(t))\end{array}}{\int_0^{\infty}\overline{D}(t)[1-F^{(N)}(t)]\mathrm{d}t+\sum_{j=N}^{\infty}\int_0^T[\int_0^{\infty}\overline{D}(t+u)\overline{F}(u)\mathrm{d}u]\mathrm{d}F^{(j)}(t)}, \tag{8.73}$$

and the expected cost rate for differential backup is

$$C_{ODL}(T,N)=$$
$$\frac{\begin{array}{l}c_F+\sum_{j=1}^{N-1}[c_K+(j+1)c_O/\omega]\int_0^{\infty}\overline{D}(t)\mathrm{d}F^{(j)}(t)\\ +\sum_{j=N}^{\infty}[c_K+(j+1)c_O/\omega]\int_0^T[\int_0^{\infty}\overline{D}(t+u)\mathrm{d}F(u)]\mathrm{d}F^{(j)}(t)\\ +\sum_{j=N}^{\infty}(c_R+jc_O/\omega)(\int_0^{\infty}[1-F^{(N)}(t)]\mathrm{d}D(t)\\ \int_0^T\{\int_0^{\infty}[D(t+u)-D(t)]\mathrm{d}F(u)\}\mathrm{d}F^{(j)}(t))\end{array}}{\int_0^{\infty}\overline{D}(t)[1-F^{(N)}(t)]\mathrm{d}t+\sum_{j=N}^{\infty}\int_0^T\{\int_0^{\infty}\overline{D}(t+u)\overline{F}(u)\mathrm{d}u\}\mathrm{d}F^{(j)}(t)}. \tag{8.74}$$

Clearly,

$$C_{OIL}(0,N)=C_{OIF}(\infty,N)=C_I(N),$$
$$C_{OIL}(T,0)=C_{OIF}(T,\infty)=C_{OI}(T)$$

in (8.36) and (8.55), respectively, and

$$C_{ODL}(0, N) = C_{ODF}(\infty, N) = C_D(N),$$
$$C_{ODL}(T, 0) = C_{ODF}(T, \infty) = C_{OD}(T)$$

in (8.44) and (8.56), respectively (Problem 8.6).

8.3 Problem 8

8.1 Derive (8.21) and (8.23).
8.2 Derive (8.54).
8.3 Prove that there exists a finite and unique T_{OI}^* which satisfies (8.58).
8.4 Prove that there exists a finite and unique T_{OD}^* which satisfies (8.62).
8.5 Derive (8.67) and (8.72).
8.6 Discuss optimum policies to minimize the expected cost rates in (8.68), (8.69), (8.73) and (8.74).

Chapter 9
Other Maintenance Models

In order to conduct safe and economical maintenance strategy, modeling and analysis of the damage due to shocks in an analytical way plays an important role in reliability theory and engineering. The damage models have been studied for decades, some of which were summarized in the literature [2] and have been discussed extensively in the above chapters. However, it is troublesome for engineers to predict reliability measures and conduct maintenance schedules exactly when the stochastic damage models are provided, as it is almost impossible to monitor the adequate number of shocks and their respective damages. Without considering the cumulative process of damage in mathematical models, the engineers can use a *periodic damage model* to formulate preventive replacement policies done at times jT $(j = 1, 2, \cdots)$ for a specified $T > 0$, which will be addressed in Sect. 9.1.

It has been widely recognized that maintenance cannot make the unit like new but younger, i.e., maintenance should be imperfect [1]. The general approach of *imperfect maintenance* modeling is to suppose that the age or failure rate after maintenance reduces in proportion to that before maintenance [62–65]. An asymptotic model for sequential imperfect maintenance, using the cumulative hazard function in a simple way, was proposed [66]. In Sect. 9.2, an improvement factor a $(0 < a < 1)$ and sequential improvement factors a_k $(k = 1, 2, \cdots, N-1)$ [67] will be introduced to reduce the total damage at periodic and sequential maintenance times.

The *continuum damage mechanics* was applied to predict crack initiations in structures subjected to heavy loadings [68]. In Sect. 9.3, we suppose that the damage stored in an operating unit increases continuously and swayingly with time at a stochastic path, and a *continuous damage model* $W(t) = A_t t + B_t$ $(A_t \geq 0)$ is given [3, 69]. Although, it is by no means certain that the damage would follow the stochastic path of the model in an exact way, we give five possible situations of models when different parameters are supposed. The reliabilities functions are obtained and the models of replacement first, last, and overtime, and sequential inspection times are surveyed.

X. Zhao and T. Nakagawa, *Advanced Maintenance Policies for Shock and Damage Models*, Springer Series in Reliability Engineering,
https://doi.org/10.1007/978-3-319-70456-2_9

We have supposed in above discussions that the damage K is a failure threshold of the mechanical strength of an operating unit, under which case, only corrective replacement that costs much can be done. When a preventive maintenance or is conducted, its damage accumulated by shocks should be less than K, e.g., Z discussed above. Taking the airframe cracks as an example [25], there exist several levels of damages for maintenances and replacement, e.g., damages Z_i $(i = 1, 2, \cdots, n-1)$ for maintenances and damage Z_n for replacement. In Sect. 9.4, a simple Markov chain model with three states of damages and a general Markov chain model with n states of damages for inspection, maintenance, and replacement are obtained.

We give compactly the above damage models and their maintenance polices in the following sections, some of which are discussed to find optimum policies analytically and numerically.

9.1 Periodic Damage Models

The damage produced by shocks is measured exactly at periodic times jT $(j = 1, 2, \cdots)$ for a specified $T > 0$, and each amount W_j of damage between $[(j-1)T, jT]$ has an identical distribution $G(x) \equiv \Pr\{W_j \leq x\}$ [2]. An operating unit, degrading with the accumulated damage, fails when the total amount of damage exceeds a failure threshold K $(0 < K < \infty)$ and corrective replacement is done immediately. If shocks which occur randomly in Chap. 2 are supposed to arrive at periodic times jT, i.e., $F(0) \equiv 0$ for $t < T$ and $\equiv 1$ for $t \geq T$, replacement policies of the damage models discussed in the above chapters can be modified to those of the periodic damage models.

9.1.1 Standard Replacement Policies

(1) Replacement First

Suppose that the unit is replaced preventively at time NT $(N = 1, 2, \cdots)$ or at damage Z $(0 < Z \leq K)$, whichever occurs first. Then, replacing μ with T in (2.30), the expected cost rate is (Problem 9.1)

$$C_F(N, Z) = \frac{\begin{array}{c} c_Z + (c_N - c_Z)G^{(N)}(Z) \\ +(c_K - c_Z)\sum_{j=0}^{N-1}\int_0^Z \overline{G}(K-x)\mathrm{d}G^{(j)}(x) \end{array}}{T\sum_{j=0}^{N-1} G^{(j)}(Z)}, \tag{9.1}$$

where c_N = replacement cost at time NT, c_Z = replacement cost at damage Z, and c_K = replacement cost at failure K, where $c_K > c_N$ and $c_K > c_Z$. Optimum policies (N_F^*, Z_F^*) to minimize $C_F(N, Z)$ have been obtained in **(3)** of Sect. 2.1.2 when $c_N = c_Z$.

(2) Replacement Last

Suppose that the unit is replaced preventively at time NT $(N = 0, 1, 2, \cdots)$ or at damage Z $(0 \le Z \le K)$, whichever occurs last. Then, replacing μ with T in (3.15), the expected cost rate is

$$C_L(N, Z) = \frac{\begin{array}{c} c_Z + (c_N - c_Z)[G^{(N)}(K) - G^{(N)}(Z)] \\ +(c_K - c_Z)\{\sum_{j=N}^{\infty} \int_0^Z \overline{G}(K - x)\mathrm{d}G^{(j)}(x) + [1 - G^{(N)}(K)]\} \end{array}}{T[\sum_{j=N}^{\infty} G^{(j)}(Z) + \sum_{j=0}^{N-1} G^{(j)}(K)]}. \tag{9.2}$$

Optimum policies (N_L^*, Z_L^*) to minimize $C_L(N, Z)$ have been obtained in **(3)** of Sect. 3.2 when $c_N = c_Z$.

(3) Replacement Overtime

Suppose that the unit is replaced preventively at the next periodic time over damage Z. Then, replacing μ with T in (4.12), the expected cost rate is

$$C_O(Z) = \frac{c_K - (c_K - c_O) \sum_{j=0}^{\infty} \int_0^Z [\int_{Z-x}^{K-x} G(K - x - y)\mathrm{d}G(y)]\mathrm{d}G^{(j)}(x)}{T[1 + \sum_{j=0}^{\infty} \int_0^Z G(K - x)\mathrm{d}G^{(j)}(x)]}, \tag{9.3}$$

where c_O = replacement cost over damage Z. Optimum policy Z_O^* to minimize $C_O(Z)$ has been obtained in **(2)** of Sect. 4.1.1.

9.1.2 Replacement with Repairs

The unit fails with probability $p(x)$ when the total damage reaches x at each periodic time jT $(j = 1, 2, \cdots)$, and can be quickly resumed to operation after minimal repair at failure, where the function $p(x)$ increases strictly with damage x from $p(0) = 0$. We modify the models in Chap. 5 as follows:

(1) Replacement First

Suppose that the unit is replaced preventively at time NT $(N = 1, 2, \cdots)$ or at damage Z $(0 < Z \le K)$, whichever occurs first. Then, putting that $T \to \infty$ and replacing μ with T in (5.6), the expected cost rate is

$$C_F(N, Z) = \frac{c_Z - (c_Z - c_N)G^{(N)}(Z) + c_M \sum_{j=0}^{N-1} \int_0^Z p(x)\mathrm{d}G^{(j)}(x)}{T \sum_{j=0}^{N-1} G^{(j)}(Z)}, \tag{9.4}$$

where c_M = cost for minimal repair at each failure. Optimum policies (N_F^*, Z_F^*) to minimize $C_F(N, Z)$ have been obtained in **(3)** of Sect. 5.1.2 when $c_N = c_Z$.

(2) Replacement Last

Suppose that the unit is replaced preventively at time NT $(N = 0, 1, 2, \cdots)$ or at damage Z $(0 \leq Z \leq K)$, whichever occurs last. Then, putting that $T \to 0$ and replacing μ with T in (5.37), the expected cost rate is

$$C_L(N, Z) = \frac{\begin{array}{l} c_N + (c_Z - c_N)G^{(N)}(Z) \\ +c_M[\sum_{j=0}^{\infty}\int_0^{\infty} p(x)\mathrm{d}G^{(j)}(x) - \sum_{j=N}^{\infty}\int_Z^{\infty} p(x)\mathrm{d}G^{(j)}(x)] \end{array}}{T[N + \sum_{j=N}^{\infty} G^{(j)}(Z)]}. \tag{9.5}$$

Optimum policies (N_L^*, Z_L^*) to minimize $C_L(N, Z)$ have been obtained in **(3)** of Sect. 5.2.1 when $c_N = c_Z$.

(3) Replacement Overtime

Suppose that the unit is replaced preventively at the next periodic time over damage Z. Then, the expected cost rate is (Problem 9.2)

$$C_O(Z) = \frac{c_O + c_M \sum_{j=0}^{\infty} \int_0^Z [\int_0^{\infty} p(x+y)\mathrm{d}G(y)]\mathrm{d}G^{(j)}(x)}{T\sum_{j=0}^{\infty} G^{(j)}(Z)}, \tag{9.6}$$

where c_O = replacement cost over damage Z.

9.1.3 Replacement with Maintenances

The unit fails when the total damage has exceeded a failure threshold K and can be quickly resumed to operation with undisturbed damage K by reactive maintenances at the following periodic times $(j+1)T$. We modify the models in Chap. 6 as follows:

(1) Replacement at Periodic Time

Suppose that the unit is replaced preventively at time NT $(N = 1, 2, \cdots)$. Then, replacing μ with T in (6.7), the expected cost rate is

$$C(N) = \frac{c_N + c_M \sum_{j=1}^{N}[1 - G^{(j)}(K)]}{NT}, \tag{9.7}$$

where c_M = cost for reactive maintenance. Optimum policy N^* to minimize $C(N)$ has been obtained in Sect. 6.1.1.

(2) Replacement at Failure Number

Suppose that the unit is replaced preventively at failure M $(M = 1, 2, \cdots)$. Then, replacing μ with T in (6.43), the expected cost rate is

$$C(M) = \frac{c_F + c_M M}{T[M + \sum_{j=1}^{\infty} G^{(j)}(K)]}, \tag{9.8}$$

where c_F = replacement cost at failure M. Optimum policy to minimize $C(M)$ is $M^* = \infty$.

9.1.4 Additive and Independent Damages

We introduce independent damage defined in Chap. 7 into periodic damage model: The independent damage occurs at a nonhomogeneous Poisson process with an intensity function $h(t)$ and a mean value function $H(t) \equiv \int_0^t h(u)\mathrm{d}t$, and the probability that a number j of independent damages occurs exactly in $[0, t]$ is $p_j(t) = [H(t)^j/j!]\mathrm{e}^{-H(t)}$ $(j = 0, 1, 2, \cdots)$, where denote that $P_j(t) \equiv \sum_{i=j}^{\infty} p_i(t)$ and $\overline{P}_j(t) \equiv 1 - P_j(t)$.

Suppose that minimal repair is made for the independent damage to let the unit return to operation, and the unit is replaced at time NT $(N = 1, 2, \cdots)$ or at failure K, whichever occurs first. Then, the probability that the unit is replaced at time NT is $G^{(N)}(K)$, and the probability that it is replaced at failure is $[1 - G^{(N)}(K)]$. Thus, the mean time to replacement is

$$NTG^{(N)}(K) + \sum_{j=0}^{N-1}(j+1)T[G^{(j)}(K) - G^{(j+1)}(K)] = T\sum_{j=0}^{N-1} G^{(j)}(K), \tag{9.9}$$

and the expected number of independent damages until replacement is

$$H(NT)G^{(N)}(K) + \sum_{j=0}^{N-1} H[(j+1)T][G^{(j)}(K) - G^{(j+1)}(K)]$$
$$= \sum_{j=0}^{N-1}\{H[(j+1)T] - H(jT)\}G^{(j)}(K). \tag{9.10}$$

Therefore, the expected cost rate is

$$C_I(N) = \frac{c_K - (c_K - c_N)G^{(N)}(K) + c_M \sum_{j=0}^{N-1}\{H[(j+1)T] - H(jT)\}G^{(j)}(K)}{T\sum_{j=0}^{N-1} G^{(j)}(K)}\}, \tag{9.11}$$

where c_M = cost for minimal repair at each independent damage.

We find optimum N^* to minimize $C_I(N)$. Forming the inequality $C_I(N+1) - C_I(N) \geq 0$,

$$(c_K - c_N)\left[r_{N+1}(K)\sum_{j=0}^{N-1} G^{(j)}(K) + G^{(N)}(K)\right]$$

$$+ c_M \sum_{j=0}^{N-1} \{H[(N+1)T] - H(NT) - H[(j+1)T] + H(jT)\} G^{(j)}(K) \geq c_K, \tag{9.12}$$

where

$$r_{N+1}(K) \equiv \frac{G^{(N)}(K) - G^{(N+1)}(K)}{G^{(N)}(K)} \quad (N = 0, 1, 2, \cdots),$$

which was defined in (2.14). Therefore, if either $r_{N+1}(x)$ or $h(t)$ increases strictly, then the left-hand side $L(N)$ of (9.12) increases strictly to $L(\infty)$. Therefore, if $L(\infty) > c_K$, then there exists a finite and unique minimum N_I^* $(1 \leq N_I^* < \infty)$ which satisfies (9.12). In particular, when $H(t) = \lambda t$, (9.12) becomes

$$r_{N+1}(K) \sum_{j=0}^{N-1} G^{(j)}(K) - [1 - G^{(N)}(K)] \geq \frac{c_N}{c_K - c_N},$$

which agrees with (2.14), and in this case, $N_I^* = N^*$ in (2.14).

9.2 Imperfect Preventive Maintenance Policies

We summarize imperfect preventive maintenance (PM) policies for cumulative damage modes [2, 67, 69].

9.2.1 *Periodic Preventive Maintenance*

Imperfect preventive maintenance is made at periodic times jT $(j = 1, 2, \cdots)$, and the total damage becomes aZ_j with an improvement factor $(0 < a < 1)$ when it was Z_j at jT before maintenance, i.e., an amount $(1 - a)Z_j$ of damage is reduced by the jth maintenance. Then, the total damage Z_j at jT before maintenance is

$$Z_j = \sum_{i=1}^{j} a^{j-i} W_i \quad (j = 1, 2, \cdots). \tag{9.13}$$

Noting that $G_j(x) \equiv \Pr\{a^{j-i}W_i \leq x\} = G(x/a^{j-i})$, Z_j has a distribution $G_a^{(j)}(x) = \Pr\{Z_j \leq x\} = G_1(x) * \cdots * G_j(x)$, where the asterisk denotes the Stieltjes convolution. In particular, when $a = 1$, there is no maintenance that should be made, and $G_a^{(j)}(x) = G^{(j)}(x)$ which represents the j-fold Stieltjes convolution of $G(x)$ with itself.

When $G(x) = 1 - \mathrm{e}^{-\omega x}$, $G_i(x) = 1 - \mathrm{e}^{-\omega x/a^{j-i}}$ $(i = 1, 2, \cdots, j)$. Thus,

$$G_a^{(j)}(x) = G_1(x) * G_2(x) * \cdots * G_j(x) = \sum_{i=1}^{j} \frac{1 - \mathrm{e}^{-\omega x/a^{i-1}}}{\prod_{k=1,k\neq i}^{j}(1 - a^{k-i})} \quad (j = 1, 2, \cdots), \tag{9.14}$$

where $\prod_{k=1,k\neq i}^{1} \equiv 1$ (Problem 9.3).

(1) Replacement First

Suppose that the unit is replaced preventively at time NT $(N = 1, 2, \cdots)$ or at damage Z $(0 < Z \leq K)$, whichever occurs first. Then, replacing $G^{(j)}(x)$ with $G_a^{(j)}(x)$ and μ with T in (9.1), the expected cost rate is

$$C_F(N, Z; a) = \frac{\begin{array}{c} c_Z + (c_N - c_Z)G_a^{(N)}(Z) \\ +(c_K - c_Z)\sum_{j=0}^{N-1}\int_0^Z \overline{G}_a(K - x)\mathrm{d}G_a^{(j)}(x) \end{array}}{T\sum_{j=0}^{N-1} G_a^{(j)}(Z)}. \tag{9.15}$$

When the unit is replaced preventively only at time NT, the expected cost rate is, putting that $Z \to K$ in (9.15),

$$C_F(N; a) = \frac{c_K - (c_K - c_N)G_a^{(N)}(K)}{T\sum_{j=0}^{N-1} G_a^{(j)}(K)} \quad (N = 1, 2, \cdots), \tag{9.16}$$

which agrees with (2.13), and optimum N_a^* to minimize (9.16) is given in (2.14) by replacing $G^{(j)}(x)$ with $G_a^{(j)}(x)$. Thus, if

$$r_{N+1}(x) = \frac{G_a^{(N)}(x) - G_a^{(N+1)}(x)}{G_a^{(N)}(x)}$$

increases strictly with N to 1 and $M_a(K) \equiv \sum_{j=1}^{\infty} G_a^{(j)}(K) > c_N/(c_F - c_N)$, then there exists a finite and unique minimum N_a^* $(1 \leq N_a^* < \infty)$ to minimize $C_F(N; a)$ in (9.16).

Table 9.1 presents optimum N_a^* and its cost rate $TC_F(N_a^*; a)/c_N$ for a and ωK when $c_K/c_N = 5$. In this case, when $a = 1$, $N_a^* = 3, 6$ for $\omega K = 5, 10$, respectively, from Table 2.2. Optimum N_a^* increases with ωK and decreases with a to N^* given in Table 2.2, which is easily understandable from the physical meanings of ωK and a.

Table 9.1 Optimum N_a^* and its cost rate $TC_F(N_a^*; a)/c_N$ when $c_K/c_N = 5$

ωK	$a = 0.8$		$a = 0.85$		$a = 0.9$		$a = 0.95$	
	N_a^*	$TC_F(N_a^*; a)/c_N$	N_a^*	$TC_F(N_a^*; a)/c_N$	N_a^*	$TC_F(N_a^*; a)/c_N$	N_a^*	$TC_F(N_a^*; a)/c_N$
1.0	3	2.311	2	2.341	2	2.368	2	2.393
2.0	3	1.272	2	1.361	2	1.347	2	1.378
3.0	4	0.760	3	0.891	3	0.872	2	0.898
4.0	∞	0.191	4	0.546	3	0.595	3	0.634
5.0	∞	0.089	6	0.377	4	0.433	3	0.480
6.0	∞	0.050	∞	0.128	5	0.326	4	0.370
7.0	∞	0.033	∞	0.072	6	0.253	5	0.298
8.0	∞	0.026	∞	0.045	8	0197	6	0.246
9.0	∞	0.023	∞	0.032	11	0.154	7	0.206
10.0	∞	0.021	∞	0.026	∞	0.078	8	0.175

Next, when the unit is replaced preventively only at damage Z, the expected cost rate is, putting that $N \to \infty$ in (9.15),

$$C_F(Z; a) = \frac{c_K - (c_K - c_Z)[G_a(K) - \int_0^Z \overline{G}_a(K-x)\mathrm{d}M_a(x)]}{T[1 + M_a(Z)]}, \tag{9.17}$$

which agrees with (2.16), and optimum Z_a^* to minimize (9.17) is given in (2.17) by replacing $G^{(j)}(x)$ with $G_a^{(j)}(x)$. Thus, if $M_a(K) > c_Z/(c_K - c_Z)$, then there exists a finite and unique Z_a^* $(0 < Z_a^* < K)$ to minimize $C_F(Z; a)$.

(2) Replacement Last

Suppose that the unit is replaced preventively at time NT $(N = 1, 2, \cdots)$ or at damage Z $(0 < Z \le K)$, whichever occurs last. Then, replacing $G^{(j)}(x)$ with $G_a^{(j)}(x)$ in (9.2), the expected cost rate is

$$C_L(N, Z; a) = \frac{\begin{array}{l} c_Z + (c_N - c_Z)[G_a^{(N)}(K) - G_a^{(N)}(Z)] \\ +(c_K - c_Z)[\sum_{j=N}^{\infty} \int_0^Z \overline{G}_a(K-x)\mathrm{d}G_a^{(j)}(x) + 1 - G_a^{(N)}(K)] \end{array}}{T[\sum_{j=N}^{\infty} G_a^{(j)}(Z) + \sum_{j=0}^{N-1} G_a^{(j)}(K)]}. \tag{9.18}$$

Clearly, $\lim_{Z \to K} C_F(N, Z; a) = \lim_{Z \to 0} C_L(N, Z; a)$ and $\lim_{N \to \infty} C_F(N, Z; a) = \lim_{N \to 0} C_L(N, Z; a)$.

(3) Replacement Overtime

Suppose that the unit is replaced preventively at the next periodic time over damage Z $(0 < Z \le K)$. Then, replacing $G^{(j)}(x)$ with $G_a^{(j)}(x)$ in (9.3), the expected cost rate is

$$C_O(Z; a) = \frac{c_K - (c_K - c_O)\sum_{j=0}^{\infty} \int_0^Z [\int_{Z-x}^{K-x} G_a(K-x-y)\mathrm{d}G_a(y)]\mathrm{d}G_a^{(j)}(x)}{T[1 + \sum_{j=0}^{\infty} \int_0^Z G_a(K-x)\mathrm{d}G_a^{(j)}(x)]}. \tag{9.19}$$

9.2.2 *Sequential Preventive Maintenance*

An operating unit is maintained at sequential times T_k $(k = 1, 2, \cdots, N-1)$ and is replaced preventively at $S_N = \sum_{k=1}^N T_k$. It is assumed that shocks occur at a Poisson process with rate λ, and random variables Y_k $(k = 1, 2, \cdots, N)$ are denoted as the number of shocks during respective times T_k. Then,

$$\Pr\{Y_k = j\} = \frac{(\lambda T_k)^j}{j!}\mathrm{e}^{-\lambda T_k} \quad (j = 0, 1, 2, \cdots). \tag{9.20}$$

In addition, let W_{kj} denote an amount of damage produced by the jth shock during T_k, where W_{kj} ($W_{k0} \equiv 0$) are nonnegative, independent, and have an identical distribution $G(x) \equiv \Pr\{W_{kj} \le x\}$ for all k and j. Then,

$$\Pr\{W_{k1} + W_{k2} + \cdots + W_{kj} \le x\} = G^{(j)}(x) \quad (j = 0, 1, 2, \cdots).$$

Suppose that minimal repair is made to fix the failure which occurs with probability $p(x)$ when the total damage is x at some shock, and the maintenance done at T_k reduces $100(1 - a_k)\%$ of the damage with improvement factor a_k. Letting Z_k be the total damage at the end of T_k, then it becomes $a_k Z_k$ after maintenance. Thus,

$$Z_k = a_{k-1} Z_{k-1} + \sum_{i=1}^{Y_k} W_{ki} \quad (k = 1, 2, \cdots, N), \tag{9.21}$$

where $Z_0 \equiv 0$ and $\sum_{i=1}^{0} \equiv 0$, and the total costs for T_k are, when $k = 1, 2, \cdots, N-1$,

$$\widetilde{C}(k) = c_T + c_M \sum_{i=1}^{Y_k} p(a_{k-1} Z_{k-1} + W_{k1} + W_{k2} + \cdots + W_{ki}), \tag{9.22}$$

and when $k = N$,

$$\widetilde{C}(N) = c_N + c_M \sum_{i=1}^{Y_N} p(a_{N-1} Z_{N-1} + W_{N1} + W_{N2} + \cdots + W_{Ni}), \tag{9.23}$$

where c_M = cost for minimal repair at failure, c_T = maintenance cost at T_k, and c_N = replacement cost at T_N with $c_N > c_T$.

When $p(x) = 1 - \mathrm{e}^{-\theta x}$,

$$E\{\exp[-\theta(W_{k1} + W_{k2} + \cdots + W_{kj})]\} = [G^*(\theta)]^j,$$

where $G^*(\theta)$ is the LS transform of $G(x)$. Thus, from (9.22),

$$\begin{aligned} E\{\widetilde{C}(k)\} &= c_T + c_M E\left\{\sum_{i=1}^{Y_k} p(a_{k-1} Z_{k-1} + W_{k1} + \cdots + W_{ki})\right\} \\ &= c_T + c_M \sum_{n=1}^{\infty} \sum_{i=1}^{n} E\{1 - \exp[-\theta(a_{k-1} Z_{k-1} + W_{k1} + \cdots + W_{ki})]\} \\ &\quad \times \Pr\{Y_k = n\}. \end{aligned}$$

Denoting $B_k^*(\theta) \equiv E\{\exp(-\theta Z_k)\}$, and noting that Z_{k-1} and W_{kj} are independent with each other, from (9.22),

$$E\{1-\exp[-\theta(a_{k-1}Z_{k-1}+W_{k1}+\cdots+W_{ki})]\}=1-B^*_{k-1}(\theta a_{k-1})[G^*(\theta)]^i.$$

From (9.20) and (9.22),

$$\begin{aligned}E\{\widetilde{C}(k)\}&=c_T+c_M\sum_{n=1}^{\infty}\frac{(\lambda T_k)^n}{n!}\mathrm{e}^{-\lambda T_k}\sum_{i=1}^{n}\{1-B^*_{k-1}(\theta a_{k-1})[G^*(\theta)]^i\}\\&=c_T+c_M\left\{\lambda T_k-\frac{G^*(\theta)}{1-G^*(\theta)}B^*_{k-1}(\theta a_{k-1})[1-\mathrm{e}^{-\lambda[1-G^*(\theta)]T_k}]\right\},\end{aligned} \tag{9.24}$$

and from (9.23),

$$E\{\widetilde{C}(N)\}=c_N+c_M\left\{\lambda T_N-\frac{G^*(\theta)}{1-G^*(\theta)}B^*_{N-1}(\theta a_{N-1})[1-\mathrm{e}^{-\lambda[1-G^*(\theta)]T_N}]\right\}. \tag{9.25}$$

Letting $A^k_r\equiv\prod_{i=r}^k a_i$ for $r\le k$ and 0 for $r>i$, from (9.21),

$$a_{k-1}Z_{k-1}=\sum_{r=1}^{k-1}A^{k-1}_r\sum_{i=1}^{Y_r}W_{ri}.$$

Recalling that W_{ri} are independent and have an identical distribution $G(x)$,

$$B_{k-1}(\theta a_{k-1})\equiv E\left\{\mathrm{e}^{-\theta a_{k-1}Z_{k-1}}\right\}=E\left\{\exp\left[-\theta\sum_{r=1}^{k-1}A^{k-1}_r\sum_{i=1}^{Y_r}W_{ri}\right]\right\}.$$

Because

$$\begin{aligned}E\left\{\exp\left[-\theta A^{k-1}_r\sum_{i=1}^{Y_r}W_{ri}\right]\right\}&=\sum_{j=0}^{\infty}\Pr\{Y_r=j\}E\left\{\exp(-\theta A^{k-1}_r\sum_{i=1}^{j}W_{ri})\right\}\\&=\sum_{j=0}^{\infty}\frac{(\lambda T_r)^j}{j!}\mathrm{e}^{-\lambda T_r}[G^*(\theta A^{k-1}_r)]^j\\&=\exp\left\{-\lambda T_r[1-G^*(\theta A^{k-1}_r)]\right\},\end{aligned}$$

we consequently have

$$B_{k-1}(\theta a_{k-1})=\exp\left\{-\sum_{i=1}^{k-1}\lambda T_i\left[1-G^*(\theta A^{k-1}_i)\right]\right\}. \tag{9.26}$$

Substituting (9.26) for (9.24) and (9.25), respectively, the expected costs for T_k $(k = 1, 2, \cdots, N-1)$ and T_N are

$$E\{\widetilde{C}(k)\} = c_T + c_M\Bigg(\lambda T_k - \frac{G^*(\theta)}{1-G^*(\theta)}\exp\Bigg\{-\sum_{i=1}^{k-1}\lambda T_i[1-G^*(\theta A_i^{k-1})]\Bigg\} \times \left\{1-\mathrm{e}^{-\lambda T_k[1-G^*(\theta)]}\right\}\Bigg) \quad (k=1,2,\cdots,N-1), \tag{9.27}$$

$$E\{\widetilde{C}(N)\} = c_N + c_M\Bigg(\lambda T_N - \frac{G^*(\theta)}{1-G^*(\theta)}\exp\Bigg\{-\sum_{i=1}^{N-1}\lambda T_i[1-G^*(\theta A_i^{N-1})]\Bigg\} \times \left\{1-\mathrm{e}^{-\lambda T_N[1-G^*(\theta)]}\right\}\Bigg). \tag{9.28}$$

Therefore, the expected cost rate until replacement is

$$\mathbf{C}(\mathbf{T}_N) = \frac{\sum_{k=1}^{N-1} E\{\widetilde{C}(k)\} + E\{\widetilde{C}(N)\}}{\sum_{k=1}^{N} T_k} = \frac{\begin{array}{c}(N-1)c_T + c_N - c_M(\{G^*(\theta)/[1-G^*(\theta)]\} \\ \times \sum_{k=1}^{N}\exp\{-\sum_{i=1}^{k-1}\lambda T_i[1-G^*(\theta A_i^{k-1})]\}\{1-\mathrm{e}^{-\lambda T_k[1-G^*(\theta)]}\})\end{array}}{\sum_{k=1}^{N} T_k} + \lambda c_M, \quad (N=1,2,\cdots), \tag{9.29}$$

where $\mathbf{T}_N = (T_1, T_2, \cdots, T_N)$.

When $a_k \equiv a$ and $G(x) = 1 - \mathrm{e}^{-\omega x}$, (9.29) is rewritten as

$$\widetilde{\mathbf{C}}_1(\mathbf{T}_N) = c_M - \frac{1}{\lambda}\mathbf{C}(\mathbf{T}_N) = \frac{\begin{array}{c}c_M(\omega/\theta)\sum_{k=1}^{N}\exp\{-\sum_{i=1}^{k-1}[\lambda T_i[\theta a^{k-i}/(\theta a^{k-i}/(\theta a^{k-i}+\omega))]\} \\ \times\{1-\mathrm{e}^{-\lambda T_k[\theta/(\theta+\omega)]} - (N-1)c_T - c_N\}\end{array}}{\lambda\sum_{k=1}^{N} T_k}, \quad (N=1,2,\cdots). \tag{9.30}$$

We find optimum T_k^* $(k = 1, 2, \cdots, N)$ to maximize $\widetilde{\mathbf{C}}_1(\mathbf{T}_N)$. When $N = 1$,

$$\widetilde{C}_1(T_1) = \frac{c_M(\omega/\theta)\{1-\mathrm{e}^{-\lambda T_1[\theta/(\theta+\omega)]}\} - c_N}{\lambda T_1}. \tag{9.31}$$

Differentiating $\widetilde{C}_1(T_1)$ with respect to T and setting it equal to zero,

$$\frac{\omega}{\theta}\left\{1-\mathrm{e}^{-\lambda T_1[\theta/(\theta+\omega)]}\right\}-\frac{\lambda\omega T_1}{\theta+\omega}\mathrm{e}^{-\lambda T_1[\theta/(\theta+\omega)]}=\frac{c_N}{c_M}, \tag{9.32}$$

whose left-hand side increases strictly with T_1 from 0 to ω/θ. Therefore, if $\omega/\theta > c_N/c_M$, then there exists a finite and unique T_1^* $(0 < T_1^* < \infty)$ which satisfies (9.32). Conversely, if $\omega/\theta \le c_N/c_M$, then $T_1^* = 0$.

Furthermore, differentiating $\widetilde{\mathbf{C}}_1(\mathbf{T}_N)$ with respect to T_k and setting it equal to zero,

$$\begin{aligned}&\frac{\theta}{\theta+\omega}\exp\left[-\sum_{i=1}^{k}\lambda T_i\left(\frac{\theta a^{k-i}}{\theta a^{k-i}+\omega}\right)\right]\\&-\sum_{j=k+1}^{N}\frac{\theta a^{j-k}}{\theta a^{j-k}+\omega}\exp\left[-\sum_{i=1}^{j-1}\lambda T_i\left(\frac{\theta a^{j-i}}{\theta a^{j-i}+\omega}\right)\right]\left\{1-\mathrm{e}^{-\lambda T_j[\theta/(\theta+\omega)]}\right\}\\&=\frac{\widetilde{\mathbf{C}}_1(\mathbf{T}_N)}{(\omega/\theta)c_M}\qquad\qquad (k=1,2,\cdots,N-1),\end{aligned} \tag{9.33}$$

$$\frac{\theta}{\theta+\omega}\exp\left[-\sum_{i=1}^{N}\lambda T_i\left(\frac{\theta a^{N-i}}{\theta a^{N-i}+\omega}\right)\right]=\frac{\widetilde{\mathbf{C}}_1(\mathbf{T}_N)}{(\omega/\theta)c_M}. \tag{9.34}$$

Solving (9.33) and (9.34) simultaneously and comparing $\widetilde{\mathbf{C}}_1(\mathbf{T}_N)$, optimum N^* and T_k^* $(k = 1, 2, \cdots, N^*)$ to maximize $\widetilde{\mathbf{C}}_1(\mathbf{T}_N)$ can be found.

Next, when the unit is operating for a finite interval $(0, S]$, i.e., it is replaced at a specified time $S = T_1 + T_2 + \cdots + T_N$, we consider the optimum policy to maximize the expected cost

$$\begin{aligned}\widetilde{\mathbf{C}}_2(\mathbf{T}_N)=&\frac{\omega c_M}{\theta}\sum_{k=1}^{N}\exp\left[-\sum_{i=1}^{k-1}\lambda T_i\left(\frac{\theta a^{k-i}}{\theta a^{k-i}+\omega}\right)\right]\\&\times\left\{1-\mathrm{e}^{-\lambda T_k[\theta/(\theta+\omega)]}\right\}-(N-1)c_T-c_N\quad (N=1,2,\cdots).\end{aligned} \tag{9.35}$$

For example, when $N = 1$,

$$\widetilde{\mathbf{C}}_2(S)=\frac{\omega c_M}{\theta}(1-\mathrm{e}^{-\lambda S D_0})-c_N, \tag{9.36}$$

where $D_j \equiv \theta a^j/(\theta a^j + \omega)$ $(j = 0, 1, 2, \cdots)$, which decreases strictly with j to 0 for $0 < a < 1$.

When $N = 2$,

$$\widetilde{C}_2(T_1)=\frac{\omega c_M}{\theta}\left(1-\mathrm{e}^{-\lambda T_1 D_0}+\mathrm{e}^{-\lambda T_1 D_1}\right)\left[1-\mathrm{e}^{-\lambda(S-T_1)D_0}\right]-c_T-c_N. \tag{9.37}$$

Differentiating $\widetilde{C}_2(T_1)$ with respect to T_1 and setting it equal to zero,

$$D_0\left[e^{-\lambda T_1(D_0-D_1)}-e^{-\lambda(S-T_1)D_0}\right]-D_1\left[1-e^{-\lambda(S-T_1)D_0}\right]=0. \tag{9.38}$$

Letting $L(T)$ denote the left-hand side of (9.38),

$$L(0)=(D_0-D_1)(1-e^{-\lambda SD_0})>0,$$
$$L(S)=-D_0[1-e^{-\lambda S(D_0-D1)}]<0,$$
$$\frac{dL(T)}{dT}=-\,D_0(D_0-D_1)[e^{-\lambda T_1(D_0-D_1)}+e^{-\lambda(S-T_1)D_0}]<0.$$

Thus, there exists optimum T_1^* $(0<T_1^*<S)$ which satisfies (9.38).

When $N=3$,

$$\widetilde{C}_2(T_1,T_2)=\frac{\omega c_M}{\theta}\left[1-e^{-\lambda T_1D_0}+e^{-\lambda T_1D_1}(1-e^{-\lambda T_2D_0})\right]$$
$$+\,e^{-\lambda(T_1D_2+T_2D_1)}\left[1-e^{-\lambda(S-T_1-T_2)D_0}\right]-2c_T-c_N. \tag{9.39}$$

Differentiating $\widetilde{C}_2(T_1,T_2)$ with respect to T_1 and T_2 and setting them equal to zero,

$$D_0\left[e^{-\lambda T_1D_0}-e^{-\lambda(T_1D_2+T_2D_1)-\lambda(S-T_1-T_2)D_0}\right]-D_1e^{-\lambda T_1D_1}(1-e^{-\lambda T_2D_0})$$
$$-\,D_2e^{-\lambda(T_1D_2+T_2D_1)}\left[1-e^{-\lambda(S-T_1-T_2)D_0}\right]=0, \tag{9.40}$$
$$D_0\left[e^{-\lambda(T_1D_1+T_2D_0)}-e^{-\lambda(T_1D_2+T_2D_1)-\lambda(S-T_1-T_2)D_0}\right]$$
$$-\,D_1e^{-\lambda(T_1D_2+T_2D_1)}\left[1-e^{-\lambda(S-T_1-T_2)D_0}\right]=0. \tag{9.41}$$

In general, differentiating $\widetilde{\mathbf{C}}_2(\mathbf{T}_N)$ with respect to T_k $(N\geq 2)$ and setting it equal to zero,

$$D_0\left[\exp\left(-\sum_{i=1}^{k}\lambda T_iD_{k-i}\right)-\exp\left(-\sum_{i=1}^{N}\lambda T_iD_{N-i}\right)\right]$$
$$-\sum_{j=k+1}^{N}D_{j-k}\exp\left(-\sum_{i=1}^{j-1}\lambda T_iD_{j-i}\right)(1-e^{-\lambda T_kD_0})=0$$
$$(k=1,2,\cdots,N-1), \tag{9.42}$$

where $T_N\equiv S-T_1-T_2-\cdots-T_{N-1}$. Therefore, solving equations (9.42) simultaneously and comparing $\widetilde{\mathbf{C}}_2(\mathbf{T}_N)$ in (9.35) for all $N\geq 1$, optimum N^* and T_k^* $(k=1,2,\cdots,N^*)$ for given S can be found.

Table 9.2 presents optimum λT_k^* and $\widetilde{\mathbf{C}}_2(\mathbf{T}_N)/c_M$ for N when $a=0.5$, $\omega/\theta=10.0$, $c_N/c_M=5.0$, $c_T/c_M=1.0$, $S=40.0$. Comparing $\widetilde{\mathbf{C}}_2(\mathbf{T}_N)/c_M$ for $N=1,2,\cdots,10$, $\widetilde{\mathbf{C}}_2(\mathbf{T}_N)/c_M$ is maximum for $N=8$, i.e., $\mathbf{C}(\mathbf{T}_N)$ in (9.29) is minimum at $N=8$. In this case, optimum maintenance times are sequentially obtained at

Table 9.2 Optimum λT_k^* and $\widetilde{C}_2(\mathbf{T}_N)/c_M$ when $a = 0.5$, $\omega/\theta = 10.0$, $c_N/c_M = 5.0$, $c_T/c_M = 1.0$, $S = 40.0$

λT_k^*	N									
	1	2	3	4	5	6	7	8	9	10
λT_1^*	40.00	13.17	12.41	11.37	10.32	9.36	8.52	7.80	7.17	6.63
λT_2^*		26.83	5.60	5.27	4.82	4.38	3.99	3.66	3.37	3.11
λT_3^*			21.99	5.23	4.87	4.45	4.06	3.72	3.42	3.17
λT_4^*				18.22	4.78	4.45	4.07	3.73	3.44	3.18
λT_5^*					15.22	4.35	4.06	3.73	3.44	3.18
λT_6^*						13.01	3.97	3.71	3.44	3.18
λT_7^*							11.33	3.64	3.42	3.18
λT_8^*								10.01	3.35	3.16
λT_9^*									8.96	3.10
λT_{10}^*										8.10
$\widetilde{C}_2(\mathbf{T}_N)/c_M$	4.74	5.86	6.87	7.70	8.34	8.78	9.05	9.17	9.16	9.03

7.80, 11.46, 15.18, 18.91, 22.64, 26.35, 29.99, 40.00. Table 9.2 also shows that λT_1^* and λT_8^* are larger than other maintenance intervals, and λT_3^*, λT_4^*, λT_5^* are almost stable at 3.73.

9.3 Continuous Damage Models

In Sect. 1.2.3, we have introduced briefly a continuous damage model in which the total damage stored in an operating unit increases continuously and swayingly with time t from $W(0) \equiv 0$ at a stochastic path $W(t) = A_t t + B_t$ for $A_t \geq 0$ and the unit fails when the total damage $W(t)$ exceeds a failure threshold K. In this case, the reliability of the unit at time t is given in (1.30).

The path of $W(t)$ and the reliability of $R(t)$ can be considered in the following cases [3]:

1. When $A_t \equiv \omega$ (constant), $K = k$ (constant), and B_t is normally distributed with mean 0 and variance $\sigma^2 t$,

$$R(t) = \Pr\{B_t \leq k - \omega t\} = \Phi\left(\frac{k - \omega t}{\sigma\sqrt{t}}\right). \tag{9.43}$$

2. When $B_t \equiv 0$, $K = k$, and A_t is normally distributed with mean ω and variance σ^2/t,

$$R(t) = \Pr\{A_t \leq \frac{k}{t}\} = \Phi\left(\frac{k - \omega t}{\sigma\sqrt{t}}\right), \tag{9.44}$$

which agrees with (9.43).

3. When $A_t \equiv \omega$, $K = k$, and B_t is exponentially distributed with mean $\sigma\sqrt{t}$, i.e., $\Pr\{B_t \le t\} = 1 - \mathrm{e}^{-t/\sigma\sqrt{t}}$,

$$R(t) = \Pr\{B_t \le k - \omega t\} = 1 - \exp\left(-\frac{k - \omega t}{\sigma\sqrt{t}}\right). \tag{9.45}$$

4. When $A_t \equiv \omega$, $B_t \equiv 0$, and K is normally distributed with mean k and variance $\sigma^2 t$, $R(t)$ agrees with (9.43) and (9.44).
5. Replacing $\alpha \equiv \sigma/\sqrt{\omega k}$ and $\beta \equiv k/\omega$ in (9.43) or (9.44),

$$R(t) = \Phi\left[\frac{1}{\alpha}\left(\sqrt{\frac{\beta}{t}} - \sqrt{\frac{t}{\beta}}\right)\right], \tag{9.46}$$

which is Birnbaum-Saunders distribution and widely used in fatigue failure resulted from stress.

9.3.1 Age Replacement Policy

Suppose that the unit is replaced preventively at planned time T $(0 < T \le \infty)$ or correctively at failure when the total damage has exceeded a failure threshold K, whichever occurs first. Then, the expected cost rate is [1]

$$C(T) = \frac{c_K - (c_K - c_T)R(T)}{\int_0^T R(t)\mathrm{d}t}, \tag{9.47}$$

where c_T = replacement cost at time T and c_K = replacement cost at failure K, where $c_K > c_T$. Optimum T^* to minimize $C(T)$ satisfies

$$h(T)\int_0^T R(t)\mathrm{d}t + R(T) = \frac{c_K}{c_K - c_T}, \tag{9.48}$$

where $h(t) \equiv -R'(t)/R(t)$.

(1) Policy I

When $R(t)$ is given in (9.43) or (9.44), the expected cost rate $C_1(T_1)$ is obtained in (9.47), and when $\omega = 1.0$ and $\sigma^2 = 1.0$, optimum T_1^* to minimize $C_1(T_1)$ satisfies

$$\frac{(K/T + 1)\phi(K/\sqrt{T} - \sqrt{T})}{2\sqrt{T}\Phi(K/\sqrt{T} - \sqrt{T})}\int_0^T \Phi\left(\frac{K}{\sqrt{t}} - \sqrt{t}\right)\mathrm{d}t$$
$$+ \Phi\left(\frac{K}{\sqrt{T}} - \sqrt{T}\right) = \frac{c_K}{c_K - c_T}, \tag{9.49}$$

Table 9.3 Optimum T_1^* and its cost rate $C_1(T_1^*)/c_T$ when $\omega = 1.0$ and $\sigma^2 = 1.0$

c_K/c_T	$K = 5.0$		$K = 10.0$		$K = 20.0$	
	T_1^*	$C(T_1^*)/c_T$	T_1^*	$C(T_1^*)/c_T$	T_1^*	$C(T_1^*)/c_T$
5	2.5	0.483	6.0	0.198	13.1	0.084
10	2.0	0.559	5.1	0.219	12.0	0.090
20	1.8	0.626	4.6	0.237	11.3	0.095
50	1.6	0.704	4.1	0.260	10.6	0.102

Table 9.4 Optimum T_2^* and its cost rate $C_2(T_2^*)/c_T$ when $\omega = 1.0$ and $\sigma^2 = 1.0$

c_K/c_T	$K = 5.0$		$K = 10.0$		$K = 20.0$	
	T_2^*	$C(T_2^*)/c_T$	T_2^*	$C(T_2^*)/c_T$	T_2^*	$C(T_2^*)/c_T$
5	2.2	0.762	4.8	0.292	10.9	0.116
10	1.6	1.103	3.8	0.365	9.2	0.137
20	1.1	1.336	3.1	0.446	7.8	0.159
50	0.8	1.813	2.3	0.568	6.5	0.192

where $\phi(x) \equiv (1/\sqrt{2\pi})\mathrm{e}^{-x^2/2}$.

Table 9.3 presents optimum T_1^* and its cost rate $C_1(T_1^*)/c_T$ for c_K/c_T and K when $\omega = 1.0$ and $\sigma^2 = 1.0$, which shows the same properties as in Table 2.1, and $C_1(T_1^*) < C(T^*)$. It is of interest that optimum T_1^* are almost equal to N^* in Table 2.2.

(2) Policy II

When $R(t)$ is given in (9.45), i.e., the total damage increases linearly with time t and also increases with some positive values according to an exponential distribution. Then, the expected cost rate is

$$C_2(T) = \frac{c_K - (c_K - c_T)\{1 - \exp[-(K - \omega T)/\sigma\sqrt{T}]\}}{\int_0^T \{1 - \exp[-(K - \omega t)/\sigma\sqrt{t}]\}\mathrm{d}t}. \tag{9.50}$$

Optimum T_2^* to minimize $C_2(T_2)$ satisfies

$$\frac{(K/T + \omega)\exp[-(K - \omega T)/\sigma\sqrt{T}]}{2\sigma\sqrt{T}\{1 - \exp[-(K - \omega T)/\sigma\sqrt{T}]\}} \int_0^T \left[1 - \exp\left(-\frac{K - \omega t}{\sigma\sqrt{t}}\right)\right]\mathrm{d}t$$
$$- \exp\left(-\frac{K - \omega T}{\sigma\sqrt{T}}\right) = \frac{c_T}{c_K - c_T}. \tag{9.51}$$

Table 9.4 presents optimum T_2^* and its cost rate $C_2(T_2^*)/c_T$ for c_K/c_T and K when $\omega = 1.0$ and $\sigma^2 = 1.0$, which also shows the same properties as in Table 9.3, however, $T_2^* < T_1^*$ and $C_2(T_2^*) > C_1(T_1^*)$.

Table 9.5 Optimum N_3^* and its cost rate $C_3(N_3^*)/c_T$ when $\omega = 1.0$, $\sigma^2 = 1.0$ and $T = 1.0$

c_K/c_T	$K = 5.0$		$K = 10.0$		$K = 20.0$	
	N_3^*	$C_3(N_3^*)/c_T$	N_3^*	$C_3(N_3^*)/c_T$	N_3^*	$C_3(N_3^*)/c_T$
5	3	0.544	6	0.214	13	0.089
10	2	0.615	5	0.235	12	0.096
20	2	0.739	4	0.263	11	0.101
50	1	1.009	4	0.281	10	0.108

(3) Policy III

Suppose that the unit is replaced at time NT ($N = 1, 2, \cdots$) or at failure K, whichever occurs first. Then, the expected cost rate is, replacing T with NT in (9.47),

$$C_3(N) = \frac{c_K - (c_K - c_T)R(NT)}{\int_0^{NT} R(t)\mathrm{d}t} \quad (N = 1, 2, \cdots), \tag{9.52}$$

and optimum N_3^* to minimize $C_3(N)$ satisfies

$$\frac{R(NT) - R[(N+1)T]}{\int_{NT}^{(N+1)T} R(t)\mathrm{d}t} \int_0^{NT} R(t)\mathrm{d}t + R(NT) \geq \frac{c_K}{c_K - c_T}. \tag{9.53}$$

When $R(t)$ is given in (9.43) or (9.44), and $\omega = \sigma^2 = 1.0$, Table 9.5 presents optimum N_3^* and its cost rate $C_3(N_3^*)/c_T$ for c_K/c_T and K when $\omega = 1.0$ and $\sigma^2 = 1.0$. It is of great interest that optimum N_3^* are almost equal to N^* in Table 2.2 and the expected cost rates $C_3(N_3^*)$ are smaller than those in Table 2.2 as K becomes larger. These numerical examples indicate that continuous damage models could be used as asymptotic damage models in real applications.

9.3.2 *Replacement First, Last and Overtime*

The unit degrading with continuous damage defined in Sect. 9.3 works for a job with cycle time Y, where Y is a random variable and has a general distribution $G(t) \equiv \Pr\{Y \leq t\}$ with finite mean $1/\theta$. In order to prevent the failure K and preserve the completion of the cycle time Y, the following replacement first, replacement last, and replacement overtime are obtained.

(1) Replacement First

Suppose that the unit is replaced at time T ($0 < T \leq \infty$), at cycle Y, or at failure K, whichever occurs first. Then, the expected cost rate is [15]

$$C_F(T) = \frac{c_K - (c_K - c_T)R(T)\overline{G}(T) - (c_K - c_R)\int_0^T R(t)\mathrm{d}G(t)}{\int_0^T R(t)\overline{G}(t)\mathrm{d}t}, \tag{9.54}$$

where c_R = replacement cost at cycle Y, and c_K and c_T are given in (9.47).

When $c_T = c_R$, optimum T_F^* to minimize $C_F(T)$ satisfies

$$\int_0^T R(t)\overline{G}(t)[h(T) - h(t)]\mathrm{d}t = \frac{c_T}{c_K - c_T}, \tag{9.55}$$

where $h(t) \equiv -R'(t)/R(t)$. When $R(t)$ is given in (9.43), and $\omega = \sigma^2 = 1.0$,

$$h(t) = \frac{(K/t + 1)\phi(K/\sqrt{t} - \sqrt{t})}{2\sqrt{t}\Phi(K/\sqrt{t} - \sqrt{t})}, \tag{9.56}$$

and when $R(t)$ is given in (9.45),

$$h(t) = \frac{K/t + \omega}{2\sigma\sqrt{t}\{\exp[(K - \omega t)/\sigma\sqrt{t}] - 1\}}. \tag{9.57}$$

Optimum T_F^* can be computed numerically when $h(t)$ in (9.56) or (9.57) is given.

(2) Replacement Last

Suppose that the unit is replaced preventively at time T $(0 < T \le \infty)$ or at cycle Y, whichever occurs last. Then, the expected cost rate is [15]

$$C_L(T) = \frac{c_K - (c_K - c_T)R(T)G(T) - (c_K - c_R)\int_T^\infty R(t)\mathrm{d}G(t)}{\int_0^T \overline{F}(t)\mathrm{d}t + \int_T^\infty R(t)\overline{G}(t)\mathrm{d}t}. \tag{9.58}$$

When $c_T = c_R$, optimum T_L^* to minimize $C_L(T)$ satisfies

$$h(T)\left[\int_0^T R(t)\mathrm{d}t + \int_T^\infty R(t)\overline{G}(t)\mathrm{d}t\right] + R(T)G(T) + \int_T^\infty R(t)\mathrm{d}G(t) = \frac{c_K}{c_K - c_T}. \tag{9.59}$$

Optimum T_L^* can be computed numerically when $R(t)$ in (9.43) or (9.45) is given.

(3) Replacement Overtime

Suppose the unit works for jobs with successive cycle times Y_j $(j = 1, 2,)$, and it is replaced preventively at the first completion of cycle time Y_j over time T. When $G(t) = 1 - \mathrm{e}^{-\theta t}$, the expected cost rate is [15]

$$C_O(T) = \frac{c_K - (c_K - c_O)\int_T^\infty R(t)\theta \mathrm{e}^{-\theta(t-T)}\mathrm{d}t}{\int_0^T R(t)\mathrm{d}t + \int_T^\infty R(t)\mathrm{e}^{-\theta(t-T)}\mathrm{d}t}. \tag{9.60}$$

Optimum T_O^* to minimize $C_O(T)$ satisfies

$$Q(T;\theta)\int_0^T R(t)\mathrm{d}t + R(T) = \frac{c_O}{c_K - c_O}, \tag{9.61}$$

where

$$Q(T;\theta) \equiv \frac{\mathrm{e}^{-\theta T}R(T)}{\int_T^\infty \mathrm{e}^{-\theta t}R(t)\mathrm{d}t} - \theta.$$

Optimum T_O^* can be computed numerically when $R(t)$ is given in (9.43) or (9.45).

(4) Replacement with Working Times

Suppose an operating unit works for jobs with successive cycle times Y_j $(j = 1, 2,)$, where Y_j are random variables and have an identical distribution $G(t) \equiv \Pr\{Y_j \le t\}$. The unit is replaced preventively at time T $(0 < T \le \infty)$ or at number N $(N = 1, 2, \cdots)$ of cycle times, whichever occurs first. Then, the expected cost rate is [15]

$$C_F(T,N) = \frac{c_K - (c_K - c_T)R(T)[1 - G^{(N)}(T)] - (c_K - c_N)\int_0^T R(t)\mathrm{d}G^{(N)}(t)}{\int_0^T R(t)[1 - G^{(N)}(t)]\mathrm{d}t}, \tag{9.62}$$

where c_N = replacement cost at cycle N.

When the unit is replaced preventively at time T $(0 < T \le \infty)$ or at number N $(N = 1, 2, \cdots)$ of cycle times, whichever occurs last. Then, the expected cost rate is [15]

$$C_L(T,N) = \frac{c_K - (c_K - c_T)R(T)G^{(N)}(T) - (c_K - c_N)\int_T^\infty R(t)\mathrm{d}G^{(N)}(t)}{\int_0^T R(t)\mathrm{d}t + \int_T^\infty R(t)[1 - G^{(N)}(t)]\mathrm{d}t}. \tag{9.63}$$

Similarly, optimum policies to minimize $C_F(T,N)$ and $C_L(T,N)$ can be computed when $R(t)$ is given.

9.4 Markov Chain Models

When we consider several levels of damages Z_i $(i = 1, 2, \cdots, n-1)$ for maintenances and damage Z_n for replacement, a Markov chain model will be formulated [25].

9.4.1 Simple Model

We firstly give three states of damages to formulate a simple Markov chain model, and a general model will be addressed in Sect. 9.4.2. That is, the following states of damages are defined:

State 0: When the damage is less than Z_1, inspection is done normally.
State 1: When the damage is between Z_1 and Z_2, the unit is maintained.
State 2: When the damage reaches Z_2, replacement is done for failure.

The total damage can be checked at periodic times kT $(k = 1, 2, \cdots)$ for given T $(0 < T < \infty)$, and the increment of damage W_k for the interval $[(k-1)T, kT]$ $(k = 1, 2, \cdots)$ is independent with each other and has an identical distribution $G(x) \equiv \Pr\{W_k \leq x\}$. It is assumed that the damage is 0 at time 0, i.e., $W_0 = 0$ when the unit starts operation, and becomes 0 during State 0 after inspections. When the unit goes into State 1, the damage becomes Z_1 after maintenances. Both States 0 and 1 have the possibilities to fall into State 2, and then, the unit is replaced with a new one. States 0, 1 and 2 and their stochastic transitions form a Markov model with an absorbing State 2.

One-step transition probability Q_{ij} $(i, j = 0, 1, 2)$ from State i to State j is given in the following matrix form:

$$(Q_{ij}) = \begin{pmatrix} G(Z_1) & G(Z_2) - G(Z_1) & 1 - G(Z_2) \\ 0 & G(Z_2 - Z_1) & 1 - G(Z_2 - Z_1) \\ 0 & 0 & 1 \end{pmatrix}. \tag{9.64}$$

The expected numbers I_i $(i = 0, 1)$ of inspections from State i to State 2 are

$$I_0 = Q_{00}(1 + I_0) + Q_{01}(1 + I_1) + Q_{02},$$
$$I_1 = Q_{11}(1 + I_1) + Q_{12}.$$

The expected numbers M_i $(i = 0, 1)$ of PMs from State i to State 2 are

$$M_0 = Q_{00}M_0 + Q_{01}(1 + M_1),$$
$$M_1 = Q_{11}(1 + M_1).$$

Solving the above equations for respective I_0 and M_0,

$$I_0 = \frac{1 + Q_{01} - Q_{11}}{(1 - Q_{00})(1 - Q_{11})} = \frac{1 - G(Z_2 - Z_1) + G(Z_2) - G(Z_1)}{\overline{G}(Z_1)\overline{G}(Z_2 - Z_1)}, \tag{9.65}$$

$$M_0 = \frac{Q_{01}}{(1 - Q_{00})(1 - Q_{11})} = \frac{G(Z_2) - G(Z_1)}{\overline{G}(Z_1)\overline{G}(Z_2 - Z_1)}. \tag{9.66}$$

Therefore, the total expected cost until failure detection is

$$C_1(Z_1) = c_I I_0 + c_M M_0 + c_F, \tag{9.67}$$

and the expected cost rate is

$$\widetilde{C}_1(Z_1) = \frac{C_1(Z_1)}{I_0 T} = \frac{1}{T}\left(c_I + \frac{c_M M_0 + c_F}{I_0}\right), \tag{9.68}$$

where c_I = inspection cost, c_M = maintenance cost, and c_F = replacement cost with $c_I < c_M < c_F$.

In particular, when $G(x) = 1 - \mathrm{e}^{-\omega x}$, I_0 and M_0 are, from (9.65) and (9.66),

$$I_0 = \mathrm{e}^{\omega(Z_2 - Z_1)} + \mathrm{e}^{\omega Z_1} - 1, \quad M_0 = \mathrm{e}^{\omega(Z_2 - Z_1)} - 1.$$

Thus, from (9.67),

$$C_1(Z_1) = (c_I + c_M)[\mathrm{e}^{\omega(Z_2 - Z_1)} - 1] + c_I \mathrm{e}^{\omega Z_1} + c_F. \tag{9.69}$$

Differentiating $C_1(Z_1)$ with respect to Z_1 and setting it equal to zero,

$$\omega Z_1^* = \frac{1}{2}\left[\omega Z_2 + \ln\left(1 + \frac{c_M}{c_I}\right)\right]. \tag{9.70}$$

If $Z_1^* \geq Z_2$, i.e.,

$$\frac{c_M}{c_I} \geq \mathrm{e}^{\omega Z_2} - 1,$$

then we should not make any PM. In other words, if the cost rate $c_M/c_I < \mathrm{e}^{\omega Z_2} - 1$, then PM should be made at periodic times for Z_1^*.

Next, letting $x_1 \equiv \mathrm{e}^{\omega Z_1}$ and $x_2 \equiv \mathrm{e}^{\omega Z_2}$, from (9.68),

$$\widetilde{C}_1(x_1; x_2) = \frac{1}{T}\left[c_I + \frac{(c_F - c_M)x_1 + c_M x_2}{x_1^2 - x_1 + x_2}\right], \tag{9.71}$$

and optimum x_1^* to minimize $\widetilde{C}_1(x_1; x_2)$ satisfies

$$x_1^2 + \frac{2c_M x_2 x_1}{c_F - c_M} - \frac{c_F x_2}{c_F - c_M} = 0,$$

i.e.,

$$x_1^* = -\frac{c_M x_2}{c_F - c_M} + \sqrt{\left(\frac{c_M x_2}{c_F - c_M}\right)^2 + \frac{c_F x_2}{c_F - c_M}}, \tag{9.72}$$

Table 9.6 Optimum ωZ_1^*, $\omega \widetilde{Z}_1^*$, and their cost rates when $\omega Z_2 = 5.0$

c_M/c_I	ωZ_1^*	$[C_1(Z_1^*) - c_F]/c_I$	$\omega \widetilde{Z}_1^*$	$[T\widetilde{C}_1(\widetilde{Z}_1^*) - c_I]/c_M$
5.0	3.396	53.682	4.339	0.076
10.0	3.699	69.810	3.621	0.321
15.0	3.886	81.460	3.299	0.617
20.0	4.022	90.654	3.116	0.909
25.0	4.129	98.238	2.999	1.186
30.0	4.217	104.659	2.919	1.449

and $\omega \widetilde{Z}_1^* = \ln x_1^*$.

Table 9.6 presents optimum ωZ_1^* and $\omega \widetilde{Z}_1^*$ to minimize $C_1(Z_1)$ and $\widetilde{C}_1(Z_1)$ for $c_M/c_I = c_F/c_M$ and c_F/c_M, respectively. Both Z_1^* and $\widetilde{Z}_1^*$ increase with c_M/c_I and c_M/c_F.

9.4.2 General Model

We next formulate a general Markov chain model with n states of damages such as $0 \equiv Z_0 < Z_1 < Z_2 < \cdots < Z_{n-1} < Z_n$ $(n = 2, 3, \cdots)$:

State 0: When the damage is less than Z_1, the unit operates normally.
State i: When the damage is between Z_i and Z_{i+1}, inspection is done.
State $n-1$: When the damage reaches Z_{n-1}, the unit is maintained.
State n: When the damage reaches Z_n, replacement is done for failure.

It is assumed that the damage is 0 at time 0 and the unit operates normally during State 0. The damage can only be checked at periodic times kT $(k = 1, 2, \cdots)$ when the unit goes into State i $(i = 1, 2, \cdots, n-1)$, i.e., the damage becomes Z_i after inspections when it is between Z_i and Z_{i+1}. When the unit goes into State $n-1$, the damage becomes Z_{n-1} after maintenances. All of States i $(i = 1, 2, \cdots, n-1)$ have the possibilities to fall into State n, and the unit is replaced with a new one.

One-step transition probability Q_{ij} is

$$Q_{ij} = G(Z_{j+1} - Z_i) - G(Z_j - Z_i) \quad (i, j = 0, 1, \cdots, n), \tag{9.73}$$

where $Z_0 \equiv 0$, $Z_{n+1} \equiv \infty$, $G(0) \equiv 0$ and $G(\infty) \equiv 1$. Thus, the expected number I_i of inspections from State i to State n is

$$I_i = \sum_{j=i}^{n-1} Q_{ij}(1 + I_j) + Q_{in} \quad (i = 0, 1, \cdots, n-1),$$

and the expected number M_i of PMs from State i to State n is

$$M_i = \sum_{j=i}^{n-1} Q_{ij} M_j + Q_{in-1} \quad (i = 0, 1, \cdots, n-1). \tag{9.74}$$

Solving the above equations for respective I_i and M_i,

$$I_i = \frac{\sum_{j=i+1}^{n-1} Q_{ij} I_j + 1}{1 - Q_{ii}}, \tag{9.75}$$

$$M_i = \frac{\sum_{j=i+1}^{n-1} Q_{ij} M_j + Q_{in-1}}{1 - Q_{ii}}, \tag{9.76}$$

where $\sum_j^{n-1} \equiv 0$ for $j \geq n$.

Therefore, the total expected cost until failure detection is

$$C_{n-1}(Z_{n-1}) = c_I I_0 + c_M M_0 + c_F, \tag{9.77}$$

and the expected cost rate is

$$\widetilde{C}_{n-1}(Z_{n-1}) = \frac{1}{T}\left(c_1 + \frac{c_M M_0 + c_F}{I_0}\right). \tag{9.78}$$

In particular, when $G(x) \equiv 1 - \mathrm{e}^{-\omega x}$, for $i = 0, 1, \cdots, n-1$,

$$Q_{ii} = 1 - \mathrm{e}^{-\omega(Z_{i+1} - Z_i)},$$
$$Q_{ij} = \mathrm{e}^{-\omega(Z_j - Z_i)} - \mathrm{e}^{-\omega(Z_{j+1} - Z_i)} \quad (j = i+1, i+2, \cdots, n-1),$$
$$Q_{in} = \mathrm{e}^{-\omega(Z_n - Z_i)}.$$

Thus, from (9.75), for $i = 0, 1, \cdots, n-1$,

$$I_i \mathrm{e}^{-\omega Z_{i+1}} = \mathrm{e}^{-\omega Z_i} + \sum_{j=i+1}^{n-1} \left(\mathrm{e}^{-\omega Z_j} - \mathrm{e}^{-\omega Z_{j+1}}\right) I_j.$$

Computing I_i for $i = n-1, n-2, \cdots$,

$$I_i = \sum_{j=i}^{n-1} \mathrm{e}^{\omega(Z_{j+1} - Z_j)} - (n - 1 - i),$$

and then, for I_0 (Problem 9.4),

$$I_0 = \sum_{j=0}^{n-1} \mathrm{e}^{\omega(Z_{j+1}-Z_j)} - (n-1). \tag{9.79}$$

From (9.76), for $i = 0, 1, \cdots, n-1$,

$$M_i \mathrm{e}^{-\omega Z_{i+1}} = \sum_{j=i+1}^{n-1} \left(\mathrm{e}^{-\omega Z_j} - \mathrm{e}^{-\omega Z_{j+1}}\right) M_j + \mathrm{e}^{-\omega Z_{n-1}} - \mathrm{e}^{-\omega Z_n}.$$

Computing M_i for $i = n-1, n-2, \cdots$,

$$M_i = \mathrm{e}^{\omega(Z_n - Z_{n-1})} - 1.$$

Therefore, from (9.77), the expected cost until failure detection is

$$C_{n-1}(Z_{n-1}) = c_I \left[\sum_{j=0}^{n-1} \mathrm{e}^{\omega(Z_{j+1}-Z_j)} - (n-1)\right] + c_M \left[\mathrm{e}^{\omega(Z_n - Z_{n-1})} - 1\right] + c_F, \tag{9.80}$$

and the expected cost rate is

$$\widetilde{C}_{n-1}(Z_{n-1}) = \frac{1}{T}\left[c_I + \frac{c_M \mathrm{e}^{\omega(Z_n - Z_{n-1})} + c_F}{\sum_{j=0}^{n-1} \mathrm{e}^{\omega(Z_{j+1}-Z_j)} - (n-1)}\right]. \tag{9.81}$$

Differentiating $C_{n-1}(Z_{n-1})$ with respect to Z_{n-1} and setting it equal to zero,

$$Z_{n-1}^* = \frac{1}{2}\left[Z_{n-2} + Z_n + \frac{1}{\omega}\ln\left(1 + \frac{c_M}{c_I}\right)\right],$$

which indicates that optimum Z_{n-1}^* a little larger than half of $Z_{n-2} + Z_n$.

9.5 Problem 9

9.1 Derive (9.1), (9.2) and (9.3).
9.2 Derive (9.6), and when $p(x) = 1 - \mathrm{e}^{-\theta x}$, obtain optimum Z_O^* to minimize $C_O(Z)$.
9.3 Derive (9.14).
9.4 Derive (9.79) and (9.80).

Appendix

A.1 Answers to Problem 1

1.1 Changing n with α, the respective mean, variance and LS transform are easily given by

$$\frac{\alpha}{\lambda}, \ \frac{\alpha}{\lambda^2}, \ \left(\frac{\lambda}{s+\lambda}\right)^{\alpha}.$$

1.2 By the mathematical induction and from (1.5),

$$\begin{aligned}
G_1(t) =& \Pr\{X_1 \le t\} = 1 - \mathrm{e}^{-H(t)}, \\
G_{n+1}(t) =& \int_0^{\infty} \Pr\{X_{n+1} \le t-u | S_n = u\} \mathrm{d}G_n(u) \\
=& \int_0^t \frac{F(t)-F(u)}{\overline{F}(u)} \frac{H(u)^{n-1}}{(n-1)!} \mathrm{d}F(u) \\
=& 1 - \sum_{j=0}^{n-1} \frac{H(t)^j}{j!} \mathrm{e}^{-H(t)} - \mathrm{e}^{-H(t)} \int_0^t \frac{H(u)^{n-1}}{(n-1)!} h(u) \mathrm{d}u \\
=& 1 - \sum_{j=0}^{n} \frac{H(t)^j}{j!} \mathrm{e}^{-H(t)} \quad (n = 0, 1, 2, \cdots).
\end{aligned}$$

Thus,

$$\begin{aligned}
E\{S_n\} =& \int_0^{\infty} \overline{G}_n(t) \mathrm{d}t = \sum_{j=0}^{n-1} \int_0^{\infty} \frac{H(t)^j}{j!} \mathrm{e}^{-H(t)} \mathrm{d}t, \\
E\{X_n\} =& E\{S_n\} - E\{S_{n-1}\} = \int_0^{\infty} \frac{H(t)^{n-1}}{(n-1)!} \mathrm{e}^{-H(t)} \mathrm{d}t.
\end{aligned}$$

X. Zhao and T. Nakagawa, *Advanced Maintenance Policies for Shock and Damage Models*, Springer Series in Reliability Engineering,
https://doi.org/10.1007/978-3-319-70456-2

1.3 Note that for $0 < x < \infty$,

$$\lim_{n\to 0} r_{n+1}(x) = \mathrm{e}^{-\omega x}, \quad \lim_{n\to\infty} r_{n+1}(x) = 1.$$

Furthermore,

$$r_{n+1}(x) - r_n(x) = \frac{[(\omega x)^{n-1}/n!]\sum_{j=n}^{\infty}[(\omega x)^j/j!](j-n)}{\sum_{j=n}^{\infty}[(\omega x)^j/j!]\sum_{j=n-1}^{\infty}[(\omega x)^j/j!]} > 0,$$

which follows that $r_{n+1}(x)$ increases strictly with n from $\mathrm{e}^{-\omega x}$ to 1. Similarly,

$$\lim_{x\to 0} r_{n+1}(x) = 1, \quad \lim_{x\to\infty} r_{n+1}(x) = 0.$$

Differentiating $r_{n+1}(x)$ with respect to x,

$$\frac{\omega[(\omega x)^{n-1}/n!]\sum_{j=n}^{\infty}[(\omega x)^j/j!](n-j)}{\{\sum_{j=n}^{\infty}[(\omega x)^j/j!]\}^2} < 0,$$

which follows that $r_{n+1}(x)$ decreases strictly with x from 1 to 0.

1.4 When $i = n-1$ and $i = n-2$, respectively, from (1.36),

$$I_{n-1} = \mathrm{e}^{\omega(Z_n - Z_{n-1})},$$
$$I_{n-2} = \mathrm{e}^{\omega(Z_n - Z_{n-1})} + \mathrm{e}^{\omega(Z_{n-1} - Z_{n-2})} - 1.$$

Generally,

$$I_i = \sum_{j=i}^{n-1} \mathrm{e}^{\omega(Z_{j+1} - Z_j)} - (n-1-i).$$

A.2 Answers to Problem 2

2.1 Summing up from (2.1) to (2.4),

$$\sum_{j=0}^{N-1}[F^{(j)}(T) - F^{(j+1)}(T)]G^{(j)}(Z) + F^{(N)}(T)G^{(N)}(Z)$$
$$+ \sum_{j=0}^{N-1} F^{(j+1)}(T)[G^{(j)}(Z) - G^{(j+1)}(Z)]$$
$$= \sum_{j=0}^{N} F^{(j)}(T)G^{(j)}(Z) - \sum_{j=0}^{N-1} F^{(j+1)}(T)G^{(j+1)}(Z) = 1.$$

Using the relation

$$\int_0^T t\mathrm{d}F^{(j+1)}(t) = TF^{(j+1)}(T) - \int_0^T F^{(j+1)}(t)\mathrm{d}t,$$

the mean time to replacement is

$$T\sum_{j=0}^{N-1}[F^{(j)}(T) - F^{(j+1)}(T)]G^{(j)}(Z) + G^{(N)}(Z)\left[TF^{(N)}(T) - \int_0^T F^{(N)}(t)\mathrm{d}t\right]$$
$$+\sum_{j=0}^{N-1}\left[TF^{(j+1)}(T) - \int_0^T F^{(j+1)}(t)\mathrm{d}t\right]\int_0^Z \overline{G}(Z-x)\mathrm{d}G^{(j)}(x)$$
$$= \sum_{j=0}^{N-1} G^{(j)}(Z)\int_0^T F^{(j)}(t)\mathrm{d}t - \sum_{j=0}^{N-1} G^{(j)}(Z)\int_0^T F^{(j+1)}(t)\mathrm{d}t$$
$$= \sum_{j=0}^{N-1} G^{(j)}(Z)\int_0^T [F^{(j)}(t) - F^{(j+1)}(t)]\mathrm{d}t.$$

2.2 Differentiating the left-hand side of (2.10) with T,

$$\frac{\mathrm{d}Q_1(T)}{\mathrm{d}T}\sum_{j=0}^{\infty} G^{(j)}(K)\int_0^T [F^{(j)}(t) - F^{(j+1)}(t)]\mathrm{d}t > 0,$$

and as $T \to \infty$, it goes to

$$Q_1(\infty)\sum_{j=0}^{\infty} G^{(j)}(K)\int_0^{\infty} [F^{(j)}(t) - F^{(j+1)}(t)]\mathrm{d}t - \sum_{j=0}^{\infty}[G^{(j)}(K) - G^{(j+1)}(K)]$$
$$= \mu Q_1(\infty)[1 + M_G(K)] - 1.$$

Next, note that when $r_{j+1}(x)$ increases strictly with j,

$$\lim_{T\to 0} \widetilde{Q}_1(T,N) = \overline{G}(K), \quad \lim_{T\to\infty} \widetilde{Q}_1(T,N) = r_N(K),$$
$$\lim_{N\to 1} \widetilde{Q}_1(T,N) = \overline{G}(K), \quad \lim_{N\to\infty} \widetilde{Q}_1(T,N) = Q_1(T)/\lambda.$$

Differentiating $[1 - \widetilde{Q}_1(T,N)]$ with respect to T,

$$\frac{\sum_{j=0}^{N}\sum_{i=0}^{N}[(\lambda T)^j/j!][(\lambda T)^i/i!]G^{(j+1)}(K)G^{(i)}(K)(j-i)}{T\{\sum_{j=0}^{\infty}[(\lambda T)^j/j!]G^{(j)}(K)\}^2}.$$

Letting $L(N)$ be the numerator of the above fraction,

$$L(N) = \sum_{j=0}^{N} \left\{ \sum_{i=0}^{j} \frac{(\lambda T)^j}{j!} \frac{(\lambda T)^i}{i!} G^{(j+1)}(K) G^{(i)}(K)(j-i) \right.$$
$$\left. + \sum_{i=j}^{N} \frac{(\lambda T)^j}{j!} \frac{(\lambda T)^i}{i!} G^{(j+1)}(K) G^{(i)}(K)(j-i) \right\}.$$

On the other hand,

$$\sum_{j=0}^{N} \sum_{i=j}^{N} \frac{(\lambda T)^j}{j!} \frac{(\lambda T)^i}{i!} G^{(j+1)}(K) G^{(i)}(K)(j-i)$$
$$= \sum_{i=0}^{N} \sum_{j=0}^{i} \frac{(\lambda T)^j}{j!} \frac{(\lambda T)^i}{i!} G^{(j+1)}(K) G^{(i)}(K)(j-i)$$
$$= \sum_{j=0}^{N} \sum_{i=0}^{j} \frac{(\lambda T)^i}{i!} \frac{(\lambda T)^j}{j!} G^{(i+1)}(K) G^{(j)}(K)(i-j).$$

Thus,

$$L(N) = \sum_{j=0}^{N} \sum_{i=0}^{j} \frac{(\lambda T)^i}{i!} \frac{(\lambda T)^j}{j!} G^{(i)}(K) G^{(j)}(K)(j-i)[r_{i+1}(K) - r_{j+1}(K)] < 0,$$

which follows that $\widetilde{Q}_1(T, N)$ increases strictly with T to $r_N(K)$ for $N \geq 2$, and $\widetilde{Q}_1(T, 1) \equiv \overline{G}(K)$ for $0 \leq T < \infty$.

Forming $\widetilde{Q}_1(T, N+1) - \widetilde{Q}_1(T, N)$,

$$\frac{\begin{array}{c}[(\lambda T)^N/N!]G^{(N)}(K)\sum_{j=0}^{N-1}[(\lambda T)^j/j!]G^{(j+1)}(K) \\ -[(\lambda T)^N/N!]G^{(N+1)}(K)\sum_{j=0}^{N-1}[(\lambda T)^j/j!]G^{(j)}(K)\end{array}}{\sum_{j=0}^{N-1}[(\lambda T)^j/j!]G^{(j)}(K)\sum_{j=0}^{N}[(\lambda T)^j/j!]G^{(j)}(K)}.$$

The numerator of the above fraction becomes

$$\frac{(\lambda T)^N}{N!} \sum_{j=0}^{N-1} \frac{(\lambda T)^j}{j!} G^{(N)}(K) G^{(j)}(K)[r_{N+1}(K) - r_{j+1}(K)] > 0,$$

which follows that $\widetilde{Q}_1(T, N)$ increases strictly with N from $\overline{G}(K)$ to $\widetilde{Q}_1(T, \infty) = Q_1(T)/\lambda$.

2.3 In order to obtain optimum N^* to minimize $C(N)$, we have to compute a finite and unique integer N^* $(1 \leq N^* < \infty)$ which satisfies

$$C(N-1) - C(N) > 0 \quad \text{and} \quad C(N+1) - C(N) \geq 0 \quad (N = 1, 2, \cdots).$$

That is, we derive a finite and unique minimum N^* which satisfies $C(N+1) - C(N) \geq 0$, or a finite and unique maximum N^* which satisfies $C(N-1) - C(N) > 0$. If there exists no finite N^* to satisfy $C(N+1) - C(N) \geq 0$, then $C(N+1) < C(N)$ for any N, i.e., $C(N)$ decreases with N and $N^* = \infty$. Conversely, if $C(N+1) - C(N) \geq 0$ for any N, then $C(N+1) \geq C(N)$ for any N, i.e., $C(N)$ increases with N and $N^* = 1$. If there exist several N_i to satisfy $C(N+1) - C(N) \geq 0$, then N^* can be decided by comparing $C(N_i)$. The same description can be given for N^* obtained from the inequality $C(N-1) - C(N) > 0$. In general, the inequality $C(N+1) - C(N) \geq 0$ is widely used to obtain optimum N^* for discrete optimization problems [1, 2].

2.4 Letting $L(N)$ be the left-hand side of (2.14),

$$\lim_{N\to\infty} L(N) = r_\infty(K)[1 + M_G(K)] - 1,$$

$$L(N+1) - L(N) = [r_{N+2}(K) - r_{N+1}(K)] \sum_{j=0}^{N} G^{(j)}(K) > 0.$$

When $G(x) = 1 - \mathrm{e}^{-\omega x}$ for $0 < x < \infty$,

$$\lim_{N\to 0} r_{N+1}(x) = \mathrm{e}^{-\omega x}, \quad \lim_{N\to\infty} r_{N+1}(x) = 1,$$

$$r_{N+1}(x) - r_N(x) = \frac{[(\omega x)^{N-1}/N!] \sum_{j=N}^{\infty}[(\omega x)^j (j-N)/j!]}{\sum_{j=N}^{\infty}[(\omega x)^j/j!] \sum_{j=N-1}^{\infty}[(\omega x)^j/j!]} > 0,$$

which follows that $r_{N+1}(x)$ increases strictly with N from $\mathrm{e}^{-\omega x}$ to 1.

Furthermore, differentiating $r_{N+1}(x)$ with x,

$$\frac{1}{N!} \sum_{j=N}^{\infty} \frac{(\omega x)^{N-1-j}}{j!}(N-j) < 0,$$

which follows that $r_{N+1}(x)$ decreases strictly with x from 1 to $r_{N+1}(K)$.

Similarly, $\widetilde{r}_{N+1}(x)$ in (2.55) decreases strictly with N from ωx to 0 and increases strictly with x from 0 to $\widetilde{r}_{N+1}(K)$.

2.5 Differentiating $C(Z)$ with respect to Z and setting it equal to zero,

$$\overline{G}(K-Z)[1 + M_G(Z)] - \overline{G}(K) - \int_0^Z \overline{G}(K-x)\mathrm{d}M_G(x) = \frac{c_Z}{c_K - c_Z},$$

i.e.,

$$\int_{K-Z}^{K} [1 + M_G(K-x)]\mathrm{d}G(x) = \frac{c_Z}{c_K - c_Z},$$

whose left-hand side increases strictly with Z from 0 to

$$M_G(K) = \int_0^K [1 + M_G(K - x)]\mathrm{d}G(x).$$

2.6 When $\int_0^Z \overline{G}(K-x)\mathrm{d}G^{(j)}(x)/G^{(j)}(Z)$ increases with j to $\overline{G}(K-Z)$, using a similar method of Problem 2.2, $\widetilde{Q}_3(T,Z)$ increases strictly with T from $\overline{G}(K)$ to $\overline{G}(K-Z)$. In particular, when $G(x) = 1 - \mathrm{e}^{-\omega x}$,

$$\frac{\int_0^Z \overline{G}(K-x)\mathrm{d}G^{(j)}(x)}{G^{(j)}(Z)} = \frac{[(\omega Z)^j/j!]\mathrm{e}^{-\omega[K-Z]}}{\sum_{i=j}^{\infty}[(\omega Z)^i/i!]}$$

increases strictly with j from $\mathrm{e}^{-\omega K}$ to $\mathrm{e}^{-\omega(K-Z)}$.

2.7 Substituting (2.31) for (2.32),

$$\frac{\sum_{j=0}^{N-1} G^{(j)}(Z)}{G^{(N)}(Z)} \int_0^Z \overline{G}(K-x)\mathrm{d}G^{(N)}(x) - \sum_{j=0}^{N-1} \int_0^Z \overline{G}(K-x)\mathrm{d}G^{(j)}(x)$$

$$\geq \sum_{j=0}^{N-1} \int_0^Z [\overline{G}(K-Z) - \overline{G}(K-x)]\mathrm{d}G^{(j)}(x),$$

which follows that

$$\int_0^Z [\overline{G}(K-Z) - \overline{G}(K-x)]\mathrm{d}G^{(N)}(x) \leq 0.$$

However, the above inequality dose not hold for any Z as $\overline{G}(K-Z) > \overline{G}(K-x)$ for $0 \leq x < Z$, which means that a finite N_F^* does not exist.

2.8 Prove that

$$Q_4(T, N+1) = \frac{[(\lambda T)^N/N!]G^{(N+1)}(K)}{\sum_{j=0}^{N}[(\lambda T)^j/j!]G^{(j)}(K)}$$

increases strictly with T from 0 to $G^{(N+1)}(K)/G^{(N)}(K)$ for $N \geq 1$.

Note that

$$\lim_{T\to 0} Q_4(T, N+1) = 0, \quad \lim_{T\to\infty} Q_4(T, N+1) = \frac{G^{(N+1)}(K)}{G^{(N)}(K)}.$$

Differentiating $Q_4(T, N+1)$ with respect to T,

$$\frac{\lambda G^{(N+1)}(K)}{N!\{\sum_{j=0}^{N}[(\lambda T)^j/j!]G^{(j)}(K)\}^2} \sum_{j=0}^{N} \frac{(\lambda T)^{N+j-1}}{j!} G^{(j)}(K)(N-j) > 0,$$

which follows that $Q_4(T, N+1)$ increases strictly with T from 0 to $G^{(N+1)}(K)/G^{(N)}(K)$. When $N = 0$, $Q_4(T, 1) = G(K)$ for $T > 0$.

2.10 Prove that $Q_6(T, Z)$ increases strictly with Z from $F(T)/F^{(2)}(T)$ for $0 < T < \infty$. Firstly, note that $F^{(j)}(T)/F^{(j+1)}(T)$ increases strictly with j when $F(t) = 1 - e^{-\lambda t}$. Differentiating $Q_6(T, Z)$ with respect to Z,

$$\frac{\begin{array}{c}\sum_{j=0}^{\infty} F^{(j+2)}(T)[(\omega Z)^j/j!] \sum_{i=0}^{\infty} F^{(i+2)}(T)[(\omega Z)^i/i!] \\ -\sum_{j=0}^{\infty} F^{(j+3)}(T)[(\omega Z)^j/j!] \sum_{i=0}^{\infty} F^{(i+1)}(T)[(\omega Z)^i/i!]\end{array}}{Z\{\sum_{j=0}^{\infty} F^{(j+2)}(T)[(\omega Z)^j/j!]\}^2}.$$

Letting $L(Z)$ be the numerator of the above fraction,

$$\begin{aligned} L(Z) &= \sum_{j=0}^{\infty} \sum_{i=0}^{\infty} F^{(j+2)}(T) \frac{(\omega Z)^j}{j!} F^{(i+1)}(T) \frac{(\omega Z)^i}{i!} (i - j) \\ &= \sum_{j=0}^{\infty} \left[\sum_{i=0}^{j} F^{(j+2)}(T) \frac{(\omega Z)^j}{j!} F^{(i+1)}(T) \frac{(\omega Z)^i}{i!} (i - j) \right. \\ &\quad \left. + \sum_{i=j}^{\infty} F^{(j+2)}(T) \frac{(\omega Z)^j}{j!} F^{(i+1)}(T) \frac{(\omega Z)^i}{i!} (i - j) \right]. \end{aligned}$$

On the other hand,

$$\begin{aligned} &\sum_{j=0}^{\infty} \sum_{i=j}^{\infty} F^{(j+2)}(T) \frac{(\omega Z)^j}{j!} F^{(i+1)}(T) \frac{(\omega Z)^i}{i!} (i - j) \\ &= \sum_{i=0}^{\infty} \sum_{j=0}^{i} F^{(j+2)}(T) \frac{(\omega Z)^j}{j!} F^{(i+1)}(T) \frac{(\omega Z)^i}{i!} (i - j) \\ &= \sum_{j=0}^{\infty} \sum_{i=0}^{j} F^{(i+2)}(T) \frac{(\omega Z)^i}{i!} F^{(j+1)}(T) \frac{(\omega Z)^j}{j!} (j - i). \end{aligned}$$

Thus,

$$\begin{aligned} L(Z) &= \sum_{j=0}^{\infty} \sum_{i=0}^{j} F^{(j+2)}(T) F^{(i+2)}(T) \frac{(\omega Z)^j}{j!} \frac{(\omega Z)^i}{i!} (i - j) \\ &\quad \times \left[\frac{F^{(i+1)}(T)}{F^{(i+2)}(T)} - \frac{F^{(j+1)}(T)}{F^{(j+2)}(T)} \right] > 0, \end{aligned}$$

which follows that $Q_6(T, Z)$ increases strictly with Z from $F(T)/F^{(2)}(T)$ to $Q_6(T, K)$.

Similarly, for $0 < Z \le K$,

$$\lim_{T\to 0} Q_6(T,Z) = \infty, \quad \lim_{T\to\infty} Q_6(T,Z) = 1.$$

Differentiating $Q_6(T,Z)$ with respect to T,

$$\sum_{j=0}^{\infty} F^{(j)}(T)\frac{(\omega Z)^j}{j!}\sum_{j=0}^{\infty} F^{(j+2)}(T)\frac{(\omega Z)^j}{j!}$$
$$-\sum_{j=0}^{\infty} F^{(j+1)}(T)\frac{(\omega Z)^j}{j!}\sum_{j=0}^{\infty} F^{(j+1)}(T)\frac{(\omega Z)^j}{j!} < 0,$$

which follows that $Q_6(T,Z)$ decreases strictly with T from ∞ to 1.

2.11 Differentiating $C_R(T;\theta)$ with respect to T and setting it equal to zero, we have (2.61), whose left-hand side increases strictly with T from 1 to

$$L(\infty) \equiv [1 - G^*(\theta)]h(\infty)\int_0^{\infty} \exp\left\{-[1 - G^*(\theta)]H(t)\right\}\mathrm{d}t.$$

Thus, if $L(\infty) > c_K/(c_K - c_T)$, then there exists a finite and unique T_R^* $(0 < T_R^* < \infty)$ which satisfies (2.61), and the resulting cost rate is given in (2.62).

2.12 From $C_R(N+1;L) - C_R(N;L) \ge 0$, we have (2.64). Letting $L(N)$ be the left-hand side of (2.64),

$$L(N+1) - L(N) = [Q_7(N+1) - Q_7(N)]\sum_{j=0}^{N}\int_0^{\infty} G^{(j)}(x)\mathrm{d}L(x).$$

Thus, if $Q_7(N)$ increases strictly with N to 1, then the left-hand side of (2.64) also increases strictly with N to $\int_0^{\infty}[1 + M_G(x)]\mathrm{d}L(x)$.

A.3 Answers to Problem 3

3.1 Summing up the above three probabilities,

$$1 - \sum_{j=N}^{\infty}[1 - F^{(j+1)}(T)]\left[G^{(j)}(K) - G^{(j+1)}(K) - \int_0^{Z}\overline{G}(K-x)\mathrm{d}G^{(j)}(x)\right]$$
$$= 1 - \sum_{j=N}^{\infty}[1 - F^{(j+1)}(T)]\int_Z^{K}\overline{G}(K-x)\mathrm{d}G^{(j)}(x).$$

3.2 It can be easily proved that when $r_{j+1}(K)$ increases strictly with j to 1 for $0 \le N < \infty$ and $0 < K < \infty$,

$$\frac{\sum_{j=N}^{\infty}[(\lambda T)^j/j!][G^{(j)}(K) - G^{(j+1)}(K)]}{\sum_{j=N}^{\infty}[(\lambda T)^j/j!]G^{(j)}(K)} > r_{N+1}(K).$$

When $1 - r_{j+1}(K) = G^{(j+1)}(K)/G^{(j)}(K)$ decreases strictly with j to 0, let

$$\begin{aligned}
L(T) &\equiv \sum_{j=N}^{\infty} j\frac{(\lambda T)^j}{j!}G^{(j+1)}(K)\sum_{i=N}^{\infty}\frac{(\lambda T)^i}{i!}G^{(i)}(K) \\
&\quad - \sum_{j=N}^{\infty}\frac{(\lambda T)^j}{j!}G^{(j+1)}(K)\sum_{i=N}^{\infty} i\frac{(\lambda T)^i}{i!}G^{(i)}(K) \\
&= \sum_{j=N}^{\infty}\frac{(\lambda T)^j}{j!}G^{(j+1)}(K)\left[\sum_{i=N}^{j}\frac{(\lambda T)^i}{i!}G^{(i)}(K)(j-i)\right. \\
&\quad \left. + \sum_{i=j}^{\infty}\frac{(\lambda T)^i}{i!}G^{(i)}(K)(j-i)\right].
\end{aligned}$$

On the other hand,

$$\begin{aligned}
&\sum_{j=N}^{\infty}\frac{(\lambda T)^j}{j!}G^{(j+1)}(K)\sum_{i=N}^{j}\frac{(\lambda T)^i}{i!}G^{(i)}(K)(j-i) \\
&= \sum_{j=N}^{\infty}\frac{(\lambda T)^j}{j!}G^{(j)}(K)\sum_{i=j}^{\infty}\frac{(\lambda T)^i}{i!}G^{(i+1)}(K)(i-j).
\end{aligned}$$

So that,

$$\begin{aligned}
L(T) &= \sum_{j=N}^{\infty}\frac{(\lambda T)^j}{j!}G^{(j)}(K)\sum_{i=j}^{\infty}\frac{(\lambda T)^i}{i!}G^{(i)}(K)(j-i) \\
&\quad \times\left[\frac{G^{(j+1)}(K)}{G^{(j)}(K)} - \frac{G^{(i+1)}(K)}{G^{(i)}(K)}\right] < 0,
\end{aligned}$$

which follows that $\widetilde{Q}_1(T, N)$ increases strictly with T. Clearly,

$$\lim_{T\to 0}\widetilde{Q}_1(T, N) = r_{N+1}(K), \quad \lim_{T\to\infty}\widetilde{Q}_1(T, N) = \lim_{N\to\infty} r_{N+1}(K) = 1.$$

Furthermore,

$$
\begin{aligned}
L(N) \equiv & \sum_{j=N+1}^{\infty} \frac{(\lambda T)^j}{j!} G^{(j+1)}(K) \sum_{j=N}^{\infty} \frac{(\lambda T)^j}{j!} G^{(j)}(K) \\
& - \sum_{j=N}^{\infty} \frac{(\lambda T)^j}{j!} G^{(j+1)}(K) \sum_{j=N+1}^{\infty} \frac{(\lambda T)^j}{j!} G^{(j)}(K) \\
= & \frac{(\lambda T)^N}{N!} \sum_{j=N+1}^{\infty} \frac{(\lambda T)^j}{j!} G^{(j)}(K) G^{(N)}(K) \\
& \times \left[\frac{G^{(j+1)}(K)}{G^{(j)}(K)} - \frac{G^{(N+1)}(K)}{G^{(N)}(K)} \right] < 0,
\end{aligned}
$$

which follows that $\widetilde{Q}_1(T, N)$ increases strictly with N. Clearly,

$$
\lim_{N \to \infty} \widetilde{Q}_1(T, N) = \lim_{N \to \infty} \frac{G^{(N)}(K) - G^{(N+1)}(K)}{G^{(N)}(K)} = 1.
$$

When $G(x) = 1 - \mathrm{e}^{-\omega x}$, note that

$$
r_{j+1}(x) = \frac{(\omega x)^j / j!}{\sum_{i=j}^{\infty} [(\omega x)^i / i!]} \quad (j = 0, 1, 2, \cdots)
$$

increases strictly with j from $\mathrm{e}^{-\omega x}$ to 1 (Problem 2.4).

3.3 Because $\overline{G}(K - x) > \overline{G}(K - Z)$ for $Z < x \le K$ and $Q_2(T, Z) = \lambda \overline{G}(K - Z)$, we have $Q_3(T, Z) > \lambda \overline{G}(K - Z)$, i.e., $Q_3(T, Z) > Q_2(T, Z)$ for $0 \le Z < K$. Similarly,

$$
\frac{\int_Z^K \overline{G}(K - x) \mathrm{d}G^{(N)}(x)}{G^{(N)}(K) - G^{(N)}(Z)} > \overline{G}(K - Z)
$$

for $0 \le Z < K$ and $N = 1, 2, \cdots$.

3.4 When $F(t) = 1 - \mathrm{e}^{-\lambda t}$, (3.19) can be obtained from (3.9), using the following relations:

$$
\int_0^T [F^{(j)}(t) - F^{(j+1)}(t)] \mathrm{d}t = \int_0^T \frac{(\lambda t)^j}{j!} \mathrm{e}^{-\lambda t} \mathrm{d}t = \frac{1}{\lambda} F^{(j+1)}(T),
$$

$$
\int_T^{\infty} [F^{(j)}(t) - F^{(j+1)}(t)] \mathrm{d}t = \int_T^{\infty} \frac{(\lambda t)^j}{j!} \mathrm{e}^{-\lambda t} \mathrm{d}t = \frac{1}{\lambda} [1 - F^{(j+1)}(T)].
$$

3.5 From (3.21) and (3.23),

$$A(N) = r_{N+1}(K)\left\{\sum_{j=0}^{N-1}[1-F^{(j+1)}(T)]G^{(j)}(K) + \sum_{j=N}^{\infty} F^{(j+1)}(T)G^{(j)}(K)\right\}$$

$$+\sum_{j=N}^{\infty}[1-F^{(j+1)}(T)][G^{(j)}(K)-G^{(j+1)}(K)] - 1$$

$$+\sum_{j=0}^{N-1} F^{(j+1)}(T)[G^{(j)}(K)-G^{(j+1)}(K)].$$

So that, using $G^{(N)}(K) - G^{(N+1)}(K) = r_{N+1}(K)G^{(N)}(K)$,

$$A(N+1) - A(N) = [r_{N+2}(K) - r_{N+1}(K)]\left\{\sum_{j=0}^{N}[1-F^{(j+1)}(T)]G^{(j)}(K)\right.$$

$$\left.+\sum_{j=N+1}^{\infty} F^{(j+1)}(T)G^{(j)}(K)\right\} > 0,$$

and

$$\lim_{N\to\infty} A(N) = \sum_{j=0}^{\infty}[1-F^{(j)}(T)]G^{(j)}(K).$$

A.4 Answers to Problem 4

4.1 The probability that the unit is replaced at failure is classified into two cases: The probability that the unit fails after time T is

$$\sum_{j=0}^{\infty}[G^{(j)}(K)-G^{(j+1)}(K)]\int_0^T \overline{F}(T-t)\mathrm{d}F^{(j)}(t),$$

and the probability that it is fails before time T is

$$\sum_{j=0}^{\infty}[G^{(j)}(K)-G^{(j+1)}(K)]\int_0^T F(T-t)\mathrm{d}F^{(j)}(t).$$

Using the following relation,

$$F^{(j+1)}(t) = \int_0^t F^{(j)}(t-u)\mathrm{d}F(u),$$

$$\int_0^T F^{(j+1)}(t)\mathrm{d}t = \int_0^T \left[\int_0^t F^{(j)}(t-u)\mathrm{d}F(u)\right]\mathrm{d}t = \int_0^T F^{(j)}(t)F(T-t)\mathrm{d}t,$$

the mean time to replacement is

$$\begin{aligned}
&\sum_{j=0}^{\infty}[G^{(j)}(K) - G^{(j+1)}(K)]\left\{\int_0^T \left[\int_{T-t}^{\infty}(t+u)\mathrm{d}F(u)\right]\mathrm{d}F^{(j)}(t)\right. \\
&\left.+\int_0^T \left[\int_0^{T-t}(t+u)\mathrm{d}F(u)\right]\mathrm{d}F^{(j)}(t)\right\} \\
&+\sum_{j=0}^{\infty} G^{(j+1)}(K)\int_0^T \left[\int_{T-t}^{\infty}(t+u)\mathrm{d}F(u)\right]\mathrm{d}F^{(j)}(t) \\
&=\sum_{j=0}^{\infty} G^{(j)}(K)\int_0^T \left[\int_{T-t}^{\infty}(t+u)\mathrm{d}F(u)\right]\mathrm{d}F^{(j)}(t) \\
&\quad+\sum_{j=0}^{\infty}[G^{(j)}(K) - G^{(j+1)}(K)]\int_0^T t\mathrm{d}F^{(j+1)}(t) \\
&=\mu\sum_{j=0}^{\infty} F^{(j)}(T)G^{(j)}(K) + \sum_{j=0}^{\infty} G^{(j)}(K)\left\{\int_0^T [F^{(j)}(t) - F^{(j+1)}(t)]\mathrm{d}t\right. \\
&\quad\left.-\int_0^T \left[\int_0^{T-t}\overline{F}(u)\mathrm{d}u\right]\mathrm{d}F^{(j)}(t)\right\} \\
&=\mu\sum_{j=0}^{\infty} F^{(j)}(T)G^{(j)}(K) + \sum_{j=0}^{\infty} G^{(j)}(K)\int_0^T [F^{(j)}(t)F(T-t) - F^{(j+1)}(t)]\mathrm{d}t \\
&=\mu\sum_{j=0}^{\infty} F^{(j)}(T)G^{(j)}(K).
\end{aligned}$$

4.2 Note that $Q_1(T, N) \le r_N(K)$ for $N \ge 2$. Let

$$
\begin{aligned}
L_1(T) &\equiv \sum_{j=0}^{N} j\frac{(\lambda T)^j}{j!}G^{(j+2)}(K)\sum_{i=0}^{N}\frac{(\lambda T)^i}{i!}G^{(i+1)}(K)\\
&\quad-\sum_{j=0}^{N}\frac{(\lambda T)^j}{j!}G^{(j+2)}(K)\sum_{i=0}^{N} i\frac{(\lambda T)^i}{i!}G^{(i+1)}(K)\\
&=\sum_{j=0}^{N}\frac{(\lambda T)^j}{j!}G^{(j+2)}(K)\left[\sum_{i=0}^{j}\frac{(\lambda T)^i}{i!}G^{(i+1)}(K)(j-i).\right.\\
&\quad\left.+\sum_{i=j}^{N}\frac{(\lambda T)^i}{i!}G^{(i+1)}(K)(j-i)\right].
\end{aligned}
$$

On the other hand,

$$
\begin{aligned}
&\sum_{j=0}^{N}\frac{(\lambda T)^j}{j!}G^{(j+2)}(K)\sum_{i=0}^{j}\frac{(\lambda T)^i}{i!}G^{(i+1)}(K)(j-i)\\
&=\sum_{j=0}^{N}\frac{(\lambda T)^j}{j!}G^{(j+1)}(K)\sum_{i=j}^{N}\frac{(\lambda T)^i}{i!}G^{(i+2)}(K)(i-j).
\end{aligned}
$$

So that,

$$
\begin{aligned}
L_1(T) &= \sum_{j=0}^{N}\frac{(\lambda T)^j}{j!}G^{(j+1)}(K)\sum_{i=j}^{N}\frac{(\lambda T)^i}{i!}G^{(i+1)}(K)(j-i)\\
&\quad\times\left[\frac{G^{(j+2)}(K)}{G^{(j+1)}(K)}-\frac{G^{(i+2)}(K)}{G^{(i+1)}(K)}\right]<0,
\end{aligned}
$$

which follows that $Q_1(T, N)$ increases strictly with T to $r_N(K)$ for $N \geq 2$. Furthermore,

$$
\begin{aligned}
L_2(N) &\equiv \sum_{j=0}^{N}\frac{(\lambda T)^j}{j!}G^{(j+2)}(K)\sum_{j=0}^{N-1}\frac{(\lambda T)^j}{j!}G^{(j+1)}(K)\\
&\quad-\sum_{j=0}^{N-1}\frac{(\lambda T)^j}{j!}G^{(j+2)}(K)\sum_{j=0}^{N}\frac{(\lambda T)^j}{j!}G^{(j+1)}(K)\\
&=\frac{(\lambda T)^N}{N!}G^{(N+1)}(K)\sum_{j=0}^{N-1}\frac{(\lambda T)^j}{j!}G^{(j+1)}(K)\\
&\quad\times\left[\frac{G^{(N+2)}(K)}{G^{(N+1)}(K)}-\frac{G^{(j+2)}(K)}{G^{(j+1)}(K)}\right]<0,
\end{aligned}
$$

which follows that $Q_1(T, N)$ increases strictly with N for $N \geq 2$.

In particular, when $G(x) = 1 - \mathrm{e}^{-\omega x}$, $Q_1(T, N)$ increases strictly with T from $r_2(K)$ to $r_N(K)$, and increases strictly with N for $N \geq 2$ from $r_2(K)$ to $Q_1(T)$. Thus, $Q_1(T)$ increases strictly with T from $r_2(K)$ to 1.

4.3 Setting that $x \equiv \omega K$, for $0 < x < \infty$,

$$x + \mathrm{e}^{-x} - 1 < \frac{x^2}{2}, \quad \frac{x^2/2}{\mathrm{e}^x - 1} < \frac{x}{2},$$

which follows that $x/2 > (x + \mathrm{e}^{-x} - 1)/(\mathrm{e}^x - 1)$.

4.4 When $r_2(K - Z)$ increases strictly with Z from $r_2(K)$ to 1, denote the left-hand side of (4.13) as $L(Z)$,

$$\frac{\mathrm{d}L(Z)}{\mathrm{d}Z} = \frac{\mathrm{d}r_2(K - Z)}{\mathrm{d}Z}\left[1 + G(K) + \int_0^Z G(K - x)\mathrm{d}M_G(x)\right] > 0,$$

$$\lim_{Z\to 0} L(Z) = \frac{G(K)^2 - G^{(2)}(K)}{G(K)} > 0, \quad \lim_{Z\to K} L(Z) = M_G(K),$$

which follows that $L(Z)$ increases strictly from $[G(K)^2 - G^{(2)}(K)]/G(K)$ to $M_G(K)$.

4.5 When $G^{(j+1)}(K)/G^{(j)}(K)$ decreases strictly with j, let

$$\begin{aligned}
L(T) \equiv & \sum_{j=0}^{\infty}\frac{(\lambda T)^j}{j!}G^{(j+1)}(K)\sum_{i=0}^{\infty}\frac{(\lambda T)^i}{i!}G^{(i+1)}(K) \\
& - \sum_{j=0}^{\infty}\frac{(\lambda T)^j}{j!}G^{(j+2)}(K)\sum_{i=0}^{\infty}\frac{(\lambda T)^i}{i!}G^{(i)}(K) \\
= & \sum_{j=0}^{\infty}\frac{(\lambda T)^j}{j!}\left\{\sum_{i=0}^{j}\frac{(\lambda T)^i}{i!}[G^{(j+1)}(K)G^{(i+1)}(K) - G^{(j+2)}(K)G^{(i)}(K)]\right. \\
& \left. + \sum_{i=j}^{\infty}\frac{(\lambda T)^i}{i!}[G^{(j+1)}(K)G^{(i+1)}(K) - G^{(j+2)}(K)G^{(i)}(K)]\right\}.
\end{aligned}$$

On the other hand,

$$\begin{aligned}
& \sum_{j=0}^{\infty}\frac{(\lambda T)^j}{j!}\sum_{i=0}^{j}\frac{(\lambda T)^i}{i!}[G^{(j+1)}(K)G^{(i+1)}(K) - G^{(j+2)}(K)G^{(i)}(K)] \\
& = \sum_{j=0}^{\infty}\frac{(\lambda T)^j}{j!}\sum_{i=j}^{\infty}\frac{(\lambda T)^i}{i!}[G^{(j+1)}(K)G^{(i+1)}(K) - G^{(j)}(K)G^{(i+2)}(K)].
\end{aligned}$$

So that,

$$L(T) = \sum_{j=0}^{\infty} \frac{(\lambda T)^j}{j!} G^{(j)}(K) \sum_{i=j}^{\infty} \frac{(\lambda T)^i}{i!} G^{(i+1)}(K) \times \left[\frac{G^{(j+1)}(K)}{G^{(j)}(K)} - \frac{G^{(i+2)}(K)}{G^{(i+1)}(K)} \right] > 0,$$

which follows that $\widetilde{Q}_1(T) < Q_1(T)$.

4.6 Summing up from (4.32) to (4.35),

$$\sum_{j=0}^{N-1} G^{(j)}(K)[F^{(j)}(T) - F^{(j+1)}(T)] + \sum_{j=N}^{\infty} G^{(j)}(Z)[F^{(j)}(T) - F^{(j+1)}(T)]$$
$$+ F^{(N)}(T)[G^{(N)}(K) - G^{(N)}(Z)] + \sum_{j=N}^{\infty} F^{(j+1)}(T)[G^{(j)}(Z) - G^{(j+1)}(Z)]$$
$$+ \sum_{j=0}^{N-1} F^{(j+1)}(T)[G^{(j)}(K) - G^{(j+1)}(K)]$$
$$= \sum_{j=0}^{N-1} G^{(j)}(K)[F^{(j)}(T) - F^{(j+1)}(T)] + \sum_{j=N}^{\infty} G^{(j)}(Z)[F^{(j)}(T) - F^{(j+1)}(T)]$$
$$+ 1 - \sum_{j=0}^{N-1} G^{(j)}(K)[F^{(j)}(T) - F^{(j+1)}(T)]$$
$$- \sum_{j=N}^{\infty} G^{(j)}(Z)[F^{(j)}(T) - F^{(j+1)}(T)] = 1.$$

The mean time to replacement is

$$T \left\{ \sum_{j=0}^{N-1} G^{(j)}(K)[F^{(j)}(T) - F^{(j+1)}(T)] + \sum_{j=N}^{\infty} G^{(j)}(Z)[F^{(j)}(T) - F^{(j+1)}(T)] \right\}$$
$$+ [G^{(N)}(K) - G^{(N)}(Z)] \int_0^T t \mathrm{d}F^{(N)}(t)$$
$$+ \sum_{j=N}^{\infty} [G^{(j)}(Z) - G^{(j+1)}(Z)] \int_0^T t \mathrm{d}F^{(j+1)}(t)$$

$$
\begin{aligned}
&+\sum_{j=0}^{N-1}[G^{(j)}(K)-G^{(j+1)}(K)]\int_0^T \mathrm{d}F^{(j+1)}(t)\\
&=\sum_{j=N}^{\infty} G^{(j)}(Z)\int_0^T [F^{(j)}(t)-F^{(j+1)}(t)]\mathrm{d}t\\
&+\sum_{j=0}^{N-1} G^{(j)}(K)\int_0^T [F^{(j)}(t)-F^{(j+1)}(t)]\mathrm{d}t.
\end{aligned}
$$

4.7 Summing up from (4.45) to (4.48),

$$
\begin{aligned}
&\sum_{j=0}^{N-1}[F^{(j)}(T)-F^{(j+1)}(T)][G^{(j)}(K)-G^{(j)}(Z)]\\
&+\sum_{j=N}^{\infty} G^{(j)}(Z)[F^{(j)}(T)-F^{(j+1)}(T)]\\
&+G^{(N)}(Z)[1-F^{(N)}(T)]+F^{(N)}(T)[G^{(N)}(K)-G^{(N)}(Z)]\\
&+\sum_{j=0}^{N-1}[1-F^{(j+1)}(T)][G^{(j)}(Z)-G^{(j+1)}(Z)]\\
&+\sum_{j=N}^{\infty} F^{(j+1)}(T)[G^{(j)}(Z)-G^{(j+1)}(Z)]\\
&+\sum_{j=0}^{N-1} F^{(j+1)}(T)[G^{(j)}(K)-G^{(j+1)}(K)]=1.
\end{aligned}
$$

The mean time to replacement is

$$
\begin{aligned}
&\sum_{j=0}^{N-1}[G^{(j)}(Z)-G^{(j+1)}(Z)]\int_T^{\infty}[1-F^{(j+1)}(t)]\mathrm{d}t\\
&+\sum_{j=0}^{N-1}[G^{(j)}(K)-G^{(j+1)}(K)]\int_0^T[1-F^{(j+1)}(t)]\mathrm{d}t\\
&+G^{(N)}(Z)\int_T^{\infty}[1-F^{(N)}(t)]\mathrm{d}t+[G^{(N)}(K)-G^{(N)}(Z)]\int_0^T[1-F^{(N)}(t)]\mathrm{d}t\\
&+\sum_{j=N}^{\infty}[G^{(j)}(Z)-G^{(j+1)}(Z)]\int_0^T[1-F^{(j+1)}(t)]\mathrm{d}t\\
&=\sum_{j=0}^{N-1} G^{(j)}(Z)\int_T^{\infty}[F^{(j)}(t)-F^{(j+1)}(t)]\mathrm{d}t
\end{aligned}
$$

$$+\sum_{j=0}^{N-1} G^{(j)}(K)\int_0^T [F^{(j)}(t)-F^{(j+1)}(t)]\mathrm{d}t$$

$$+\sum_{j=N}^{\infty} G^{(j)}(Z)\int_0^T [F^{(j)}(t)-F^{(j+1)}(t)]\mathrm{d}t.$$

A.5 Answers to Problem 5

5.1 The expected number of shocks before replacement is

$$\sum_{j=0}^{N-1} j[F^{(j)}(T)-F^{(j+1)}(T)]G^{(j)}(Z)+(N-1)F^{(N)}(T)G^{(N)}(Z)$$
$$+\sum_{j=0}^{N-1} jF^{(j+1)}(T)[G^{(j)}(Z)-G^{(j+1)}(Z)]=\sum_{j=1}^{N-1}F^{(j)}(T)G^{(j)}(Z). \quad \text{(A.1)}$$

The expected number of shocks until replacement, including the shock at N or at Z, is

$$\sum_{j=0}^{N-1} j[F^{(j)}(T)-F^{(j+1)}(T)]G^{(j)}(Z)+NF^{(N)}(T)G^{(N)}(Z)$$
$$+\sum_{j=0}^{N-1}(j+1)F^{(j+1)}(T)[G^{(j)}(Z)-G^{(j+1)}(Z)]=\sum_{j=0}^{N-1}F^{(j+1)}(T)G^{(j)}(Z). \quad \text{(A.2)}$$

The expected number of failures before replacement is

$$\sum_{j=0}^{N-1}[F^{(j)}(T)-F^{(j+1)}(T)]\sum_{i=1}^{j}\int_0^Z p(x)\mathrm{d}G^{(i)}(x)$$
$$+\sum_{j=0}^{N-1}F^{(j+1)}(T)\sum_{i=1}^{j}\left[\int_0^Z p(x)\mathrm{d}G^{(i)}(x)-\int_0^Z p(x)\mathrm{d}G^{(i+1)}(x)\right]$$
$$+F^{(N)}(T)\sum_{j=1}^{N-1}\int_0^Z p(x)\mathrm{d}G^{(j)}(x)=\sum_{j=1}^{N-1}F^{(j)}(T)\int_0^Z p(x)\mathrm{d}G^{(j)}(x), \quad \text{(A.3)}$$

which agrees with (A.1) when $p(x)\equiv 1$. The expected number of failures until replacement is

$$\sum_{j=0}^{N-1}[F^{(j)}(T)-F^{(j+1)}(T)]\sum_{i=1}^{j}\int_0^Z p(x)\mathrm{d}G^{(i)}(x)$$
$$+F^{(N)}(T)\sum_{j=1}^{N}\int_0^Z p(x)\mathrm{d}G^{(j)}(x)$$
$$+\sum_{j=0}^{N-1}F^{(j+1)}(T)\sum_{i=1}^{j+1}\left[\int_0^Z p(x)\mathrm{d}G^{(i)}(x)-\int_0^Z p(x)\mathrm{d}G^{(i+1)}(x)\right]$$
$$=\sum_{j=0}^{N}F^{(j)}(T)\int_0^Z p(x)\mathrm{d}G^{(j)}(x)+\sum_{j=0}^{N-1}F^{(j+1)}(T)\int_0^Z p(x)\mathrm{d}G^{(j)}(x)$$
$$-\sum_{j=1}^{N}F^{(j)}(T)\int_0^Z p(x)\mathrm{d}G^{(j)}(x)=\sum_{j=0}^{N-1}F^{(j+1)}(T)\int_0^Z p(x)\mathrm{d}G^{(j)}(x),\tag{A.4}$$

which agrees with (A.2) when $p(x)\equiv 1$. Using this method, we can easily obtain the expected number of failures from the expected number of shocks.

5.2 The expected number of failures until replacement is

$$\sum_{j=1}^{N-1}[F^{(j)}(T)-F^{(j+1)}(T)]\sum_{i=1}^{j}p_iG^{(i)}(Z)+F^{(N)}(T)\sum_{j=1}^{N}p_jG^{(j)}(Z)$$
$$+\sum_{j=1}^{N-1}F^{(j+1)}(T)\left[p_jG^{(j)}(Z)-p_{j+1}G^{(j+1)}(Z)\right]=\sum_{j=1}^{N-1}F^{(j+1)}(T)p_jG^{(j)}(Z),$$

which agrees with (A.2) when $p_j=p(x)=1$. Thus, the expected cost rate is

$$\widehat{C}_F(T,N,Z)=\frac{\begin{array}{l}c_Z-(c_Z-c_T)\sum_{j=0}^{N-1}[F^{(j)}(T)-F^{(j+1)}(T)]G^{(j)}(Z)\\ -(c_Z-c_N)F^{(N)}(T)G^{(N)}(Z)\\ +c_M\sum_{j=1}^{N-1}F^{(j+1)}(T)p_jG^{(j)}(Z)\end{array}}{\sum_{j=0}^{N-1}G^{(j)}(Z)\int_0^T[F^{(j)}(t)-F^{(j+1)}(t)]\mathrm{d}t}.$$

In particular, when $Z\to\infty$, $p_j=1-q^j$ and $p(x)=1-\mathrm{e}^{-\theta x}$, the expected cost rate $C_F(T,N,\infty)$ in (5.6) is equal to $\widehat{C}_F(T,N,\infty)$ in (5.8), by setting that $q=G^*(\theta)$.

5.3 When the inequality (5.23) does not hold, from (5.21) and (5.22),

$$\frac{\sum_{j=0}^{N-1}[(\lambda T)^j/j!]\{1-[G^*(\theta)]^j\}}{\sum_{j=0}^{N-1}[(\lambda T)^j/j!]}\sum_{j=0}^{N-1}F^{(j+1)}(T)$$
$$-\sum_{j=0}^{N-1}F^{(j+1)}(T)\{1-[G^*(\theta)]^j\}-\frac{c_T}{c_M}$$
$$\leq\sum_{j=0}^{N-1}F^{(j+1)}(T)\{[G^*(\theta)]^j-[G^*(\theta)]^{N-1}\}-\frac{c_T}{c_M}<0,$$

which follows that $\mathrm{d}C_F(T,N)/\mathrm{d}T<0$, and $T_F^*=\infty$.

5.4 First, we prove that when $\int_0^Z p(x)\mathrm{d}G^{(j)}(x)/G^{(j)}(Z)$ increases strictly with j to $p(Z)$,

$$Q(T,N)=\frac{\sum_{j=0}^{N}[(\lambda T)^j/j!]\int_0^Z p(x)\mathrm{d}G^{(j)}(x)}{\sum_{j=0}^{N}[(\lambda T)^j/j!]G^{(j)}(Z)}$$

increases strictly with T to $\int_0^Z p(x)\mathrm{d}G^{(N)}(x)/G^{(N)}(Z)$. Letting

$$L_1(N)\equiv\sum_{j=0}^{N}j\frac{(\lambda T)^j}{j!}\int_0^Z p(x)\mathrm{d}G^{(j)}(x)\sum_{i=0}^{N}\frac{(\lambda T)^i}{i!}G^{(i)}(Z)$$
$$-\sum_{j=0}^{N}\frac{(\lambda T)^j}{j!}\int_0^Z p(x)\mathrm{d}G^{(j)}(x)\sum_{i=0}^{N}i\frac{(\lambda T)^i}{i!}G^{(i)}(Z)$$
$$=\sum_{j=0}^{N}\frac{(\lambda T)^j}{j!}\int_0^Z p(x)\mathrm{d}G^{(j)}(x)\sum_{i=0}^{N}\frac{(\lambda T)^i}{i!}G^{(i)}(Z)(j-i)$$
$$=\sum_{j=0}^{N}\frac{(\lambda T)^j}{j!}\int_0^Z p(x)\mathrm{d}G^{(j)}(x)\left\{\sum_{i=0}^{j}\frac{(\lambda T)^i}{i!}G^{(i)}(Z)(j-i)\right.$$
$$\left.+\sum_{i=j}^{N}\frac{(\lambda T)^i}{i!}G^{(i)}(Z)(j-i)\right\}.$$

On the other hand,

$$\sum_{j=0}^{N}\frac{(\lambda T)^j}{j!}\int_0^Z p(x)\mathrm{d}G^{(j)}(x)\sum_{i=j}^{N}\frac{(\lambda T)^i}{i!}G^{(i)}(Z)(j-i)$$
$$=\sum_{j=0}^{N}\frac{(\lambda T)^j}{j!}G^{(j)}(Z)\sum_{i=0}^{j}\frac{(\lambda T)^i}{i!}\int_0^Z p(x)\mathrm{d}G^{(i)}(x)(i-j).$$

Thus,

$$L_1(N) = \sum_{j=0}^{N} \frac{(\lambda T)^j}{j!} G^{(j)}(Z) \sum_{i=0}^{j} \frac{(\lambda T)^i}{i!} G^{(i)}(Z)$$
$$\times \left[\frac{\int_0^Z p(x)\mathrm{d}G^{(j)}(x)}{G^{(j)}(x)} - \frac{\int_0^Z p(x)\mathrm{d}G^{(i)}(x)}{G^{(i)}(Z)} \right] (j - i) > 0,$$

which follows that $Q(T, N)$ increases strictly with T, and

$$\lim_{T\to\infty} Q(T, N) \equiv \frac{\int_0^Z p(x)\mathrm{d}G^{(N)}(x)}{G^{(N)}(Z)}.$$

Furthermore, from the assumption that $\int_0^Z p(x)\mathrm{d}G^{(N)}(x)/G^{(N)}(Z)$ goes to $p(Z)$ as $N \to \infty$, $Q(T, \infty)$ increases strictly with T to $p(Z)$.

In particular, when $G(x) = 1 - \mathrm{e}^{-\omega x}$, letting

$$L_2(Z) \equiv \int_0^Z p(x)\mathrm{d}G^{(j+1)}(x)G^{(j)}(Z) - \int_0^Z p(x)\mathrm{d}G^{(j)}(x)G^{(j+1)}(Z),$$

we have $\lim_{Z\to 0} L_2(Z) = 0$, and

$$\frac{\mathrm{d}L_2(Z)}{\mathrm{d}Z} = \frac{\omega(\omega Z)^{j-1}}{(j-1)!}\mathrm{e}^{-\omega Z} \int_0^Z \frac{\omega(\omega x)^{j-1}}{j!}\mathrm{e}^{-\omega x}[p(Z) - p(x)](\omega Z - \omega x)\mathrm{d}x > 0,$$

which follows that $\int_0^Z p(x)\mathrm{d}G^{(j)}(x)/G^{(j)}(Z)$ increases strictly with j. Next, because $\lim_{j\to\infty} G^{(j)}(x)/G^{(j)}(Z) = 0$ for $0 < x < Z$,

$$\lim_{j\to\infty} \frac{\int_0^Z p(x)\mathrm{d}G^{(j)}(x)}{G^{(j)}(Z)} = p(Z) - \lim_{j\to\infty} \frac{\int_0^Z G^{(j)}(x)\mathrm{d}p(x)}{G^{(j)}(Z)} = p(Z).$$

Thus, $\int_0^Z p(x)\mathrm{d}G^{(j)}(x)/G^{(j)}(Z)$ increases strictly with j to $p(Z)$. Therefore, the left-hand side of (5.25) increases strictly with T to that of (5.17), i.e., Z_F^* decreases with T to Z^*.

5.5 The expected number of shocks until replacement is

$$\sum_{j=N}^{\infty} j[F^{(j)}(T) - F^{(j+1)}(T)][1 - G^{(j)}(Z)] + N[1 - F^{(N)}(T)][1 - G^{(N)}(Z)]$$
$$+ \sum_{j=N}^{\infty} (j+1)[1 - F^{(j+1)}(T)][G^{(j)}(Z) - G^{(j+1)}(Z)]$$
$$= \sum_{j=0}^{\infty} G^{(j)}(Z) + \sum_{j=0}^{N-1} [1 - G^{(j)}(Z)] + \sum_{j=N}^{\infty} F^{(j+1)}(T)[1 - G^{(j)}(Z)].$$

Thus, the expected number of failures until replacement is

$$\sum_{j=0}^{\infty}\int_0^Z p(x)\mathrm{d}G^{(j)}(x)+\sum_{j=0}^{N-1}\int_Z^{\infty} p(x)\mathrm{d}G^{(j)}(x)$$
$$+\sum_{j=N}^{\infty}F^{(j+1)}(T)\int_Z^{\infty} p(x)\mathrm{d}G^{(j)}(x)$$
$$=\sum_{j=0}^{\infty}\int_0^{\infty} p(x)\mathrm{d}G^{(j)}(x)-\sum_{j=N}^{\infty}[1-F^{(j+1)}(T)]\int_Z^{\infty} p(x)\mathrm{d}G^{(j)}(x).$$

5.6 Letting $L(T)$ denote the left-hand side of (5.40),

$$\begin{aligned}\frac{\mathrm{d}L(T)}{\mathrm{d}T}=&\{1-[G^*(\theta)]^N\}F^{(N)}(T)\\ &-\sum_{j=N}^{\infty}[F^{(j)}(T)-F^{(j+1)}(T)]\{1-[G^*(\theta)]^j\}\\ =&-\sum_{j=N}^{\infty}F^{(j+1)}(T)\{[G^*(\theta)]^j-[G^*(\theta)]^{j+1}\}<0,\end{aligned}$$

$$\begin{aligned}\lim_{T\to 0}L(T)=&\{1-[G^*(\theta)]^N\}N-\sum_{j=0}^{N-1}\{1-[G^*(\theta)]^j\}\\ =&\sum_{j=0}^{N-1}\{[G^*(\theta)]^j-[G^*(\theta)]^N\},\end{aligned}$$

which agrees with that of (5.14).

5.7 When the inequality does not hold, from (5.40) and (5.41),

$$\frac{\sum_{j=N}^{\infty}[(\lambda T)^j/j!]\{1-G^*(\theta)]^j\}}{\sum_{j=N}^{\infty}[(\lambda T)^j/j!]}\left\{\lambda T+\sum_{j=0}^{N-1}[1-F^{(j+1)}(T)]\right\}$$

$$-\sum_{j=0}^{\infty}F^{(j+1)}(T)\{1-[G^*(\theta)]^j\}-\sum_{j=0}^{N-1}[1-F^{(j+1)}(T)]\{1-[G^*(\theta)]^j\}-\frac{c_T}{c_M}$$

$$>\{1-[G^*(\theta)]^N\}\left\{\lambda T+\sum_{j=0}^{N-1}[1-F^{(j+1)}(T)]\right\}$$

$$-\sum_{j=0}^{\infty}F^{(j+1)}(T)\{1-[G^*(\theta)]^j\}-\sum_{j=0}^{N-1}[1-F^{(j+1)}(T)]\{1-[G^*(\theta)]^j\}$$

$$-\frac{c_T}{c_M}\geq 0,$$

which follows that $\mathrm{d}C_L(T,N)/\mathrm{d}T>0$, and $T_L^*=0$.

5.8 Letting $L(T)$ denote the left-hand side of (5.44),

$$\begin{aligned}\frac{\mathrm{d}L(T)}{\mathrm{d}T}=&(1-\mathrm{e}^{-\theta Z})\sum_{j=0}^{\infty}\frac{(\lambda T)^j}{j!}\mathrm{e}^{-\lambda T}[1-G^{(j)}(Z)]\\&-\sum_{j=0}^{\infty}\frac{(\lambda T)^j}{j!}\mathrm{e}^{-\lambda T}\int_Z^{\infty}(1-\mathrm{e}^{-\theta x})\mathrm{d}G^{(j)}(x)\\=&\sum_{j=0}^{\infty}\frac{(\lambda T)^j}{j!}\mathrm{e}^{-\lambda T}\int_Z^{\infty}(\mathrm{e}^{-\theta x}-\mathrm{e}^{-\theta Z})\mathrm{d}G^{(j)}(x)<0,\end{aligned}$$

$$\begin{aligned}\lim_{T\to 0}L(T)=&(1-\mathrm{e}^{-\theta Z})[1+M_G(Z)]-\int_0^Z(1-\mathrm{e}^{-\theta x})\mathrm{d}M_G(x)\\=&\int_0^Z[1+M_G(x)]\theta\mathrm{e}^{-\theta x}\mathrm{d}x,\end{aligned}$$

which agrees with the left-hand side of (5.17).

5.9 From (5.32), (5.33) and (5.34),

$$\sum_{j=N}^{\infty}[1-G^{(j)}(Z)]\int_0^T\left[\int_{T-t}^{\infty}(t+u)\mathrm{d}F(u)\right]\mathrm{d}F^{(j)}(t)$$

$$+[1-G^{(N)}(Z)]\int_T^{\infty}t\mathrm{d}F^{(N)}(t)+\sum_{j=N}^{\infty}[G^{(j)}(Z)-G^{(j+1)}(Z)]\int_T^{\infty}t\mathrm{d}F^{(j+1)}(t)$$

$$=\sum_{j=N}^{\infty}[1-G^{(j)}(Z)]\left\{\mu F^{(j)}(T)+\int_0^T t\mathrm{d}[F^{(j)}(t)-F^{(j+1)}(t)]\right\}$$

$$+\mu\left[N+\sum_{j=N}^{\infty}G^{(j)}(Z)\right]-\sum_{j=N}^{\infty}[1-G^{(j)}(Z)]\int_0^T t\mathrm{d}[F^{(j)}(t)-F^{(j+1)}(t)]$$

$$=\mu\left\{N+\sum_{j=N}^{\infty}[F^{(j)}(T)+G^{(j)}(Z)-F^{(j)}(T)G^{(j)}(Z)]\right\}.$$

A.6 Answers to Problem 6

6.1 The expected number of failures until replacement is

$$\sum_{j=0}^{N-1}[F^{(j)}(T)-F^{(j+1)}(T)]\sum_{i=0}^{j}(j-i)[G^{(i)}(K)-G^{(i+1)}(K)]$$

$$+F^{(N)}(T)\sum_{j=0}^{N}(N-j)[G^{(j)}(K)-G^{(j+1)}(K)]$$

$$=\sum_{j=0}^{N-1}[F^{(j)}(T)-F^{(j+1)}(T)]\sum_{i=1}^{j}[1-G^{(i)}(K)]+F^{(N)}(T)\sum_{j=0}^{N}[1-G^{(j)}(K)]$$

$$=\sum_{j=1}^{N}F^{(j)}(T)[1-G^{(j)}(K)].$$

6.2 Letting $L(T)$ be the left-hand side of (6.8),

$$
\begin{aligned}
L(T) &\equiv \sum_{j=1}^{\infty}[1-G^{(j)}(K)]\left[jp_j(T)-\sum_{i=j}^{\infty}p_i(T)\right] \\
&= \sum_{j=1}^{\infty}p_j(T)\left\{j[1-G^{(j)}(K)]-\sum_{i=1}^{j}[1-G^{(i)}(K)]\right\} \\
&= \sum_{j=1}^{\infty}p_j(T)\sum_{i=1}^{j}[G^{(i)}(K)-G^{(j)}(K)].
\end{aligned}
$$

We obtain

$$
\begin{aligned}
\lim_{T\to 0} L(T) &= 0, \\
\lim_{T\to\infty} L(T) &= \lim_{T\to\infty}\sum_{i=1}^{\infty}\sum_{j=i}^{\infty}[G^{(i)}(K)-G^{(j)}(K)]p_j(T) \\
&= \lim_{T\to\infty}\sum_{i=1}^{\infty}G^{(i)}(K)\sum_{j=i}^{\infty}p_j(T) = M_G(K), \\
\frac{\mathrm{d}L(T)}{\mathrm{d}T} &= \lambda\sum_{j=1}^{\infty}jp_j(T)[G^{(j)}(K)-G^{(j+1)}(K)] > 0,
\end{aligned}
$$

which follows that $L(T)$ increases strictly with T from 0 to $M_G(K)$.

6.3 Note that

$$
\lim_{T\to 0} Q_1(T,N) = \overline{G}(K), \quad \lim_{T\to\infty} Q_1(T,N) = 1-G^{(N)}(K).
$$

Differentiating $Q_1(T, N+1)$ with respect to T,

$$
\begin{aligned}
L(T) &\equiv \sum_{j=0}^{N} j\frac{(\lambda T)^j}{j!}[1-G^{(j+1)}(K)]\sum_{i=0}^{N}\frac{(\lambda T)^i}{i!} \\
&\quad - \sum_{j=0}^{N}\frac{(\lambda T)^j}{j!}[1-G^{(j+1)}(K)]\sum_{i=0}^{N}i\frac{(\lambda T)^i}{i!} \\
&= \sum_{j=0}^{N}\frac{(\lambda T)^j}{j!}[1-G^{(j+1)}(K)]\left[\sum_{i=0}^{j}\frac{(\lambda T)^i}{i!}(j-i)+\sum_{i=j}^{N}\frac{(\lambda T)^i}{i!}(j-i)\right].
\end{aligned}
$$

On the other hand,

$$\sum_{j=0}^{N}\frac{(\lambda T)^j}{j!}[1-G^{(j+1)}(K)]\sum_{i=j}^{N}\frac{(\lambda T)^i}{i!}(j-i)$$
$$=\sum_{j=0}^{N}\frac{(\lambda T)^j}{j!}\sum_{i=0}^{j}\frac{(\lambda T)^i}{i!}[1-G^{(i+1)}(K)](i-j).$$

So that,

$$L(T)=\sum_{j=0}^{N}\frac{(\lambda T)^j}{j!}\sum_{i=0}^{j}\frac{(\lambda T)^i}{i!}(j-i)[G^{(i+1)}(K)-G^{(j+1)}(K)]>0,$$

which follows that $Q_1(T,N)$ increases strictly with T from $\overline{G}(K)$ to $1-G^{(N)}(K)$. Thus, the left-hand side of (6.16) increases strictly with T from 0 to $\sum_{j=1}^{N}[G^{(j)}(K)-G^{(N)}(K)]$.

6.4 The expected number of failures until replacement is

$$\sum_{j=0}^{N-1}[F^{(j)}(T)-F^{(j+1)}(T)]\sum_{i=0}^{j}(j+1-i)[G^{(i)}(K)-G^{(i+1)}(K)]$$
$$+F^{(N)}(T)\sum_{j=0}^{N}(N-j)[G^{(j)}(K)-G^{(j+1)}(K)]$$
$$=\sum_{j=0}^{N-1}F^{(j)}(T)\sum_{i=1}^{j+1}[1-G^{(i)}(K)]-\sum_{j=1}^{N}F^{(j)}(T)\sum_{i=1}^{j}[1-G^{(i)}(K)]$$
$$+F^{(N)}(T)\sum_{j=1}^{N}[1-G^{(j)}(K)]=\sum_{j=0}^{N-1}F^{(j)}(T)[1-G^{(j+1)}(K)].$$

6.5 Differentiating (6.21) with respect to T and setting it equal to zero,

$$\sum_{j=0}^{\infty}F^{(j)}(T)G^{(j+1)}(K)-(1+\lambda T)\sum_{j=0}^{\infty}p_j(T)G^{(j+2)}(K)=\frac{c_O}{c_M},$$

whose left-hand side becomes

$$\sum_{j=0}^{\infty}F^{(j)}(T)G^{(j+1)}(K)-\sum_{j=0}^{\infty}p_j(T)G^{(j+2)}(K)-\sum_{j=0}^{\infty}jp_j(T)G^{(j+1)}(K)$$
$$=\sum_{j=0}^{\infty}p_j(T)\left\{\sum_{i=1}^{j}[G^{(i)}(K)-G^{(j+1)}(K)]+[G^{(j+1)}(K)-G^{(j+2)}(K)]\right\}.$$

Letting $L(T)$ be the left-hand side of (6.22),

$$L(T) = \sum_{i=0}^{\infty} G^{(i+1)}(K) \sum_{j=i}^{\infty} p_j(T) - \sum_{j=0}^{\infty} p_j(T)[jG^{(j+1)}(K) + G^{(j+2)}(K)],$$

$$\lim_{T\to 0} L(T) = G(K) - G^{(2)}(K), \quad \lim_{T\to\infty} L(T) = \sum_{j=1}^{\infty} G^{(j)}(K) = M_G(K),$$

$$\frac{\mathrm{d}L(T)}{\mathrm{d}T} = \lambda \sum_{j=0}^{\infty} p_j(T)[jG^{(j+1)}(K) - (j-1)G^{(j+2)}(K) - G^{(j+3)}(K)] > 0,$$

which follows that $L(T)$ increases strictly with T from $G(K) - G^{(2)}(K)$ to $M_G(K)$.

6.6 The expected number of failures until replacement for replacement last is

$$\sum_{j=N}^{\infty}[F^{(j)}(T) - F^{(j+1)}(T)] \sum_{i=0}^{j}(j-i)[G^{(i)}(K) - G^{(i+1)}(K)]$$
$$+ [1 - F^{(N)}(T)] \sum_{j=0}^{N}(N-j)[G^{(j)}(K) - G^{(j+1)}(K)]$$
$$= \sum_{j=N+1}^{\infty} F^{(j)}(T)[1 - G^{(j)}(K)] + \sum_{j=0}^{N}(N-j)[G^{(j)}(K) - G^{(j+1)}(K)]$$
$$= \sum_{j=1}^{N}[1 - G^{(j)}(K)] + \sum_{j=N+1}^{\infty} F^{(j)}(T)[1 - G^{(j)}(K)].$$

The expected number of failures until replacement for replacement overtime last is

$$\sum_{j=N}^{\infty}[F^{(j)}(T) - F^{(j+1)}(T)] \sum_{i=0}^{j}(j+1-i)[G^{(i)}(K) - G^{(i+1)}(K)]$$
$$+ [1 - F^{(N)}(T)] \sum_{j=0}^{N}(N-j)[G^{(j)}(K) - G^{(j+1)}(K)]$$
$$= \sum_{j=1}^{N}[1 - G^{(j)}(K)] + \sum_{j=N}^{\infty} F^{(j)}(T)[1 - G^{(j+1)}(K)].$$

6.7 Forming $C(M+1) - C(M) \geq 0$, we obtain $M_G(K) \geq c_F/c_M$. Thus, if $M_G(K) \geq c_F/c_M$, then $M^* = 1$, and conversely, if $M_G(K) < c_F/c_M$, then $M^* = \infty$. This means that from the assumption that shocks occur at a renewal process with mean μ, both expected cost in the numerator and mean replacement time in the denominator increase constantly with M. Thus, if the cost rate of c_F and c_M is equal to or less than $M_G(K)$, the unit should be replaced at the first failure, and if the cost rate is more than $M_G(K)$, then it always undergoes maintenance and replacement should not be done.

6.8 Prove that when $F(t) = 1 - \mathrm{e}^{-\lambda t}$, $Q_1(T, M)$ increases strictly with T from 0 to 1, i.e., putting that $q_j \equiv G^{(j)}(K) - G^{(j+1)}(K)$, prove that

$$A_1(T, M) \equiv \frac{\sum_{j=0}^{\infty} q_{j+1} \sum_{i=0}^{j}[(\lambda T)^i / i!]}{\sum_{j=0}^{\infty} q_{j+1} \sum_{i=0}^{M+j}[(\lambda T)^i / i!]} < 1 \quad (M = 1, 2, \cdots)$$

decreases strictly with T from 1 to 0. Firstly, note that

$$\lim_{T \to 0} A_1(T, M) = 1, \quad \lim_{T \to \infty} A_1(T, M) = 0.$$

Differentiating $A_1(T, M)$ with respect to T,

$$\sum_{j=0}^{\infty} q_{j+2} \frac{(\lambda T)^j}{j!} \sum_{j=0}^{\infty} q_{j+1} \sum_{i=0}^{M+j} \frac{(\lambda T)^i}{i!}$$
$$- \sum_{j=0}^{\infty} q_{j+1} \sum_{i=0}^{j} \frac{(\lambda T)^i}{i!} \sum_{j=0}^{\infty} q_{j+2} \frac{(\lambda T)^{M+j}}{(M+j)!}.$$

Next, prove that

$$A_2(T, M) \equiv \frac{\sum_{j=0}^{\infty} q_{j+2}[(\lambda T)^{M+j} / (M+j)!]}{\sum_{j=0}^{\infty} q_{j+1} \sum_{i=0}^{M+j}[(\lambda T)^i / i!]} \quad (M = 0, 1, 2, \cdots)$$

decreases strictly with M. Forming $A_2(T, M) - A_2(T, M+1)$ by using the method in Problem 2.2,

$$\sum_{j=0}^{\infty} q_{j+2} \frac{(\lambda T)^{M+j}}{(M+j)!} \sum_{k=0}^{\infty} q_{k+1} \sum_{i=0}^{M+k+1} \frac{(\lambda T)^i}{i!}$$
$$- \sum_{j=0}^{\infty} q_{j+2} \frac{(\lambda T)^{M+j+1}}{(M+j+1)!} \sum_{k=0}^{\infty} q_{k+1} \sum_{i=0}^{M+k} \frac{(\lambda T)^i}{i!}$$
$$= \sum_{j=0}^{\infty} q_{j+2} \frac{(\lambda T)^{M+j}}{(M+j+1)!} \sum_{k=0}^{\infty} q_{k+1} \sum_{i=0}^{M+k+1} \frac{(\lambda T)^i}{i!} (M+j+1-i) > 0.$$

Thus, $A_2(T, M)$ decreases strictly with M, which follows that $\mathrm{d}A_1(T, M)/\mathrm{d}T < 0$, i.e., $A_1(T, M)$ decreases strictly with T from 1 to 0. It can be easily seen that $Q_1(T, M)$ increases strictly with M to

$$\sum_{j=0}^{\infty}[G^{(j)}(K) - G^{(j+1)}(K)]F^{(j)}(T).$$

When $M \to \infty$, (6.44) becomes

$$\begin{aligned}
&\lambda T \sum_{j=0}^{\infty}[G^{(j)}(K) - G^{(j+1)}(K)]F^{(j)}(T) \\
&\quad - \sum_{j=0}^{\infty}[G^{(j)}(K) - G^{(j+1)}(K)] \sum_{i=j+1}^{\infty} F^{(i)}(T) \\
&= \sum_{j=0}^{\infty}[G^{(j)}(K) - G^{(j+1)}(K)] \sum_{i=j+1}^{\infty} i p_i(T) \\
&\quad - \sum_{j=0}^{\infty}[G^{(j)}(K) - G^{(j+1)}(K)] \sum_{i=j}^{\infty}(i - j) p_i(T) \\
&= \sum_{j=0}^{\infty}[G^{(j)}(K) - G^{(j+1)}(K)] j \sum_{i=j+1}^{\infty} p_i(T) \\
&= \sum_{j=1}^{\infty} p_j(T) \sum_{i=1}^{j}[G^{(i)}(K) - G^{(j)}(K)],
\end{aligned}$$

which agrees with the left-hand side of (6.10). The same results can be proved for $Q_2(T, M)$ by using the method in the above proof.

6.9 When $F(t) = 1 - e^{-\lambda t}$, the denominator of (6.50) is

$$\begin{aligned}
&\lambda T + \sum_{j=0}^{\infty}[G^{(j)}(K) - G^{(j+1)}(K)] \int_T^{\infty} \lambda[1 - F^{(M+j)}(t)]\mathrm{d}t \\
&= \sum_{j=0}^{\infty}[G^{(j)}(K) - G^{(j+1)}(K)] \left\{ \sum_{i=1}^{\infty} F^{(i)}(T) + \sum_{i=1}^{M+j}[1 - F^{(i)}(T)] \right\} \\
&= \sum_{j=1}^{\infty} G^{(j)}(K) + M + \sum_{j=0}^{\infty}[G^{(j)}(K) - G^{(j+1)}(K)] \sum_{i=M+j+1}^{\infty} F^{(i)}(T) \\
&= M_G(K) + M + \sum_{j=0}^{\infty}[1 - G^{(j)}(K)]F^{(M+j)}(T).
\end{aligned}$$

6.10 Suppose that the unit is replaced at shock N $(N = 1, 2, \cdots)$ or at failure M $(M = 1, 2, \cdots)$, whichever occurs first. The probability that the unit is replaced at shock N before failure M is

$$\sum_{j=N-M}^{\infty}[G^{(j)}(K)-G^{(j+1)}(K)]=G^{(N-M)}(K),$$

and the probability that it is replaced at failure M before shock N is

$$\sum_{j=0}^{N-M-1}[G^{(j)}(K)-G^{(j+1)}(K)]=1-G^{(N-M)}(K).$$

The expected number shocks until replacement is

$$NG^{(N-M)}(K)+\sum_{j=0}^{N-M-1}(j+M)[G^{(j)}(K)-G^{(j+1)}(K)]$$
$$=M+\sum_{j=1}^{N-M}G^{(j)}(K),$$

and the expected number of failures until replacement is

$$\sum_{j=N-M}^{N}(N-j)[G^{(j)}(K)-G^{(j+1)}(K)]+M[1-G^{(N-M)}(K)]$$
$$=\sum_{j=N-M+1}^{N}[1-G^{(j)}(K)].$$

Therefore, the expected cost rate is

$$C_F(N,M)=\frac{c_F-(c_F-c_N)G^{(N-M)}(K)+c_M\sum_{j=N-M+1}^{N}[1-G^{(j)}(K)]}{\mu[M+\sum_{j=1}^{N-M}G^{(j)}(K)]}.$$

Clearly, $C_F(N,N)=C(N)$ in (6.7).

Next, suppose that the unit is replaced at shock N ($N=0,1,2,\cdots$) or at failure M ($M=0,1,2,\cdots$), whichever occurs last. The probability that the unit is replaced at shock N after failure M is

$$\sum_{j=0}^{N-M-1}[G^{(j)}(K)-G^{(j+1)}(K)]=1-G^{(N-M)}(K),$$

and the probability that it is replaced at failure M after shock N is

$$\sum_{j=N-M}^{\infty}[G^{(j)}(K) - G^{(j+1)}(K)] = G^{(N-M)}(K).$$

The expected number of shocks until replacement is

$$\sum_{j=N-M}^{\infty}(j+M)[G^{(j)}(K) - G^{(j+1)}(K)] + N[1 - G^{(N-M)}(K)]$$
$$= N + \sum_{j=N-M+1}^{\infty} G^{(j)}(K),$$

and the expected number of failures until replacement is

$$MG^{(N-M)}(K) + \sum_{j=0}^{N-M-1}(N-j)[G^{(j)}(K) - G^{(j+1)}(K)] = N - \sum_{j=1}^{N-M} G^{(j)}(K).$$

Therefore, the expected cost rate is

$$C_L(N, M) = \frac{c_N + (c_F - c_N)G^{(N-M)}(K) + c_M[N - \sum_{j=1}^{N-M} G^{(j)}(K)]}{\mu[N + \sum_{j=N-M+1}^{\infty} G^{(j)}(K)]}.$$

Clearly, $C_F(\infty, M) = C_L(M, M) = C(M)$ in (6.43).

6.11 The mean time to replacement is

$$\sum_{j=0}^{\infty}[G^{(j)}(K) - G^{(j+1)}(K)]$$
$$\times \sum_{i=M-1}^{\infty}\int_0^T\left\{\int_0^{T-t}\left[\int_{T-t-u}^{\infty}(t+u+y)\mathrm{d}F(y)\right]\mathrm{d}F^{(i)}(u)\right\}\mathrm{d}F^{(j+1)}(t)$$
$$+\sum_{j=0}^{\infty}[G^{(j)}(K) - G^{(j+1)}(K)]\int_T^{\infty}\left[\int_0^{\infty}(t+u)\mathrm{d}F^{(M-1)}(u)\right]\mathrm{d}F^{(j+1)}(t)$$
$$+\sum_{j=0}^{\infty}[G^{(j)}(K) - G^{(j+1)}(K)]\int_0^{T}\left[\int_{T-t}^{\infty}(t+u)\mathrm{d}F^{(M-1)}(u)\right]\mathrm{d}F^{(j+1)}(t)$$
$$= \mu\sum_{j=0}^{\infty}[G^{(j)}(K) - G^{(j+1)}(K)]\left[M + j + \sum_{i=M+j}^{\infty} F^{(i)}(T)\right]$$
$$= \mu\left\{M + M_G(K) + \sum_{j=0}^{\infty}[1 - G^{(j+1)}(K)]F^{(M+j)}(T)\right\}.$$

The expected number of failures until replacement is

$$M\sum_{j=0}^{\infty}[G^{(j)}(K)-G^{(j+1)}(K)][1-F^{(M+j)}(T)]+\sum_{j=0}^{\infty}[G^{(j)}(K)-G^{(j+1)}(K)]$$
$$\times\sum_{i=M-1}^{\infty}(i+2)\int_0^T\left[\int_0^{T-t}\overline{F}(T-t-u)\mathrm{d}F^{(i)}(u)\right]\mathrm{d}F^{(j+1)}(t)$$
$$=M+\sum_{j=0}^{\infty}[G^{(j)}(K)-G^{(j+1)}(K)]\sum_{i=M+j}^{\infty}F^{(i)}(T)$$
$$=M+\sum_{j=0}^{\infty}[1-G^{(j+1)}(K)]F^{(M+j)}(T).$$

6.12 The mean time to replacement for overtime first is, from (6.52),

$$\sum_{j=0}^{\infty}G^{(j)}(K)\int_0^T\left[\int_T^{\infty}u\mathrm{e}^{-[H(u)-H(t)]}h(u)\mathrm{d}u\right]\mathrm{d}P_j(t)$$
$$+\sum_{j=0}^{\infty}[G^{(j)}(K)-G^{(j+1)}(K)]\int_0^T t\mathrm{d}P_{M+j}(t)+\sum_{j=0}^{\infty}[G^{(j)}(K)-G^{(j+1)}(K)]$$
$$\times\sum_{i=0}^{M-2}\int_0^T\left\{\int_T^{\infty}u\mathrm{e}^{-[H(u)-H(t)]}h(u)\mathrm{d}u\right\}\mathrm{d}P_{i+j+1}(t)$$
$$=\sum_{j=0}^{\infty}G^{(j)}(K)\int_0^T\left[\int_T^{\infty}\mathrm{e}^{-[H(u)-H(t)]}\mathrm{d}u\right]\mathrm{d}P_j(t)$$
$$+\sum_{j=0}^{\infty}[G^{(j)}(K)-G^{(j+1)}(K)]\int_0^T\overline{P}_{M+j}(t)\mathrm{d}t+\sum_{j=0}^{\infty}[G^{(j)}(K)-G^{(j+1)}(K)]$$
$$\times\sum_{i=0}^{M-2}\int_0^T\left\{\int_T^{\infty}\mathrm{e}^{-[H(u)-H(t)]}\mathrm{d}u\right\}\mathrm{d}P_{i+j+1}(t)$$
$$=\left[\sum_{j=0}^{M-1}P_j(T)+\sum_{j=0}^{\infty}G^{(j+1)}(K)P_{M+j}(T)\right]\int_T^{\infty}\mathrm{e}^{-[H(t)-H(T)]}\mathrm{d}t$$
$$+\sum_{j=0}^{\infty}[G^{(j)}(K)-G^{(j+1)}(K)]\int_0^T\overline{P}_{M+j}(t)\mathrm{d}t$$
$$=\sum_{j=0}^{\infty}[G^{(j)}(K)-G^{(j+1)}(K)]$$

$$\times \left\{ \int_0^T \overline{P}_{M+j}(t)\mathrm{d}t + \sum_{i=0}^{M+j-1} P_i(T) \int_T^\infty \mathrm{e}^{-[H(t)-H(T)]}\mathrm{d}t \right\}.$$

The mean time to replacement for overtime last is, from Problem 6.10,

$$\sum_{j=0}^{\infty}[G^{(j)}(K) - G^{(j+1)}(K)]$$
$$\times \left(\sum_{i=M-1}^{\infty} \int_0^T \left\{ \int_T^\infty u\mathrm{e}^{-[H(u)-H(t)]}h(u)\mathrm{d}u \right\} \mathrm{d}P_{i+j+1}(t). \right.$$
$$+ \int_T^\infty \left[\int_t^\infty u\mathrm{e}^{H(t)}\mathrm{d}P_{M-1}(u) \right] \mathrm{d}P_{j+1}(t)$$
$$\left. + \int_0^T \left[\int_T^\infty u\mathrm{e}^{H(t)}\mathrm{d}P_{M-1}(u) \right] \mathrm{d}P_{j+1}(t) \right)$$
$$= \sum_{j=0}^{\infty}[G^{(j)}(K) - G^{(j+1)}(K)]$$
$$\times \left\{ \int_T^\infty \overline{P}_{M+j}(t)\mathrm{d}t + \sum_{i=M+j}^{\infty} P_i(T) \int_T^\infty \mathrm{e}^{-[H(t)-H(T)]}\mathrm{d}t \right\}.$$

A.7 Answers to Problem 7

7.1 Letting $L(N)$ denote the second term of (7.5) and noting that $L(0) = 0$ and

$$\sum_{j=0}^{\infty}[G^{(j)}(K) - G^{(j+1)}(K)] \int_0^T \overline{P}_N(t)\mathrm{d}F^{(j+1)}(t)$$
$$= 1 - \overline{P}_N(T) \sum_{j=0}^{\infty} G^{(j)}(K)[F^{(j)}(T) - F^{(j+1)}(T)]$$
$$- \sum_{j=0}^{\infty} G^{(j)}(K) \int_0^T [F^{(j)}(t) - F^{(j+1)}(t)]p_{N-1}(t)h(t)\mathrm{d}t,$$

we obtain

$$L(N)-L(N-1)=1-\overline{P}_N(T)\sum_{j=0}^{\infty}G^{(j)}(K)[F^{(j)}(T)-F^{(j+1)}(T)]$$
$$-\sum_{j=0}^{\infty}[G^{(j)}(K)-G^{(j+1)}(K)]\int_0^T\overline{P}_N(t)\mathrm{d}F^{(j+1)}(t)$$
$$=\sum_{j=0}^{\infty}G^{(j)}(K)\int_0^T[F^{(j)}(t)-F^{(j+1)}(t)]p_{N-1}(t)h(t)\mathrm{d}t.$$

Thus,

$$L(N)=\sum_{j=1}^{N}[L(j)-L(j-1)]$$
$$=\sum_{j=0}^{\infty}G^{(j)}(K)\int_0^T[F^{(j)}(t)-F^{(j+1)}(t)]\overline{P}_N(t)h(t)\mathrm{d}t.$$

7.2 From $Q_1(T)>Q_1(t)$ and $h(T)>h(t)$ for $0<t<T$, we have $Q_2(T,N)<Q_1(T)$ and $Q_3(T,N)<h(T)$ for any N. Similarly, from $Q_1(T)<Q_1(t)$ and $h(T)<h(t)$ for $T<t<\infty$, we have $Q_4(T,N)>Q_1(T)$ and $Q_5(T,N)>h(T)$ for any N.

7.3 Suppose that the jth ($j=0,1,2,\cdots,N-1$) independent damage occurs at time t ($0<t<T$), then the $(j+1)$th one occurs at time u after time T with probability

$$\int_0^T\left\{\int_T^{\infty}\mathrm{e}^{H(t)}\mathrm{d}[1-\mathrm{e}^{-H(u)}]\right\}\mathrm{d}P_j(t)=\frac{H(T)^j}{j!}\int_T^{\infty}h(t)\mathrm{e}^{-H(t)}\mathrm{d}t.$$

Thus, the probability that the unit is replaced over time T is

$$\overline{P}_N(T)\sum_{j=0}^{\infty}G^{(j)}(K)\int_T^{\infty}[F^{(j)}(t)-F^{(j+1)}(t)]h(t)\mathrm{e}^{-[H(t)-H(T)]}\mathrm{d}t.$$

7.4 Note that when $F(t)=1-\mathrm{e}^{-\lambda t}$ and $r_{j+1}(K)$ increases with j to 1,

$$Q_2(T)=\frac{\lambda\sum_{j=0}^{\infty}[G^{(j)}(K)-G^{(j+1)}(K)]\int_T^{\infty}[(\lambda t)^j/j!]\mathrm{e}^{-\lambda t-H(t)}\mathrm{d}t}{\sum_{j=0}^{\infty}G^{(j)}(K)\int_T^{\infty}[(\lambda t)^j/j!]\mathrm{e}^{-\lambda t-H(t)}\mathrm{d}t},$$
$$Q_3(T)=\frac{\sum_{j=0}^{\infty}G^{(j)}(K)\int_T^{\infty}[(\lambda t)^j/j!]\mathrm{e}^{-\lambda t-H(t)}h(t)\mathrm{d}t}{\sum_{j=0}^{\infty}G^{(j)}(K)\int_T^{\infty}[(\lambda t)^j/j!]\mathrm{e}^{-\lambda t-H(t)}\mathrm{d}t},$$

$\lim_{T\to\infty}Q_2(T)=\lambda$ and $\lim_{T\to\infty}Q_3(T)=h(\infty)$. Differentiating $Q_2(T)$ with respect to T,

$$\lambda \mathrm{e}^{-H(T)}\left\{\sum_{j=0}^{\infty} G^{(j)}(K)\frac{(\lambda T)^j}{j!}\mathrm{e}^{-\lambda T}\sum_{i=0}^{\infty}[G^{(i)}(K)-G^{(i+1)}(K)]\right.$$
$$\times\int_T^{\infty}\frac{(\lambda t)^i}{i!}\mathrm{e}^{-\lambda t}\mathrm{e}^{-H(t)}\mathrm{d}t-\sum_{j=0}^{\infty}[G^{(j)}(K)-G^{(j+1)}(K)]\frac{(\lambda T)^j}{j!}\mathrm{e}^{-\lambda T}$$
$$\left.\times\sum_{i=0}^{\infty}G^{(i)}(K)\int_T^{\infty}\frac{(\lambda t)^i}{i!}\mathrm{e}^{-\lambda t-H(t)}\mathrm{d}t\right\}$$
$$=\lambda \mathrm{e}^{-H(T)}\sum_{j=0}^{\infty}G^{(j)}(K)\frac{(\lambda T)^j}{j!}\mathrm{e}^{-\lambda T}\sum_{i=0}^{\infty}G^{(i)}(K)\int_T^{\infty}\frac{(\lambda t)^i}{i!}\mathrm{e}^{-\lambda t-H(t)}\mathrm{d}t$$
$$\times[r_{i+1}(K)-r_{j+1}(K)]>0,$$

as $r_{j+1}(K)$ increases strictly with j (Problem 2.3). Furthermore, $Q_2(T)$ is

$$Q_2(T)=\frac{\lambda\sum_{j=0}^{\infty}[G^{(j)}(K)-G^{(j+1)}(K)]\int_T^{\infty}[(\lambda t)^j/j!]\mathrm{e}^{-\lambda t-H(t)}\mathrm{d}t}{\sum_{j=0}^{\infty}[G^{(j)}(K)-G^{(j+1)}(K)]\int_T^{\infty}[1-F^{(j+1)}(t)]\mathrm{e}^{-\lambda t-H(t)}\mathrm{d}t}.$$

Because $[(\lambda t)^j/j!]/[1-F^{(j+1)}(t)]$ increases strictly with t, $Q_2(T)>Q_1(T)$ for $0\le T<\infty$.

Similarly, differentiating $Q_3(T)$ with respect to T,

$$\mathrm{e}^{-H(T)}\sum_{j=0}^{\infty}G^{(j)}(K)\frac{(\lambda T)^j}{j!}\mathrm{e}^{-\lambda T}$$
$$\times\sum_{i=0}^{\infty}G^{(i)}(K)\int_T^{\infty}\frac{(\lambda t)^i}{i!}\mathrm{e}^{-\lambda t-H(t)}[h(t)-h(T)]>0.$$

Thus, because both $Q_2(T)$ and $Q_3(T)$ increase strictly with T and $Q_3(T)>h(T)$, the left-hand side of (7.38) increases strictly with T to $L_4(\infty)$.

7.5 Lettting $L_1(T,N)$ denote the first term of (7.43),

$$L_1(T,N)=\sum_{j=0}^{\infty}G^{(j)}(K)\left\{\int_0^T[F^{(j)}(t)-F^{(j+1)}(t)]\overline{P}_N(t)\mathrm{d}t\right.$$
$$\left.+\int_0^T\left[\int_{T-t}^{\infty}\overline{F}(u)\overline{P}_N(t+u)\mathrm{d}u\right]\mathrm{d}F^{(j)}(t)\right\}$$
$$=\sum_{j=0}^{\infty}G^{(j)}(K)\int_0^T\left[\int_0^{\infty}\overline{F}(u)\overline{P}_N(t+u)\mathrm{d}u\right]\mathrm{d}F^{(j)}(t),$$

as

$$\int_0^T\left[\int_{T-t}^{\infty}\overline{F}(u)\overline{P}_N(t+u)\mathrm{d}u\right]\mathrm{d}F^{(j)}(t)$$
$$=\int_0^T\left[\int_0^{\infty}\overline{F}(u)\overline{P}_N(t+u)\mathrm{d}u\right]\mathrm{d}F^{(j)}(t)$$
$$-\int_0^T\left[\int_0^{T-t}\overline{F}(u)\overline{P}_N(t+u)\mathrm{d}u\right]\mathrm{d}F^{(j)}(t),$$

and

$$\int_0^T\left[\int_0^{T-t}\overline{F}(u)\overline{P}_N(t+u)\mathrm{d}u\right]\mathrm{d}F^{(j)}(t)$$
$$=\int_0^T\left[\int_t^{T}\overline{F}(u-t)\overline{P}_N(u)\mathrm{d}u\right]\mathrm{d}F^{(j)}(t)$$
$$=\int_0^T\overline{P}_N(u)\left[\int_0^{u}\overline{F}(u-t)\mathrm{d}F^{(j)}(t)\right]\mathrm{d}u$$
$$=\int_0^T[F^{(j)}(t)-F^{(j+1)}(t)]\overline{P}_N(t)\mathrm{d}t.$$

Similarly, letting $L_2(T,N)$ denote the first term of (7.44),

$$L_2(T,N)=\sum_{j=0}^{\infty}G^{(j)}(K)\left\{\int_0^T[F^{(j)}(t)-F^{(j+1)}(t)]\overline{P}_N(t)h(t)\mathrm{d}t\right.$$
$$\left.+\int_0^T\left[\int_{T-u}^{\infty}\overline{F}(u)\overline{P}_N(t+u)h(t+u)\mathrm{d}u\right]\mathrm{d}F^{(j)}(t)\right\}$$
$$=\sum_{j=0}^{\infty}G^{(j)}(K)\int_0^T\left[\int_0^{\infty}\overline{F}(u)\overline{P}_N(t+u)h(t+u)\mathrm{d}u\right]\mathrm{d}F^{(j)}(t).$$

7.6 When $c_O=c_T$, $F(t)=1-\mathrm{e}^{-\lambda t}$ and $r_{j+1}(x)$ increases strictly with j, subtracting (7.27) from (7.47),

$$(c_K-c_O)\left\{\sum_{j=0}^{\infty}G^{(j)}(K)[Q_4(T)F^{(j)}(T)-\widetilde{Q}_1(T)F^{(j+1)}(T)]\right.$$
$$\left.-\sum_{j=0}^{\infty}[G^{(j)}(K)-G^{(j+1)}(K)]\frac{(\lambda T)^j}{j!}\mathrm{e}^{-\lambda T}\right\}$$
$$+c_M\left(\sum_{j=0}^{\infty}G^{(j)}(K)\left[F^{(j)}(T)\int_0^{\infty}\mathrm{e}^{-\lambda t}h(t+T)\mathrm{d}t-\frac{h(T)}{\lambda}F^{(j+1)}(T)\right]\right.$$

$$
-\sum_{j=0}^{\infty}G^{(j)}(K)\left\{\int_0^T\left[\int_0^{\infty}\mathrm{e}^{-\lambda u}h(t+u)\mathrm{d}u\right]\mathrm{d}F^{(j)}(t)\right.
$$

$$
\left.\left.-\int_0^T\frac{(\lambda t)^j}{j!}\mathrm{e}^{-\lambda t}h(t)\mathrm{d}t\right\}\right)>(c_K-c_O)\left\{\widetilde{Q}_1(T)\sum_{j=0}^{\infty}G^{(j)}(K)\frac{(\lambda T)^j}{j!}\mathrm{e}^{-\lambda T}\right.
$$

$$
\left.-\sum_{j=0}^{\infty}[G^{(j)}(K)-G^{(j+1)}(K)]\frac{(\lambda T)^j}{j!}\mathrm{e}^{-\lambda T}\right\}
$$

$$
+\frac{c_M}{\lambda}\left\{h(T)\sum_{j=0}^{\infty}G^{(j)}(K)\frac{(\lambda T)^j}{j!}\mathrm{e}^{-\lambda T}-\sum_{j=0}^{\infty}G^{(j)}(K)\left[\int_0^T h(t)\mathrm{d}F^{(j)}(t)\right.\right.
$$

$$
\left.\left.-\int_0^T h(t)\mathrm{d}F^{(j+1)}(t)\right]\right\}>0,
$$

as $Q_4(T)>\widetilde{Q}_1(T)$ and $\lambda\int_0^{\infty}\mathrm{e}^{-\lambda t}h(t+T)\mathrm{d}t>h(T)$, where $\widetilde{Q}_1(T)$ is given in (7.27), which follows that $T_O^*<T^*$.

A.8 Answers to Problem 8

8.1 Equation (8.21) is

$$
\sum_{j=0}^{\infty}G^{(j)}(K)\left[Tp_j(T)+p_j(T)\int_T^{\infty}\mathrm{e}^{-H(t)+H(T)}\mathrm{d}t\right]
$$

$$
+\sum_{j=0}^{\infty}[G^{(j)}(K)-G^{(j+1)}(K)]\left[TP_{j+1}(T)-\int_0^T P_{j+1}(t)\mathrm{d}t\right]
$$

$$
=\sum_{j=0}^{\infty}G^{(j)}(K)\left[\int_0^T p_j(t)\mathrm{d}t+p_j(T)\int_T^{\infty}\mathrm{e}^{-H(t)+H(T)}\mathrm{d}t\right].
$$

8.2 Noting that

$$
\sum_{j=0}^{\infty}\int_0^T\left[\int_0^{T-t}\overline{D}(t+u)\overline{F}(u)\mathrm{d}u\right]\mathrm{d}F^{(j)}(t)
$$

$$
=\sum_{j=0}^{\infty}\int_0^T\overline{D}(t)[F^{(j)}(t)-F^{(j+1)}(t)]\mathrm{d}t=\int_0^T\overline{D}(t)\mathrm{d}t,
$$

Equation (8.54) is

$$\int_0^T t\mathrm{d}D(t) + T\overline{D}(T)\sum_{j=0}^{\infty}\int_0^T \overline{F}(T-t)\mathrm{d}F^{(j)}(t)$$

$$+\sum_{j=0}^{\infty}\int_0^T \left[\int_{T-t}^{\infty} \overline{D}(t+u)\overline{F}(u)\mathrm{d}u\right]\mathrm{d}F^{(j)}(t)$$

$$=\sum_{j=0}^{\infty}\int_0^T \left[\int_{T-t}^{\infty} \overline{D}(t+u)\overline{F}(u)\mathrm{d}u\right]\mathrm{d}F^{(j)}(t) + \int_0^T \overline{D}(t)\mathrm{d}t$$

$$=\sum_{j=0}^{\infty}\int_0^T \left[\int_0^{\infty} \overline{D}(t+u)\overline{F}(u)\mathrm{d}u\right]\mathrm{d}F^{(j)}(t).$$

8.3 Letting $Q_1(T) \equiv \int_T^{\infty} \mathrm{e}^{-\lambda t}\mathrm{d}D(t)/\int_T^{\infty} \mathrm{e}^{-\lambda t}\overline{D}(t)\mathrm{d}t$ for $0 \le T < \infty$,

$$Q_1(T) > r(T) \quad \text{and} \quad \lim_{T\to\infty} Q_1(T) = r(\infty).$$

Differentiating $Q_1(T)$ with respect to T,

$$\mathrm{e}^{-\lambda T}\overline{D}(T)\int_T^{\infty} \mathrm{e}^{-\lambda t}\overline{D}(t)[r(t) - r(T)] > 0,$$

which follows that the left-hand side of (8.58) increases strictly with T to ∞.

8.4 Prove that $Q_2(T) \equiv \overline{D}(T)\mathrm{e}^{-\lambda T} / \int_T^{\infty} \overline{D}(t)\mathrm{e}^{-\lambda t}\mathrm{d}t$ increases strictly with T to $\lambda + r(\infty)$. Note that for $0 \le T < \infty$,

$$\lim_{T\to\infty} Q_2(T) = \lambda + r(\infty), \quad Q_2(T) > \lambda + r(T).$$

Differentiating $Q_2(T)$ with respect to T,

$$\overline{D}(T)\mathrm{e}^{-\lambda T}\left\{\overline{D}(T)\mathrm{e}^{-\lambda T} - [\lambda + r(T)]\int_T^{\infty} \overline{D}(t)\mathrm{e}^{-\lambda t}\mathrm{d}t\right\} > 0,$$

which follows that $Q_2(T)$ increases strictly with T to $\lambda + r(\infty)$. Thus, the left-hand side increase strictly with T to ∞, and there exists a finite and unique T_{OD}^* which satisfies (8.62).

8.5 The mean time to full backup for backup overtime first is

$$\sum_{j=0}^{N-1}\int_0^T \left[\int_{T-t}^{\infty} (t+u)\overline{D}(t+u)\mathrm{d}F(u)\right]\mathrm{d}F^{(j)}(t) + \int_0^T t\overline{D}(t)\mathrm{d}F^{(N)}(t)$$

$$+\int_0^T t[1 - F^{(N)}(t)]\mathrm{d}D(t) + \sum_{j=0}^{N-1}\int_0^T \left[\int_{T-t}^{\infty} (t+u)\overline{F}(u)\mathrm{d}D(t+u)\right]\mathrm{d}F^{(j)}(t).$$

Using the method of 8.2, we derive (8.67). Similarly, the mean time to full backup for backup overtime last is

$$\sum_{j=N}^{\infty}\int_0^T\left[\int_{T-t}^{\infty}(t+u)\overline{D}(t+u)\mathrm{d}F(u)\right]\mathrm{d}F^{(j)}(t)+\int_T^{\infty}t\overline{D}(t)\mathrm{d}F^{(N)}(t)$$
$$+\int_0^T t\mathrm{d}D(t)+\int_T^{\infty}t[1-F^{(N)}(t)]\mathrm{d}D(t)$$
$$+\sum_{j=N}^{\infty}\int_0^T\left[\int_{T-t}^{\infty}(t+u)\overline{F}(u)\mathrm{d}D(t+u)\right]\mathrm{d}F^{(j)}(t),$$

which follows (8.72).

A.9 Answers to Problem 9

9.1 Expected cost rates $C_F(N,Z)$, $C_L(N,Z)$ and $C_O(Z)$ are obtained respectively from (2.30), (3.15) and (4.12), putting that $T\to\infty$ and replacing μ with T.

9.2 The mean time to replacement is

$$\sum_{j=0}^{\infty}(j+1)T[G^{(j)}(Z)-G^{(j+1)}(Z)]=T\sum_{j=0}^{\infty}G^{(j)}(Z),$$

and the expected number of failures until replacement is

$$\sum_{j=0}^{\infty}\int_0^Z p(x)\mathrm{d}G^{(j)}(x)+\int_0^Z\left[\int_{Z-x}^{\infty}p(x+y)\mathrm{d}G(y)\right]\mathrm{d}G^{(j)}(x)$$
$$=\sum_{j=0}^{\infty}\int_0^Z\left[\int_0^{\infty}p(x+y)\mathrm{d}G(y)\right]\mathrm{d}G^{(j)}(x).$$

In particular, when $p(x)=1-\mathrm{e}^{-\theta x}$, the expected cost rate is

$$C_O(Z)=\frac{c_O+c_M\{1-G^*(\theta)+\int_0^Z[1-\mathrm{e}^{-\theta x}G^*(\theta)]\mathrm{d}M_G(x)\}}{T[1+M_G(Z)]},$$

where $G^*(\theta)$ is LS transform of $G(x)$ and $M_G(x)\equiv\sum_{j=1}^{\infty}G^{(j)}(x)$. Differentiating $C_O(Z)$ with respect to Z and setting it equal to zero,

$$G^*(\theta)\left[1-\mathrm{e}^{-\theta Z}+\int_0^Z(\mathrm{e}^{-\theta x}-\mathrm{e}^{-\theta Z})\mathrm{d}M_G(x)\right]=\frac{c_O}{c_M},$$

i.e.,

$$G^*(\theta)\int_0^Z [1+M_G(x)]\theta e^{-\theta x}dx = \frac{c_O}{c_M},$$

whose left-hand increases strictly with Z from 0 to $M_G^*(\theta) = G^*(\theta)/[1-G^*(\theta)]$. Thus, if $M_G^*(\theta) > c_O/c_M$, then a finite Z_O^* $(0 < Z_O^* < \infty)$ exists, and the resulting cost rate is

$$TC_O(Z_O^*) = C_M[1 - e^{-\theta Z_O^*}G^*(\theta)].$$

9.3 When $G_i(x) = 1 - e^{-\omega x/a^{j-i}}$ $(i = 1, 2, \cdots, j)$,

$$\begin{aligned}
G^{(1)}(x) =& G_1(x) = 1 - e^{-\omega x},\\
G^{(2)}(x) =& \int_0^x G^{(1)}(x-y)dG_1(y) = \frac{1-e^{-\omega x}}{1-a} + \frac{1-e^{-\omega x/a}}{1-a^{-1}},\\
G^{(3)}(x) =& \int_0^x G^{(2)}(x-y)dG_1(y)\\
=& \frac{1-e^{-\omega x}}{(1-a)(1-a^2)} + \frac{1-e^{-\omega x/a}}{(1-a^{-1})(1-a)} + \frac{1-e^{-\omega x/a^2}}{(1-a^{-2})(1-a^{-1})},
\end{aligned}$$

and generally,

$$\begin{aligned}
G^{(j)}(x) =& \int_0^x G^{(j-1)}(x-y)dG_1(y)\\
=& \sum_{i=1}^{j}\left[\frac{1-e^{-\omega x/a^{i-1}}}{\prod_{k=1,k\neq i}^{j}(1-a^{k-i})}\right] \quad (j = 1, 2, \cdots),
\end{aligned}$$

where $\prod_{k=1,k\neq i}^{1} \equiv 1$.

9.4 (9.79) is derived in Problem 1.4. From

$$M_i e^{-\omega Z_{i+1}} = \sum_{j=i+1}^{n-1}\left(e^{-\omega Z_j} - e^{-\omega Z_{j+1}}\right)M_j + e^{-\omega Z_{n-1}} - e^{-\omega Z_n},$$

we have

$$\begin{aligned}
M_{n-1} &= e^{\omega(Z_n - Z_{n-1})} - 1,\\
M_{n-2} &= e^{\omega(Z_n - Z_{n-1})} - 1.
\end{aligned}$$

Generally,

$$M_i = \mathrm{e}^{\omega(Z_n - Z_{n-1})} - 1 \quad (i = 0, 1, \cdots, n-1),$$

which follows (9.80).

References

1. Nakagawa T (2005) Maintenance theory of reliability. Springer, London
2. Nakagawa T (2007) Shock and damage models in reliability theory. Springer, London
3. Nakagawa T (2011) Stochastic process with applications to reliability theory. Springer, London
4. Chen M, Mizutani S, Nakagawa T (2010) Random and age replacement policies. Int J Reliab Qual Saf Eng 17:27–39
5. Chen M, Nakamura S, Nakagawa T (2010) Replacement and preventive maintenance models with random working times. IEICE Trans Fund Electron Commun Comput Sci E93-A:500–507
6. Nakagawa T, Zhao X, Yun W (2011) Optimal age replacement and inspection with random failure and replacement times. Int J Reliab Qual Saf Eng 18:1–12
7. Zhao X, Nakagawa T (2012) Optimization problems of replacement first or last in reliability theory. Eur J Oper Res 223:141–149
8. Zhao X, Qian C, Nakamura S (2014) Age and periodic replacement models with overtime policies. Int J Reliab Qual Saf Eng 21:1450016 (14 pp)
9. Zhao X, Nakagawa T, Zuo M (2014) Optimal replacement last with continuous and discrete policies. IEEE Trans Reliab 63:868–880
10. Zhao X, Mizutani S, Nakagawa T (2015) Which is better for replacement policies with continuous or discrete scheduled times? Eur J Oper Res 242:477–486
11. Mizutani S, Zhao X, Nakagawa T (2015) Overtime replacement policies with finite operating interval and number. IEICE Trans Fund Electron Commun Comput Sci E98-A:2069–2076
12. Zhao X, Liu H, Nakagawa T (2015) Where does “whichever occurs first” hold for preventive maintenance modelings? Reliab Eng Syst Saf 142:203–211
13. Zhao X, Al-Khalifa KN, Hamouda AMS, Nakagawa T (2015) What is middle maintenance policy? Qual Reliab Eng Int 32:2403–2414
14. Zhao X, Al-Khalifa KN, Hamouda AMS, Nakagawa T (2016) First and last triggering event approaches for replacement with minimal repairs. IEEE Trans Reliab 65:197–207
15. Nakagawa T (2014) Random maintenance policies. Springer, London
16. Nakagawa T, Zhao X (2015) Maintenance overtime policies in reliability theory. Springer, Switzerland
17. Zhao X, Nakagawa T (2010) Optimal replacement policies for damage models with the limit number of shocks. Int J Reliab Qual Perform 2:13–20
18. Zhao X, Zhang H, Qian C, Nakagawa T, Nakamura S (2012) Replacement models for combining additive independent damages. Int J Perform Eng 8:91–100
19. Zhao X, Qian C, Nakagawa T (2013) Optimal policies for cumulative damage models with maintenance last and first. Reliab Eng Syst Saf 110:50–59
20. Zhao X, Nakamura S, Nakagawa T (2013) Optimal maintenance policies for cumulative damage models with random working times. J Qual Mainten Eng 19:25–37

X. Zhao and T. Nakagawa, *Advanced Maintenance Policies for Shock and Damage Models*, Springer Series in Reliability Engineering,
https://doi.org/10.1007/978-3-319-70456-2

21. Zhao X, Qian C, Sheu S (2014) Chapter 5: Cumulative damage models with random working times. In: Qian C, Chen M (eds) Nakamura S. Reliability modeling with applications, World Scientific, pp 79–98
22. Barlow RE, Proschan F (1965) Mathematical theory of reliability. Wiley, New York
23. Osaki S (1992) Applied stochastic system modeling. Springer, Berlin
24. Esary JD, Marshall AW, Proschan F (1973) Shock models and wear processes. Ann Probab 1:627–649
25. Ito K, Nakagawa T (2014) Optimal maintenance of airframe cracks. Int J Reliab Qual Saf Eng 21:1450014 (16 pp)
26. Endharta AJ, Yun WY (2014) A comparison study of replacement policies for a cumulative damage model. Int J Reliab Qual Saf Eng 21:1450021 (12 pp)
27. Scarf PA, Wang W, Laycock PJ (1996) A stochastic model of crack growth under periodic inspections. Reliab Eng Syst Saf 51:331–339
28. Hopp WJ, Kuo YL (1998) An optimal structured policy for maintenance of partially observable aircraft engine components. Naval Res Logist 45:335–352
29. Schijve J (1995) Multiple-site damage in aircraft fuselage structure. Fatigue Fract Eng Mater Struct 18:329–344
30. Nakagawa T (2008) Advanced reliability models and maintenance policies. Springer, London
31. Duchesne T, Lawless J (2000) Alternative time scales and failure time models. Lifetime Data Anal 6:157–179
32. Yuan F, Kumar U (2012) A general imperfect repair model considering time-dependent repair effectiveness. IEEE Trans Reliab 61:95–100
33. Pulcini G (2003) Mechanical reliability and maintenance models. In: Pham H (ed) Handbook of reliability engineering. Springer, London, pp 317–348
34. Wang W (2013) Optimum production and inspection modeling with minimal repair and rework considerations. Appl Math Modell 37:1618–1626
35. Park M, Jung K, Park D (2013) Optimal post-warranty maintenance policy with repair time threshold for minimal repair. Reliab Eng Syst Saf 111:147–153
36. Chang C (2014) Optimum preventive maintenance policies for systems subject to random working times, replacement, and minimal repair. Comput Ind Eng 67:185–194
37. Chien Y, Sheu S (2006) Extended optimal age-replacement policy with minimal repair of a system subject to shocks. Eur J Oper Res 174:169–181
38. Huynh K, Castro I, Barros A, Bérenguer C (2012) Modeling age-based maintenance strategies with minimal repairs for systems subject to competing failure modes due to degradation and shocks. Eur J Oper Res 218:140–151
39. Sim SH, Endrenyi J (2002) A failure-repair model with minimal and major maintenance. IEEE Trans Reliab 42:134–140
40. Gramopadhye AK, Drury CG (2000) Human factors in aviation maintenance: how we got to where we are. Int J Ind Ergon 26:125–131
41. Krausa DC, Gramopadhye A, AK, (2001) Effect of team training on aircraft maintenance technicians: computer-based training versus instructor-based training. Int J Ind Ergon 27:141–157
42. Grall A, Dieulle L, Berenguer C, Roussignol M (2002) Continuous-time predictive-maintenance scheduling for a deteriorating system. IEEE Trans Reliab 51:141–150
43. Zhou X, Xi L, Lee J (2007) Reliability-centered predictive maintenance scheduling for a continuously monitored system subject to degradation. Reliab Eng Syst Saf 92:530–534
44. Wang W (2012) An overview of the recent advances in delay-time-based maintenance modelling. Reliab Eng Syst Saf 106:165–178
45. Ahmad R, Kamaruddin S (2012) An overview of time-based and condition-based maintenance in industrial application. Comput Ind Eng 63:135–149
46. Chen D, Trivedi K (2005) Optimization for condition-based maintenance with semi-Markov decision process. Reliab Eng Syst Saf 90:25–29
47. Arunraj NS, Maiti J (2007) Risk-based maintenance-techniques and applications. J Hazard Mater 142:653–661

48. Swanson L (2001) Linking maintenance strategies to performance. Int J Prod Econ 70:237–244
49. Pluvinage G, Elwany MH (2008) Safety. Reliability and risks associated with water, oil and gas pipelines. Springer, Netherlands
50. Finkelstein MS, Zarudnij VI (2001) A shock process with a non-cumulative damage. Reliab Eng Syst Saf 71:103–107
51. Silberschatz A, Korth HF, Sudarshan S (2010) Database system concepts, 6th edn. McGraw-Hill Education
52. Kumar A, Segev A (1993) Cost and availability tradeoffs in replicated concurrency control. ACM Trans Database Syst 18:102–131
53. Fong Y, Manley S (2007) Efficient true image recovery of data from full, differential, and incremental backups. US Patent 7,251,749
54. Scott JA, Hamilton EC (2006) Method and apparatus for defragmentation. US Patent App.11/528,984
55. Douceur JR, Bolosky WJ (1999) A large-scale study of file-system contents. ACM SIGMETRICS Perform Eval Rev 27:59–70
56. Qian C, Nakamura S, Nakagawa T (2002) Optimal backup policies for a database system with incremental backup. Electron Commun Japan, Part 3 85:1–9
57. Nakamura S, Qian C, Fukumoto S, Nakagawa T (2003) Optimal backup policy for a database system with incremental and full backup. Math Comput Modell 11:1373–1379
58. Microsoft Support (2012) Description of full, incremental, and differential backups. Accessed 21 Aug 2012
59. Symantec Enterprise Technical Support (2012) What are the differences between Differential and Incremental backups?. Article: TECH7665. Created: 2000-01-27, Updated: 2012-05-12. Accessed 21 Aug 2012
60. NovaStor (2014) Differential and Incremental backups: why should you care?. Accessed 31 Oct 2014
61. Nakamura S, Arafua M, Iwata K (2016) Incremental and differential random backup policies. In: Pham H (ed) The 22nd ISSAT international conference on reliability and quality in design, pp 213–217
62. Nakagawa T (1979) Optimal policies when preventive maintenance is imperfect. IEEE Trans Reliab R-28:331–332
63. Nakagawa T, Yasui K (1987) Optimum policies for a system with imperfect maintenance. IEEE Trans Reliab R-36:631–633
64. Wang H, Pham H (2003) Optimal imperfect maintenance models. In: Pham H (ed) Handbook of reliability engineering. Springer, London, pp 397–414
65. Sheu SH, Chang CC (2010) Extended periodic imperfect preventive maintenance model of a system subjected to shocks. Int J Syst Sci 41:1145–1153
66. Zhao X, Al-Khalifa KN, Nakagawa T (2015) Approximate methods for optimal replacement, maintenance, and inspection policies. Reliab Eng Syst Saf 144:68–73
67. Kijima M, Nakagawa T (1992) Replacement policies of a shock model with imperfect preventive maintenance. Reliab Eng Syst Saf 57:100–110
68. Lemaitre J, Desmorat R (2005) Engineering damage mechanics. Springer, Berlin
69. Nakagawa T, Mizutani S (2008) Periodic and sequential imperfect preventive maintenance policies for cumulative damage models. In: Pham H (ed) Recent advances in reliability and quality in design. Springer, London, pp 85–99

Printed in the United States
By Bookmasters